Advances in Comparative and Environmental Physiology 22

Volumes already published

Vol. 10: Comparative Aspects of Mechanoreceptor Systems
Edited by *F. Ito* (1992)

Vol. 11: Mechanics of Animal Locomotion
Edited by *R. McN. Alexander* (1992)

Vol. 12: Muscle Contraction and Cell Motility: Molecular and Cellular Aspects.
Edited by *H. Sugi* (1992)

Vol. 13: Blood and Tissue Oxygen Carriers
Edited by *Ch.P. Mangum* (1993)

Vol. 14: Interaction of Cell Volume and Cell Function
Edited by *F. Lang* and *D. Häussinger* (1993)

Vol. 15: From the Contents: Salivary Gland Secretion – Nematocyst Discharge – Adaptations in Decapodan Crustaceans – Steroid-Sensitive Areas Mediating Reproductive Behaviors (1993)

Vol. 16: Ion Transport in Vertebrate Colon
Edited by *W. Clauss* (1993)

Vol. 17: Effects of High Pressure on Biological Systems
Edited by *A.G. Macdonald* (1993)

Vol. 18: Biomechanics of Feeding in Vertebrates
Edited by *V.L. Bels, M. Chardon* and *P. Vandewalle* (1994)

Vol. 19: Electrogenic Cl^- Transporters in Biological Membranes
Edited by *G.A. Gerencser* (1994)

Vol. 20: From the Contents: Motile Activities of Fish Chromatophores – Epithelial Transport of Heavy Metals – Heavy Metal Cytotoxicity in Marine Organisms – Comparative Pulmonary Morphology and Morphometry – Molecular Adaptations in Resistance to Penicillins (1994)

Vol. 21: Mechanisms of Systemic Regulation: Respiration and Circulation
Edited by *N. Heisler* (1995)

Vol. 22: Mechanisms of Systemic Regulation: Acid-Base Regulation, Ion Transfer and Metabolism
Edited by *N. Heisler* (1995)

Volumes in preparation

Comparative Immunology: Invertebrate Immune Responses.
Vol. 23: Cells and Molecular Products.
Vol. 24: Cell Activities and the Environment.
Edited by *E.L. Cooper*

Advances in

Comparative and Environmental Physiology 22

Mechanisms of Systemic Regulation: Acid–Base Regulation, Ion-Transfer and Metabolism

Guest Editor: N. Heisler

With Contributions by

D.E. Atkinson · E. Bourke · W.H. Dantzler · D.H. Evans
M.K. Grieshaber · B.J. Harvey · N. Heisler · P. Laurent
S.F. Perry · H.O. Pörtner · T.J. Shuttleworth · V. Urbach
S. Völkel · S.C. Wood

With 67 Figures

Springer

Guest Editor:

Prof. Dr. Norbert Heisler

Lehrstuhl für Tierphysiologie
Math. Naturwiss. Fak. I
Humboldt Universität zu Berlin
Abderhaldenhaus
Philippstr. 13
10115 Berlin
Germany

ISSN 0938-2673

ISBN 978-3-642-52365-6 ISBN 978-3-642-52363-2 (eBook)
DOI 10.1007/978-3-642-52363-2

CIP data applied for

Die Deutsche Bibliothek – CIP-Einheitsaufnahme
Mechanisms of systemic regulation/guest ed.: Heisler.-
Berlin; Heidelberg; New York: Springer
NE: Heisler, Norbert [Hrsg.]
Acid base regulation, ion transfer and metabolism/with
contributions by D. E. Atkinson ... – 1995
(Advances in comparative and environmental physiology; 22)

NE: Atkinson, D. E.; GT

Softcover reprint of the harcover 1st edition 1995

Cover design: Springer-Verlag, Design & Productions

Typesetting: Thomson Press (India) Ltd., New Delhi

SPIN: 10051809 31/3137/SPS – 5 4 3 2 1 0 – Printed on acid-free paper

Foreword to the Series

The aim of the series is to provide comprehensive, integrated reviews giving sound, critical and provocative summaries of our present knowledge in environmental and comparative physiology, from the molecular to the organismic level.

Living organisms have evolved a widespread range of basic solutions to cope with the different problems, both organismal and environmental, with which they are faced. A clear understanding of these solutions is of course of fundamental interest for all biologists, zoologically or medically oriented. They can be best comprehended in the framework of the environmental and/or comparative approaches. These approaches demand either wide surveys of animal forms or a knowledge of the specific adaptive features of the species considered. This diversity of requirements, both at the conceptual and technological level, together with the fact that physiology and biochemistry have long been mainly devoted to the service of medicine, can account for the fact that these approaches emerged only slow amongst the other new, more rapidly growing disciplines of the biological sciences.

The field has now gained the international status it deserves and the organization of a series devoted to it appeared timely to me in view of its actual rapid development and of the interest it arouses for a growing number of biologists, physiologists, and biochemists, independently of their basic, major orientation.

Raymond Gilles

List of Editors

Preface

Various endogenous and environmental challenges of homoiostasis have resulted in the evolution of apparently quite different mechanisms for the same or similar functions in individual representatives of the animal kingdom. One of the prominent achievements of comparative physiology over the last few decades has been the description of regulatory features common to many studied species beyond the extreme diversity of their morphological forms. Delineation of functional principles universally applicable to the physiology and biochemistry of living systems became often possible through technical advances in the development of numerous new techniques, in many cases modified and adopted from other fields of science, but also by approaching certain problems using multifactorial analysis. The advance in technology has facilitated studies of minute functional details of mechanisms, which finally lead to better understanding of generally similar functions, covered by the multiple developments of Nature as a response to an extreme variety of different conditions. Improved understanding of specific mechanisms, however, has presented new problems at the level of system integration. The importance of the integrative aspect became particularly apparent during an international symposium on 'Mechanisms of Systemic Regulation in Lower Vertebrates: Respiration, Circulation, Ion Transfer and Metabolism' (organized in 1990 by Norbert Heisler and Johannes Piiper at the Max-Planck-Institut für experimentelle Medizin at Göttingen/Germany). Interaction, discussion and scientific cooperation among participants of the symposium initiated further progress in the understanding of the role of specific mechanisms in the light of organismic function, but at the same time indicated still largely fragmentary knowledge and existing deficits. The resulting progress in understanding homoiostatic control prompted the idea of publishing a number of concise reviews on selected aspects of this scientific field, which are presented in volumes 21 and 22 of this book series.

These reviews do not attempt to present encyclopedic completeness, but are intended to selectively delineate recent progress in some fields of comparative animal physiology, qualified by new approaches, interesting new data, novel interpretation or particular importance in terms of integrative merits. The large weight of the last point is reflected in the title of the two volumes. The general aim is to provide a synthesis of established ideas, models and the state of knowledge with new (sometimes controversial) ideas, approaches, results and interpretations.

The various chapters present aspects of cellular mechanisms and generalized functions related to import and export of substance, intracorporeal transfer and maintenance of the overall status of animals. Ammonia is the energetically most efficient nitrogenous metabolic end product, but there are problems of toxicity and effects on acid-base status. Metabolic elimination of ammonia by the synthesis of urea is one way to deal with these problems. The interpretation of the regulatory logic of the involved metabolic processes and of the interaction with the acid-base status is still controversial (Chaps. 1 and 3). Elimination of urate as another less toxic and acid-base relevant nitrogenous waste and of other organic acids in the kidney in general is another mechanism to reduce metabolic effects on homoiostatic control (Chap. 2). Direct elimination of ammonia to the environment is limited to aquatic species because of its extreme toxicity. The form in which ammonia is transferred through membranes and epithelia has been the subject of intense study and discussion over the last few years, with recent data clearly suggesting transfer by nonionic diffusion (Chap. 4). Acid-base and ionic regulation in fish takes place mainly at the branchial epithelium, the morphological structure of which may be modified considerably in response to various external and internal homoiostatic conditions (Chap. 5). Natriuretic peptide hormones, well known from mammals, are also found in heart muscle, brain and plasma of various fish groups. Changes in activity of natriuretic peptide hormones initiate osmoregulatory and haemodynamic effects in fish, which may be closely related to systemic regulation (Chap. 6). Changes in intracellular pH regulate Na^+ and K^+ channel activity, indirectly affect the Na^+/K^+ pump in the principal cells of high resistance epithelia and will also affect the activity of the proton pump of mitochondria-rich cells in order to reestablish equilibrium after acid loading (Chap. 7). Extensive studies on nasal gland cells, particularly involved in the osmoregulation of marine birds, have provided new information with respect to the character of intracellular signals controlling ionic and acid-base regulation (Chap. 8). The amount of oxygen available for metabolic energy production is limited in hypoxia. Recent studies indicate that in both ectotherms and endotherms an alternative and energetically advantageous approach to increasing oxygen supply under such conditions is the reduction of metabolic demand by behaviorally decreasing body temperature (Chap. 9). In spite of the extreme cellular toxicity of hydrogen sulfide numerous invertebrate species as well as a few vertebrates have developed various types of effective internal and external defense mechanisms in order to cope with hydrogen sulfide as an environmental challenge (Chap. 10).

All of the chapters have been prepared by authorities in the respective fields and as such represent the state of the art of our present knowledge, sometimes spiced with controversial aspects delineated from various sides. With this high standard of the individual contributions, it is still hoped that this collection of reviews represents more than the sum of the parts. The volume is designed to attract readers from a variety of disciplines in biology and medicine, in addition to specialized fields in physiology, biochemistry and biophysics, providing solid information for established scientists as well as for students and postgraduates.

The editor would like to express his gratitude to the authors, who have provided an excellent selection of chapters and have made a great effort to make this volume comprehensible and wide ranging in a limited amount of print space. Numerous other people involved in the preparation of the manuscript and in the printing process, in particular my assistant Dr. Sabine Heinz and the editorial staff of Springer, in particular Ursula Gramm and Theodora Krammer, facilitated the publication process.

April 7, 1995 *Norbert Heisler*

Contents

Acid–Base Regulation and Ammonia Elimination

Chapter 1

pH Homeostasis in Terrestrial Vertebrates;
Ammonium Ion as a Proton Source
D.E. Atkinson and *E. Bourke*

Chapter 2

Renal Transport of Organic Acids and Bases
in Non-Mammalian Vertebrates
W.H. Dantzler

Chapter 3

pH Homeostasis in Terrestrial Vertebrates:
A Comparison of Traditional and New Concepts
H.O. Pörtner

Chapter 7

Regulation of Ion and Water Transport
by Hydrogen Ions in High Resistance Epithelia
B.J. Harvey and *V. Urbach*

Chapter 8

Intracellular Signals Controlling Ionic
and Acid–Base Regulation in Avian Nasal Gland Cells
T.J. Shuttleworth

Hypoxia and Metabolic Adaptation

Chapter 9

Interrelationships Between Hypoxia
and Thermoregulation in Vertebrates
S.C. Wood

Chapter 10

Sulfide Tolerance in Marine Invertebrates
S. Völkel and *M.K. Grieshaber*

Contents for Volume 21

Gas Exchange and Regulation of Respiration

Central Vescular Circulation

Chapter 6

Mechanisms of Intracardiac Shunting in Reptiles: Pressure vs Washout Shunting
J. W. Hicks and *G. M. Malvin*

Chapter 7

Central Cardiovascular Dynamics in Reptiles
S. Nilsson

Chapter 8

Central Cardiovascular Function in Amphibians: Qualitative Influences of Phylogeny, Ontogeny, and Season
W. W. Burggren

Veno-Lymphatic System

Chapter 9

The Role of the Lymphatic System for Water Balance and Acid–Base Regulation in the Amphibia
D. P. Toews and *L.A. Wentzell*

Chapter 10

Physiological Roles of the Secondary Circulatory System in fish
A. Ishimatsu, G. K. Iwama and *N. Heisler*

Acid–Base Regulation and Ammonia Elimination

Chapter 1

pH Homeostasis in Terrestrial Vertebrates; Ammonium Ion as a Proton Source

D.E. Atkinson[1] and *E. Bourke*[2]

Contents

1 Introduction

Homeostasis and functional regulation are among the most characteristic and essential features of organisms. Intracellular pH is especially critical, and organisms of all types have evolved elaborate and metabolically costly mechanisms for its stabilization. Of necessity, those mechanisms vary widely in order to be effective in types as diverse as bacteria and mammals, living in habitats that range from aqueous to xeric and from well-aerated to anoxic.

In most metazoans the regulation of pH, ancestrally a cellular function, has been taken over by systemic homeostatic systems. This conversion is nearly total

[1] Department of Chemistry and Biochemistry, University of California, Los Angeles, CA 90024, USA
[2] Veterans Administration Medical Center, 800 Poly Place, Brooklyn, NY 11209, USA

Advances in Comparative and Environmental Physiology, Vol. 22

in land vertebrates, in which the pH of blood and interstitial fluid is tightly controlled. Thus cells and tissues are bathed by media of nearly constant pH, which greatly facilitates the stabilization of their internal pH. There are, of course, pH gradients across membranes of various types, which participate importantly in physiological functions such as energy transduction and the movement of other ions and molecules across membranes, but those gradients and the functions that they serve are dependent on the ambient homeostasis provided by the systems that regulate blood pH.

The HCO_3^-/CO_2 system is the primary buffer of blood and interstitial fluid; thus pH homeostasis depends on regulation of the concentration ratio of its conjugate acid, CO_2, and conjugate base, HCO_3^-. It has conventionally been believed that normal metabolism produces only acid, and that blood pH is stabilized by excretion of the volatile acid, CO_2, via the lungs and generation of HCO_3^- in the kidney, at the expense of excretion of NH_4^+ in the urine, to counter the small amount of non-volatile acid that was thought to arise in the metabolism of protein. Recent reconsideration of the basic chemical events involved in protein catabolism, ureagenesis, renal metabolism, and NH_4^+ excretion indicates that those views require modification. Both CO_2 and HCO_3^- are produced in metabolism, and their concentration ratio in the blood is regulated by independent modulation of the rates of disposal of both of them in response to pH.

This chapter focuses on metabolic aspects of pH homeostasis in mammals.

2 Renal Glutamine Metabolism

It has been known for more than a century that urinary urea decreases in metabolic acidosis and that urinary ammonium (NH_4^+) increases. Until recently, studies on the causal relationships have focused on the latter observation (Bourke and Atkinson 1988). Glutamine was demonstrated to be the major source of urinary ammonium by Van Slyke et al. (1943). Pitts (1948) concluded that NH_3 derived from glutamine in renal tubular cells moved into the tubular urine, where it combined with hydrogen ion to form NH_4^+, thereby eliminating both ammonia and protons from the body. He envisioned the hydrogen ions as being derived from the hydration of CO_2, forming H_2CO_3, which dissociated into HCO_3^- and H^+. The HCO_3^- was thought to return to the circulation to replenish the depleted body stores of bicarbonate, seen as the cause of the adaptation in the first place (Fig. 1A). This concept stimulated much experimental work, and many experimental observations seem compatible with it. Thus in experimental metabolic acidosis in the dog the rate of glutamine utilization by the kidney increases, as does total renal ammoniagenesis, predominantly from the sequential deamidation and deamination of glutamine (Bourke et al. 1971).

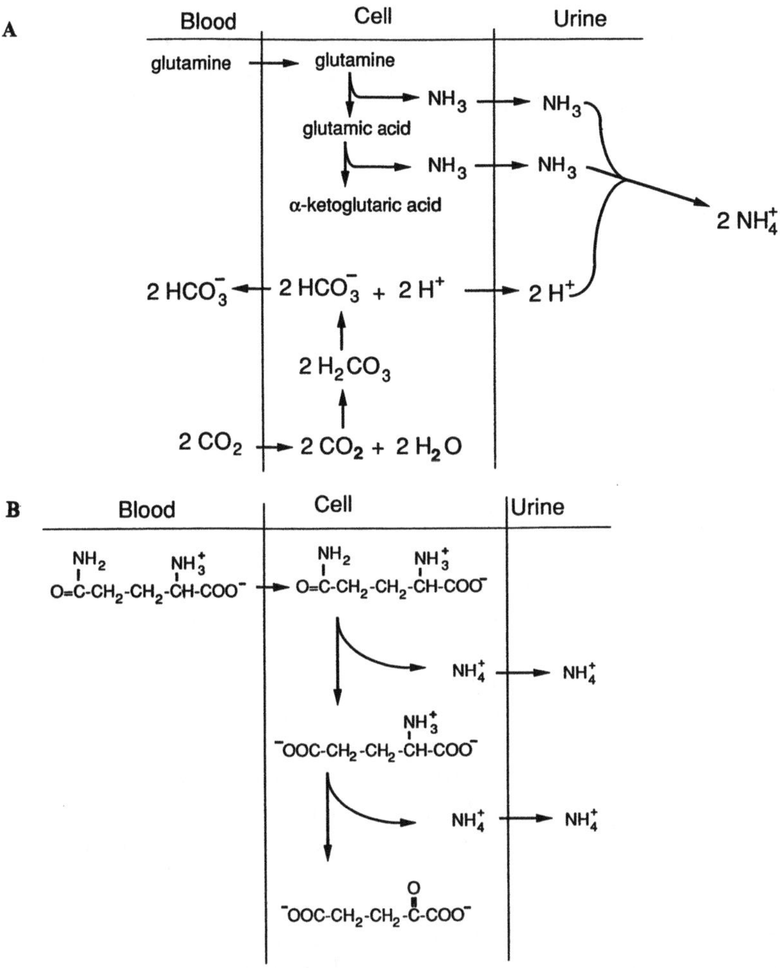

Fig. 1. Deamidation and deamination of glutamine in the kidney. **A** The conventional Pitts formulation, in which NH_3, derived from glutamine, moves into the lumen and combines with H^+ that was obtained from the body buffer, with generation of an equimolar amount of HCO_3^-, that moves into the blood. **B** The chemically valid formulation, taking ionization into account. NH_4^+, rather than NH_3, is the product of deamidation and deamination of glutamine, and excretion of NH_4^+ has no effect on the body buffer

2.1 Urinary Ammonium Is Not a Form of Acid Excretion

A theory is not proved by the fact that some observations fit it. Although it dominated thinking in the field for decades and is still encountered (e.g. Carroll and Oh 1989; Knepper et al. 1989; Toto 1990; Andreoli 1992), the Pitts concept that metabolism generates NH_3, which then serves as a vehicle for urinary excretion of proton, was incompatible with elementary chemical considerations. Metabolism of protein does not produce NH_3. Many of the misconceptions resulted from ignoring the ionic natures of the metabolites involved. Glutamine, like other amino acids, is a dipolar ion, containing an anionic carboxylate group ($-COO^-$) and a cationic substituted ammonium group ($-NH_3^+$). The deamidation and deamination of glutamine give rise to NH_4^+, not NH_3. Conversion of the α-NH_3^+ group of glutamine to NH_4^+ and movement to the urine does not result in any uptake or release of hydrogen ions from the body's stores; the nitrogen was protonated from the beginning (Fig. 1B).

Although the amide nitrogen of glutamine is not protonated, hydrolysis of an amide yields NH_4^+ and a carboxylate anion with no uptake of protons (Fig. 1B). Glutamine thus does not generate NH_3 which can mop up hydrogen ions from the dissociation of carbonic acid, thereby leaving behind a newly generated bicarbonate ion to return to the body.

The mechanisms whereby NH_4^+ enters the tubular urine are varied (DuBose et al. 1991). There is considerable direct transport of NH_4^+ across the luminal membrane in addition to the conventional route, which involves dissociation to H^+ and NH_3, which move into the lumen and recombine. The pathway is not relevant to the stoichiometry; movement of NH_4^+ from kidney cells to urine does not entail net removal of protons from the body buffers.

The prevailing view has been that urinary acid excretion is the sum of urinary titratable acidity plus the amount of NH_4^+ in the urine. In that view, the positive charge of the ammonium ion represents a proton that is somehow different from the protons of other conjugate acids. However, that distinction is not valid. If urine is titrated to the same pH as blood, the amount of base required for the titration is an accurate measure of the amount of acid that was excreted in the urine. That conclusion necessarily follows from the generalization that when two solutions are at the same pH, all of the buffer systems in the two solutions are at pH equilibrium. That is as true for the NH_4^+/NH_3 system as for other urinary systems, such as $H_2PO_4^-/HPO_4^{2-}$. In fact, in spite of its negative charge the $H_2PO_4^-$ ion is approximately 100 times as strong an acid as is NH_4^+. Thus the common practice of adding the concentration of NH_4^+ to the titratable acidity when estimating urinary acid excretion is not chemically or physiologically valid. The ammonia system, being a very weak acid, is approximately 98.5% protonated in blood at pH 7.4 (i.e. 98.5% NH_4^+ and 1.5% NH_3), so that it can accept only a negligible amount of additional proton for excretion no matter how acid the urine. In contrast, the conjugate acid of the phosphate system, $H_2PO_4^-$, makes up about 20% of the total phosphate pool at pH 7.4 in blood and about 86% in urine at pH 6. Because of that sizeable difference in protons held at the two pH values, the phosphate system is the main vehicle for excretion of acid in urine.

2.2 Bicarbonate Generation in the Kidney

Most of the α-ketoglutarate that is generated from glutamine in the kidney is metabolized there, predominantly to glucose or to CO_2 plus HCO_3^-. α-Ketoglutarate contains two negatively-charged carboxylate groups and, whether it is oxidized to CO_2 or metabolized to glucose, conservation of charge demands that two HCO_3^- ions be produced from each molecule of α-ketoglutarate (Eqs. 1 and 2). These equations, like others in this review, show overall conversions, not mechanisms.

$$^{-}OOC{-}CH_2{-}CH_2{-}\overset{\overset{\displaystyle O}{\|}}{C}{-}COO^- + 4\,O_2 \longrightarrow 3\,CO_2 + 2\,HCO_3^- + H_2O \quad (1)$$

$$^{-}OOC{-}CH_2{-}CH_2{-}\overset{\overset{\displaystyle O}{\|}}{C}{-}COO^- + O_2 + 2\,H_2O \longrightarrow \tfrac{1}{2}\,C_6H_{12}O_6 + 2\,HCO_3^-. \quad (2)$$

This renal production of bicarbonate is conventionally considered to play a role in acid-base regulation. Even if the excretion of NH_4^+ does not represent actual excretion of acid, it is often maintained that metabolism of the α-ketoglutarate that is left behind produces "new" bicarbonate. Since the rate of glutamine metabolism in the kidney increases in acidosis, the conventional view is that a resulting increase in renal production of bicarbonate acts to counter the acidosis. This proposal was initially forwarded as an alternate interpretation of Pitts' hypothesis (Oliver and Bourke 1975; Bourke 1977; Halperin and Jungus 1983; Walser 1986). Two questions arise when the validity of this proposal is considered. First, is there an increase in the rate of production of bicarbonate in acidosis? Second, does the rate of production of bicarbonate in the kidney have any effect on systemic pH homeostasis?

2.2.1 Response to Systemic Acid–Base Change

With regard to the first question, the shortcomings of the "new bicarbonate" interpretation were already implicit in the results of Pitts' own experiments in vivo (Pitts 1975) on the renal utilization of lactate in metabolic acidosis and alkalosis in the dog. The effect of acidosis on the rate at which lactate is used is virtually a mirror image of the effect on glutamine metabolism; that is, in acidosis the increase in the rate of glutamine utilization is balanced by a decrease in the rate of utilization of lactate. A paper from Pitts' laboratory (Leal-Pinto et al. 1973) reported that total energy turnover (hence bicarbonate generation) does not differ significantly in acidosis and alkalosis, although the amounts of lactate and glutamine used (hence the rate of generation of NH_4^+) are different.

The interrelationships between metabolism of glutamine and of lactate and other energy sources are complex and have been elaborated on in rat kidney in vitro (Preuss et al. 1987), but in the context of acid–base relationships the important point is that an increase in the rate of production of bicarbonate from α-ketoglutarate is compensated for by a decrease in the rate of production of bicarbonate from lactate. The rate of citrate metabolism responds to acid–base status in the same way as lactate (Cohen and Kamm 1981). Presumably the rate of total substrate metabolism is regulated by the energy demand of the kidney, which does not change appreciably in response to changes in blood pH, whereas the rate of glutamine hydrolysis is regulated in part by acid–base status. When the rate of glutamine utilization increases in acidosis, the energy-modulated metabolic regulatory system causes a decrease in the rate of oxidation of other substrates. That response is not consistent with the view that the physiological function of the increased glutamine metabolism in acidosis is to increase the rate of generation of bicarbonate, since that rate does not in fact increase.

2.2.2 Effect on Systemic Acid–Base Status

The second question is a more fundamental one: does the rate of production of bicarbonate in the kidney affect systemic acid–base relationships? The answer is that no such effect is possible. Even if the rate of production of bicarbonate in the kidney were to increase (or decrease) in acidosis, the change could have no effect on systemic acid–base status. If it is not utilized in the kidney, glutamine will be metabolized elsewhere. Metabolic acidosis, in addition to increasing the renal utilization of glutamine, decreases its hepatic utilization (Lueck and Miller 1970; Häussinger et al. 1986; Almond et al. 1991). But whether it is utilized in the liver or transported to the kidney and utilized there, two bicarbonate ions are generated from each molecule of glutamine, and the alkalinizing effect on the blood is identical in both cases. Thus the metabolism of glutamine in the kidney and the excretion of ammonium ion in the urine can have no direct role in the regulation of blood pH.

3 Catabolism of Protein Generates Bicarbonate

Glutamine has a unique physiological function as the primary carrier of nitrogen between organs and tissues. However, in chemical terms there is nothing special about glutamine. The amide bond of glutamine resembles the substituted amide bonds (peptide bonds) that join amino acids in proteins. Hydrolysis of a peptide bond yields two charged products: a negatively-charged carboxylate group and a positively-charged substituted ammonium group. The carboxylate or carboxamide ($-CONH_2$) groups or positively charged nitrogen-containing groups in the side chains of certain amino acids must also be considered in a detailed accounting of the consequences of protein metabolism (Atkinson and Camien 1982). However, for our present purposes it is enough to point out that hydrolysis of proteins yields dipolar amino acids with no uptake or donation of protons. In the oxidation of an amino acid an HCO_3^- anion is produced from each $-COO^-$ group. Each $-NH_3^+$ group yields an NH_4^+ ion, again with no gain or loss of protons. If the amino acid is used as material for gluconeogenesis the ionic yield is the same as for oxidation. Those relationships are shown for alanine in

Eqs. (3) and (4):

$$CH_3\text{—}\underset{\displaystyle |}{\overset{\displaystyle NH_3^+}{CH}}\text{—}COO^- + 3\,O_2 \longrightarrow 2\,CO_2 + HCO_3^- + NH_4^+ + H_2O \qquad (3)$$

$$2\,CH_3\text{—}\overset{\displaystyle NH_3^+}{CH}\text{—}COO^- + 4\,H_2O + 2\,CO_2 \longrightarrow$$

$$C_6H_{12}O_6 + 2\,NH_4^+ + 2\,HCO_3^-. \qquad (4)$$

Equation (4) can be considered to represent the sum of two processes. In one, two molecules of alanine are converted to glucose plus NH_4^+. This conversion consumes two molecules of water and two protons. The protons are generated from the body buffers, with the associated conversion of $2CO_2$ (plus $2H_2O$) to $2\,HCO_3^-$. The important point is that when amino acids are metabolized, whether they are oxidized to yield energy or used as starting materials for gluconeogenesis or the biosynthesis of other uncharged products, one NH_4^+ is produced for each $-NH_3^+$ group and one HCO_3^- for each $-COO^-$ group.

Because NH_4^+ is a very weak acid (its pK_a value is about 9.2), it does not interact significantly with the body buffer systems. Since the ammonium ion remains more than 98% protonated at all pH values below about 7.5 (thus in blood and in urine), its ability to either donate or accept protons in vivo is extremely limited, and the metabolic production of NH_4^+ has virtually no effect on acid–base status. Bicarbonate, on the other hand, is the conjugate base of the predominant buffer of blood and interstitial fluid, so the production of HCO_3^- is alkalinizing. One HCO_3^- ion is produced for each peptide bond in a protein (that is, for each carboxylate group in the component amino acids) and one for each glutamate, glutamine, aspartate, and asparagine side chain. It follows that the metabolism of protein is strongly alkalinizing.

The metabolism of 1g of a protein of typical amino acid composition produces about 10 mmol of HCO_3^- (Atkinson and Camien 1982; Atkinson and Bourke 1987a). Thus a daily dietary intake of 100 to 150 grams of protein generates about 1000 to 1500 mmol bicarbonate. Unless disposed of, the accumulating bicarbonate would soon cause lethal alkalosis. An approximately equal amount of ammonium ion is also generated. However, as noted above, ammonium is such a weak acid that it has no appreciable effect on blood pH. In fact, the generation of ammonium bicarbonate in protein metabolism is almost exactly equivalent, in direct effect on acid–base status, to the addition of an equimolar amount of sodium bicarbonate. This is illustrated in Table 1. Three features of the table should be noted:

a) The buffer strength increases steadily with addition of the conjugate base HCO_3^-, in spite of the fact that the pH of the system is far above the pK_a value of the CO_2/HCO_3^- buffer system. This is because the concentration of the conjugate acid is fixed and the total concentration of the buffer system increases when HCO_3^- is added.

Table 1. Calculated effects of adding $NaHCO_3$, NH_4HCO_3, or NH_4Cl to a CO_2/HCO_3^- buffer system with the concentration of CO_2 fixed at 1.2 m 2M

Addn	pH			$[HCO_3^-]$			Buffer strength		
mmol/l	$NaHCO_3$	NH_4HCO_3	NH_4Cl	$NaHCO_3$	NH_4HCO_3	NH_4Cl	$NaHCO_3$	NH_4HCO_3	NH_4Cl
0	7.40	7.40	7.40	24	24	24	0.56	0.56	0.56
5	7.48	7.48	7.40	29	29	24	0.67	0.67	0.56
10	7.55	7.55	7.40	34	34	24	0.79	0.79	0.56
20	7.66	7.66	7.40	44	44	24	1.02	1.01	0.56
100	8.11	8.09	7.38	124	118	23	2.89	2.87	0.56

The initial concentration of HCO_3^- was 24 mM, and the initial pH was 7.40. The column headings identify the salt that was added and the left column indicates the amount added (m mol/l). The entries show the resulting values of pH, the concentration of HCO_3^-, and the buffer strength (defined as the number of millimoles of OH^- ion that would be required per liter to raise the pH by 0.01 unit). Effects of ionic strength, which would be very small in the physiological range, are ignored.

b) The increases in pH and in buffer strength caused by addition of HCO_3^- are the same whether the accompanying cation is Na^+ or NH_4^+.
c) In keeping with the demonstration (point b) that NH_4^+ is equivalent to Na^+ in terms of pH, the addition of NH_4Cl is equivalent to addition of NaCl, and has no appreciable effect of pH or buffer strength. Even at 100 mM, well over 100 times normal blood concentrations, the effect of NH_4^+ on pH-related properties would be minimal. It follows that removal of NH_4^+ into the urine is equivalent to removal of Na^+ and does not affect pH.

Table 1 was calculated for a buffer system equivalent to the predominant bicarbonate buffer of blood. Because of the presence of other buffers, pH changes in actual blood would be slightly smaller than those shown. Those small differences would not affect the comparisons in Table 1 or points a, b, and c, which apply equally to blood.

4 Disposal of Metabolic Bicarbonate

4.1 Ureagenesis

Once it is recognized that urinary ammonium cannot serve to transport an appreciable amount of acid (Sect. 2.1; Fig. 1B and Table 1), the conventional picture is seen to require revision. Since urinary ammonium is not directly related to acid excretion, the decrease in urinary NH_4^+ in alkalosis and its increase in acidosis cannot contribute directly to pH homeostasis. This conclusion follows directly from Henderson's (1911) discussion of urinary titratable acidity alone as the valid measure of urinary acid excretion, and might have suggested at any time since 1911 that it is the change in amount of urea synthesized (and excreted) that is related to pH homeostasis.

That possibility might have led to consideration of the balanced equation for urea synthesis and to realization that the assumption that synthesis of urea has no effect on acid–base status could not be correct. The fact that two positively-charged NH_4^+ ions (one actually an $-NH_3^+$ group of aspartate) are incorporated into the uncharged urea molecule indicates directly that two protons must be liberated in the synthesis. Thus an increase in the rate of urea synthesis will cause the rate of liberation of protons to increase and tend to lower the blood pH. Conversely, a decrease in the rate of urea synthesis is the appropriate response to acidosis, because the rate of proton generation is decreased. In blood and interstitial fluid the $(CO_2+H_2O)/HCO_3^-$ system is the primary buffer. This system will react with nearly all protons that are liberated. Thus the terms "generation of two protons" and "consumption of two bicarbonate ions" are effectively synonymous. Whichever term is applied to ureagenesis, the effect is to contribute to a decrease in pH. The liberation of protons and their reaction with the blood buffer are represented in summary form in Eq. (5):

$$2NH_4^+ + 2HCO_3^- \longrightarrow NH_2{-}CO{-}NH_2 + 3H_2O + CO_2 \quad (5)$$

It has been generally assumed that metabolism produces a threat of acidosis, and most experimental work on acid–base relationships has dealt with experimentally induced acidosis. As noted above, however, the metabolic production of bicarbonate greatly exceeds that of all acids combined (except for carbon dioxide, which is not relevant here, because it is excreted via the lungs). The 1000 mmol or more of bicarbonate that arises in metabolism of a typical daily human diet poses a threat of alkalosis that if not countered would be fatal. Thus the physiological function of the metabolic component of the pH homeostatic mechanism is to guard against alkalosis, not acidosis. The only feasible route of disposal of bicarbonate in large amount is titration, and the only possible source of 1000 mmol protons/day is the NH_4^+ that arises, together with the HCO_3^- that causes the problem, from protein catabolism. However, since NH_4^+ is far too weak an acid to titrate HCO_3^- directly, the only means by which the protons of NH_4^+ can be obtained for titration of HCO_3^- is to incorporate the nitrogen into an uncharged group of an organic molecule, with liberation of protons.

In mammals, two protons are liberated in the synthesis of each molecule of urea. This generation of protons appears to be the metabolic function of the synthesis. Four molecules of ATP are converted to ADP for each molecule of urea synthesized (two ATPs for each proton liberated). Ureagenesis uses metabolic energy to force the titration of HCO_3^- by an acid that is much too weak to effect the titration directly.

4.1.1 Ureagenesis and pH

Mammals, including humans, maintain a stable systemic pH by individually regulating the disposal of the conjugate acid and conjugate base, CO_2 and HCO_3^-, of the major systemic buffer, both in response to pH. Humans on a typical diet excrete from 10 to 15 mol acid through the lungs each day. The typical daily synthesis of about 500 mmol (30g) of urea liberates the 1000 mmol proton that is required for titration of the daily HCO_3^- production (Sect. 3).

When the early experiments on pH and nitrogen excretion were subsequently repeated under more controlled conditions, the finding of a pH-induced shift in partitioning between urea and glutamine was essentially confirmed (Oliver and Bourke 1975). For instance, during chronic hydrochloric acidosis in the rat, the rise in urinary ammonium was accompanied by an equimolar decrease in urinary urea. Similar findings have been reported in man (Fine et al. 1977) where a fall in blood urea nitrogen was also observed. Furthermore, in the isolated perfused rat liver, lowering of medium pH led to a significant decrease in the rate of ureagenesis (Lueck and Miller 1970; Oliver et al. 1977; Bean and Atkinson 1984), accounting for the decreased excretion observed in vivo. The pattern of response of glutamine metabolism in the liver to change in pH is opposite to that of urea, with an increase in net glutamine release at lower medium pH (Lueck and Miller 1970; Oliver et al. 1977; Häussinger 1990; Häussinger et al. 1988). The relevance of these observations to acid–base homeostasis is apparent from consideration of the metabolic cycle of urea synthesis (Fig. 2). Ammonium, derived partly from the portal blood and partly from the action of pH-dependent hepatic glutaminase, reacts with bicarbonate, forming carbamoylphosphate. This reaction requires energy (ATP), and is in effect a forced (energy-driven) titration of HCO_3^- by NH_4^+. The second nitrogen (and the second proton) are incorporated into the cycle from aspartate. When ornithine reacts with carbamoylphosphate to form citrulline, the second proton is released and titrates a second bicarbonate. Thus each turn of the cycle disposes of two bicarbonate ions. Hydrolysis of arginine produces urea and ornithine, regenerating a key intermediate and completing the cycle.

Although the rate of synthesis of urea has been shown to be modulated by pH in intact animals, perfused liver, and hepatocytes, the mechanism of the regulation is at present not well understood. In contrast to typical biosynthetic pathways, in which regulatory effects are exerted on only the first enzyme of the sequence, the control of ureagenesis seems to be more diffuse. Each of the enzymes of the cycle itself responds positively to increase in pH across the physiological range in experiments in vitro, and accessory enzymes also seem to participate. Hepatic glutaminase responds to both pH and the concentration of NH_4^+ (Szweda and Atkinson 1990), and N-acetylglutamate synthetase responds to pH, arginine, and glutamate (Kamemoto and Atkinson 1985). In both cases the responses appear to involve complex interdependencies, rather than being simply additive.

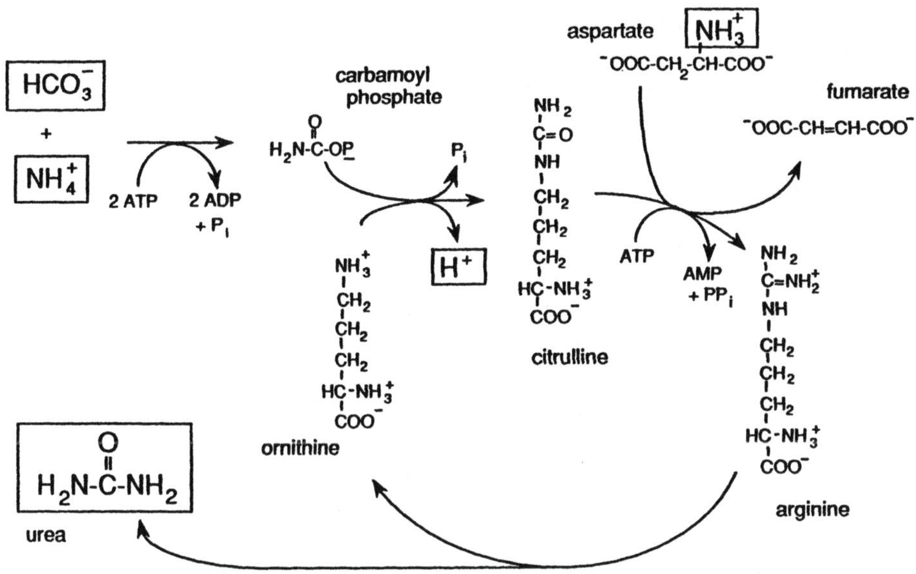

Fig. 2. The pathway of urea synthesis. Compounds and ions that are consumed or produced are enclosed in boxes. In each turn of the cycle two bicarbonate ions (one of which is retained in the product urea) are titrated with protons derived from two ammonium ions. This titration, which would be thermodynamically impossible as a direct reaction, is made energetically favorable by being coupled to the conversion of four molecules of ATP to ADP. Three ATPs are used directly and the fourth is needed for the conversion of AMP to ADP. One NH_4^+ ion in effect provides a proton for titration of HCO_3^- in its conversion to the carbamoyl group of citrulline. The proton provided by the substituted ammonium group of asparate is retained in arginine and ornithine and is released in the reaction of ornithine with carbamoyl phosphate. Argininosuccinate, an intermediate between citrulline and arginine, is omitted from this figure. The production of three molecules of water is not shown. Gains or losses of protons from phosphates, including ATP, ADP, and AMP, are not shown because they are balanced in the regeneration of ATP and so have no effect on systemic pH homeostasis. *P* Phosphoryl group; P_i inorganic orthophosphate; PP_i, inorganic pyrophosphate

A role for hepatic ureagenesis in acid-base homeostasis is gaining increasing acceptance (Cohen and Woods 1988; Meijer et al. 1990). However, certain clinical and experimental observations have been interpreted as not favoring such a role. The authors of one paper concluded, on the basis of their interpretation of the consequences of a large ammonium chloride load, that urea formation is not regulated primarily by acid–base balance in vivo (Halperin et al. 1986). During sustained administration of NH_4Cl, a constant rate of ureagenesis was observed over the time interval studied and systemic pH was noted to fall. This, from our analysis, is scarcely surprising. Systemic pH was driven down by enhanced ureagenesis caused by a massive NH_4^+ overload that overwhelmed the normal pH-based controls. As the administration continued, the inappropriately high rate of urea synthesis was sustained, with a steady decline in pH. The fall in pH was the *consequence* of the rate of urea synthesis under non-physiological conditions. The authors' interpretation of the data, however, was quite different.

By plotting urea synthesis against blood pH, and titling the resulting figure "Effect of pH on urea synthesis," they implied that pH had been experimentally varied, and that the variation had not affected the rate of urea synthesis. This non-standard mode of presentation (in which the abscissa variable is dependent on the ordinate parameter, which is itself a dependent variable) has misled several readers into citing the paper as showing that variation of pH does not affect the rate of urea synthesis.

Reports as to the consequences of functional liver impairment on acid–base status vary. For example, Almdal et al. (1989) reported that partial hepatectomy (85%) did not influence acid–base homeostasis, despite a significant decrease in urinary excretion of urea. That finding led the authors to conclude that a major role for the liver in the regulation of acid–base status was unlikely. However, other interpretations of the results are possible. The need for urea synthesis depends on the rate of metabolic bicarbonate generation, which in turn depends mainly on the net rate of protein breakdown. The net rate of protein catabolism may have been lower than normal in the experimental animals, in which liver regeneration was probably occurring, with a concomitant high rate of protein synthesis. If so, a lower than normal rate of ureagenesis would be required for maintenance of normal blood pH, and the reported results may merely indicate that the residual liver tissue was sufficient to meet the reduced needs.

Intensive studies on a series of patients with stratified liver disease (Häussinger et al. 1990) provided cogent evidence regarding the relationship between liver function and acid–base status in humans. The concentration of plasma bicarbonate increased progressively with loss of urea cycle capacity as measured in vitro. In these patients other causes of metabolic alkalosis such as diuretics or antacids, vomiting, hyperaldosteronism, or renal disease were rigorously excluded. Because alkalosis is a potent stimulus for urea synthesis, a reasonable hypothesis as to the situation in the patients with impaired liver function is as follows: a decrease of urea cycle capacity leads to hyperbicarbonatemia and alkalosis which, in turn, activates urea synthesis and restores a normal urea flux, albeit at a somewhat higher plasma bicarbonate level, despite diminished total ureagenetic capacity. This is an illustration of the generalization that when the capacity of a regulated system is diminished, a larger error signal will be required for any given level of response. Such generalizations apply to biological regulatory systems as well as to technological ones. Thus, when ureagenetic capacity is impaired, 80% of the remaining capacity might be needed to dispose of the bicarbonate load, while the same absolute rate of disposal might correspond to only 10% of normal capacity. It is obvious that a larger pH error signal would be required to obtain that rate in the impaired liver.

The metabolic pH-stabilizing machinery has evolved to protect the organism against alkalosis. Acidosis is rarely, if ever, encountered in nature except in various pathological states (and transient exercise-induced lactic acidosis, which is automatically corrected when the lactate is metabolized), and cannot have been a selective factor of any significance in evolution. Thus, there presumably is no primary regulatory response to acidosis. These considerations suggest that the responses to experimentally induced acidosis that have been extensively reported are not primary protective measures against acidity, but diminutions of the activity of the system that protects against alkalinity. The increased amount of NH_4^+ that is excreted after administration of HCl, for example, is not carrying acid away; rather it reflects the fact that administered protons have titrated part of the metabolic HCO_3^- and thus less NH_4^+ is required in urea synthesis to generate protons. In experimental or pathological metabolic acidosis some of the substituted ammonium of amino acids that would ordinarily be synthesized into urea, liberating protons for the titration of bicarbonate, is diverted via glutamine to be excreted by the kidney into the urine as NH_4^+. It is the reduction in rate of urea synthesis that mitigates the acidotic challenge, not the increase in NH_4^+ excretion.

Changes in nitrogen metabolism that are related to pH homeostasis thus occur both in the liver and the kidney. So we should no longer think of a single organ, the kidney, as the metabolic regulator of systemic pH but of the coordinated action of at least two organs in the overall response to metabolic acid–base disturbances. Adaptations in other organs, including muscle, have also been referred to in the literature but are outside the scope of this review (Oliver et al. 1977; Atkinson and Camien 1982; Guppy 1989; Atkinson and Bourke 1987a; Foley and Hume 1987; Welbourne 1988; see review by Pörtner 1989). A likely participation of the gut is discussed in Sect. 4.1.4. Although ureagenesis is the primary component of the metabolic pH homeostatic machinery, the kidney plays an essential secondary role through the elimination of ammonium that is not required for ureagenesis. The renal metabolic pathways of glutamine deamidation and deamination to ammonium have been extensively studied and remain a field for productive future investigation since the enhanced glutamine ammoniagenesis induced by metabolic acidosis is still incompletely understood (Schoolwerth 1991). One of the harmful effects of failure of the kidney to eliminate ammonium would be overriding, by an abnormally high concentration of ammonium, of the pH-sensitive regulation of ureagenesis, with resultant bicarbonate consumption and metabolic acidosis (Häussinger et al. 1986; Atkinson and Bourke 1987a). The experimental acidosis induced by ammonium chloride administration is a classic example of such override.

Those relationships can be illustrated in the rat in vivo by comparing dietary supplementation with ammonium chloride and with ammonium bicarbonate. Administration of NH_4HCO_3 causes a marked increase in urinary urea with negligible change in urinary NH_4^+ or in blood pH (Oliver and Bourke 1975). This experimental treatment mimics a normal physiological situation – the generation of NH_4^+ and HCO_3^- in virtually equal amounts in protein metabolism. The liver responds to the resulting increase in pH by increasing the rate of urea synthesis and thus the rate of generation of proton for titration of HCO_3^-. Since there is as much NH_4^+ as HCO_3^-, an adequate supply of protons can be obtained; the regulatory system functions normally and adequately, and the pH change is resisted. Thus, an addition that would significantly raise the pH of a passive buffer system similar to that of blood (Table 1) has virtually no effect in an intact normal animal because of pH homeostatic mechanisms.

The situation is very different when NH_4Cl is administered. The addition of the administered ammonium ion to that produced in metabolism produces a situation that has probably never occurred outside the laboratory – a large molar excess of NH_4^+ over HCO_3^-. It seems impossible to conceive of any natural diet whose metabolism would lead to that result. Faced with a non-physiological situation for which no defense has evolved (because it has not been encountered in nature), the pH homeostatic mechanism is unable to respond in a totally appropriate manner. After administration of ammonium chloride, urinary ammonium excretion typically increases by about 90% of the amount administered. This part of the NH_4^+ load, appearing in urine in the same form as that in which it was taken up, obviously has no effect whatever on acid–base status. However, the excess supply of NH_4^+ over that of HCO_3^- causes a relatively slight increase in the rate of urea production compared to the rate that would have occurred in the absence of the administration. The resulting oversupply of protons leads to titration of more than the appropriate amount HCO_3^-, so that pH falls. Thus, an addition that would have no significant effect in a passive buffer (Table 1) causes the pH to decrease in vivo, because it perturbs the normal operation of the homeostatic system.

It is important to recognize that even in this non-physiological case the rate of urea synthesis responds primarily to pH. The regulatory system responds to a slight fall in pH by changing the parameters of the system in the direction that limits the rate of ureagenesis. This control is sufficiently effective that it holds the rate of urea synthesis near its normal value even in the face of non-physiological excess of the substrate NH_4^+. It is because of this pH-modulated regulation that most of the administered NH_4^+ is sent to the kidney for excretion. In fact, the fraction of total NH_4^+ that is used in urea synthesis is much lower than before the administration (Oliver and Bourke 1975). If the primary function of ureagenesis were removal of NH_4^+ the rate of urea synthesis would be expected to rise enough to dispose of most or all the extra NH_4^+ load.

Dry land is not the only environment in which excretion of HCO_3^- as such may be impossible. An interesting example is seen in the adaptation of a teleost fish, *Oreochromis alcalicus grahami* (Randall et al. 1989; Wood et al. 1989), to the extremely alkaline environment of Lake Magadi in the Great Rift Valley in Kenya, in which pH is about 10 and the concentration of HCO_3^- is about 180 millimolar. This species not only cannot excrete its metabolic HCO_3^-, but must cope with HCO_3^- that diffuses in, since the concentration of that ion in lake water is about 20 times as high as in the plasma of the fish. The expedient that has evolved is ureagenesis. This is the only known instance of complete ureotelism in an entirely aquatic teleost fish. The rate of urea synthesis is regulated in response to need for protons, and fell virtually to zero when fish were transferred to water at pH 7. A closely related species of fish without the capacity for ureagenesis dies within an hour in water from the lake.

4.1.2 Renal NH_4^+ Elimination in Relation to Acid–Base Balance

After ureagenesis has generated the requisite number of protons for titration of HCO_3^-, some NH_4^+ may remain to be incorporated into glutamine for transport to other organs or tissues where needed or to the kidney for urinary excretion. The complexity of the mechanisms of urinary elimination of renally generated ammonium (for review, see DuBose et al. 1991) attests to the importance of this renal function.

Ammonium generation and elimination are complex renal functions of substantial importance to the metabolic regulation of the homeostatic response to metabolic acidosis, but do not result in direct excretion of acid or generation of "new" bicarbonate. Under normal conditions, the rate of renal ammonium excretion is determined by the amount of NH_4^+ not required for urea synthesis. Thus, an enhanced rate of excretion may be a reflection either of an increased rate of metabolic generation of NH_4^+ or of a reduced rate of bicarbonate titration in the liver.

The conventional view has been that glutamine synthesis is regulated and that urea synthesis is merely a means of disposal of excess ammonium, which was seen merely as a toxic waste product. According to the concepts that are accepted in this review (Atkinson and Camien 1982; Atkinson and Bourke 1984, 1987a,b; Häussinger et al. 1984; Häussinger 1990), the regulatory relationships are essentially the reverse: urea synthesis is regulated in response to pH in order to supply the number of protons necessary for titration of metabolic bicarbonate and maintain pH stability, and any remaining NH_4^+ is packaged into glutamine for export. It is clear that the ammonium available to the liver is partitioned between two main pathways in response to blood pH. The two views differ as to which pathway has regulatory priority. If, as proposed here, the main metabolic function of ureagenesis is its contribution to pH homeostasis, the regulation of ureagenesis must necessarily take precedence over that of glutamine synthesis.

4.1.3 Histological Localization of Ureagenesis and Glutamine Synthesis

The liver is composed of functional units, or acini. Blood flow in each acinus runs from a portal venule along a sinusoid to a terminal hepatic or central venule (for review, see Lamers et al. 1989). The hepatocytes near the sinusoidal inflow are termed periportal and those near the sinusoidal outflow are termed perivenous or pericentral. These zones differ in enzymatic capacities (for example, the capacities for glycolysis and gluconeogenesis vary reciprocally along the sinusoid). Pronounced differences in localization of enzymes that participate in nitrogen metabolism have more recently been discovered. These differences appear to be crucially important to pH homeostasis.

Comparative studies of antegrade and retrograde perfusion of the structurally and metabolically intact rat liver (Häussinger 1983; Häussinger and Gerok 1983) suggested metabolic zonation of ureagenesis and glutamine synthesis attributable to a differential distribution of the key enzymes between the periportal (glutaminase and urea cycle enzymes) and perivenous (glutamine synthetase) hepatocytes of the hepatic acinus. This subacinar localization has now been amply confirmed (for review, see Häussinger 1990), in experiments involving differential zonal liver damage and by immunohistochemical visualization of glutamine synthetase in a single layer of cells around the hepatic venule and of carbamoylphosphate synthetase, argininosuccinate synthetase, and arginase (enzymes of the urea cycle) in most or all other hepatocytes. More recent studies on the specific activity of hepatic phosphate-activated glutaminase and the presence of the mRNA for this enzyme confirmed previous functional evidence that the capacity for hydrolysis of glutamine accompanies the enzymes for urea synthesis (Watford and Smith 1990); that is, the majority of hepatocytes are capable of hydrolyzing glutamine and of using ammonium in the synthesis of urea.

Functional consequences of this differential localization of enzymes are illustrated schematically in Fig. 3. Changes in the widths of the bars reflect

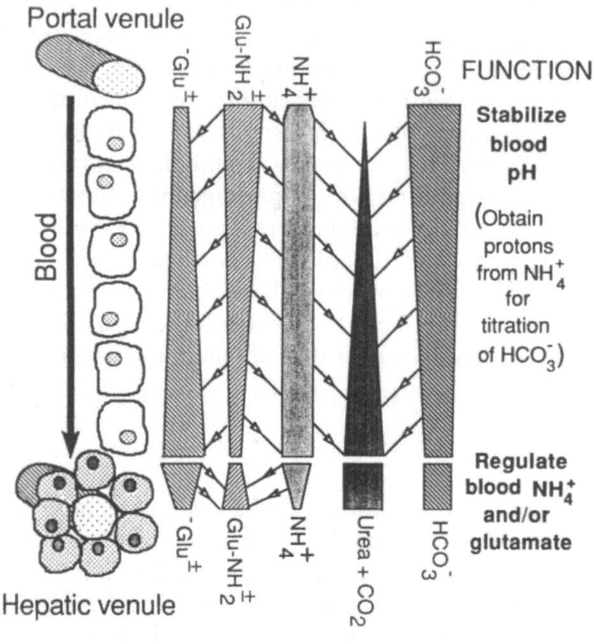

Fig. 3. Schematic representation of nitrogen metabolism in the liver. As blood flows from the portal venule through the sinusoid, it first passes cells that contain glutaminase and the enzymes of the urea cycle. Glutamine is hydrolyzed to a greater or less extent; NH_4^+ is generated from that hydrolysis and from the metabolism of amino acids (not shown); and incorporating of NH_4^+ into urea generates protons that serve to titrate HCO_3^-, half of which is incorporated into urea while the other half is converted to CO_2. The last rank of cells around the hepatic or central venule takes up glutamate and NH_4^+ and synthesizes glutamine. Changes in widths of the bars represent changes in concentrations of the corresponding substances. $-Glu^{\pm}$ Glutamate; *$-Glu\text{-}NH_2^{\pm}$* glutamine

processes in which the corresponding materials are consumed or generated, but they are not drawn to scale. The major points illustrated are that the decrease in the concentration of HCO_3^-, which results from urea synthesis and hence is linked to an increase in urea concentration, occurs in the great majority of hepatocytes, from the portal venule to very near the hepatic venule, and that there is no change in HCO_3^- and urea concentrations in the last rank of cells. Hydrolysis of glutamine and other metabolic processes contribute to maintaining the concentration of NH_4^+ in the large periportal zone sufficiently high that it will not limit ureagenesis, but this concentration probably varies more widely than is shown. The intercellular glutamine cycle demonstrated by Häussinger (1983) is reflected in the decrease in glutamine concentration in the urea-synthesizing periportal zone and its increase in the last rank of cells, which uniquely contain glutamine synthetase. This cycling, with hydrolysis of glutamine being followed by its synthesis, probably occurs under all conditions, but the relative magnitudes of the two processes, and thus whether the liver is a net consumer or producer of glutamine, depends on such factors as blood pH and the level of NH_4^+ in portal blood.

A zonal distribution of plasma membrane transport systems complementary to that of the enzymes has been reported. Thus, plasma glutamate and α-ketoglutarate are transported into the perivenous hepatocytes, which contain glutamine synthetase, whereas glutamine is transported predominantly into the periportal hepatocytes containing glutaminase and enzymes of the urea cycle (see Häussinger 1990).

To summarize the conclusions that appear to follow from the observations cited: During most of its travel from portal to hepatic venule an element of blood passes cells that are able to take up glutamine and hydrolyze it and that possess the enzymes for the rections of the urea cycle. That cycle supplies protons for titration of metabolically-produced bicarbonate. In order to maintain pH homeostasis, the rate at which urea is made in these cells is controlled primarily by the pH of the blood. A human or other mammal in good nutritional balance obtains the ammonium needed for ureagenesis mainly from catabolism of amino acids derived from dietary protein and by ammonium taken up by the gut (part of which arises from hydrolysis of urea in the colon; see Sect. 4.1.4) and exported to the portal blood. When necessary, this supply can be supplemented by the action of glutaminase on glutamine from the portal blood. At the end of its passage through the sinusoid, the blood passes a ring of cells that are able to take up glutamate and to synthesize glutamine; that synthesis packages residual ammonium for transport to the kidney or other tissues and presumably protects against high blood levels of ammonium and perhaps glutamate. Regulation of urea synthesis has total priority, since urea synthesis has been completed before the blood reaches cells capable of synthesizing glutamine.

In acidosis less urea synthesis is needed. As a consequence the amount of ammonium in blood reaching the perihepatic cells will increase. This residual ammonium is taken up and incorporated into glutamine. Thus in acidosis the liver sends more glutamine to the kidney than under more normal conditions.

4.1.4 Recycling of NH_4^+ Through the Gut

Catabolism of protein is the main metabolic source of NH_4^+, but metabolism of carboxylates in the diet, especially in fruits and leaves, leads to additional HCO_3^-. It is thus likely that total metabolic production of HCO_3^- exceeds total NH_4^+ production much of the time. This imbalance raises the question of how the extra HCO_3^- can be titrated, and why a small amount of NH_4^+ is generally excreted in the urine.

A possible answer is supplied by the observation that in humans up to 30% of the urea synthesized in the liver moves into the gut instead of the urine. The urea is hydrolyzed by intestinal bacteria and the NH_4^+ is almost quantitatively reabsorbed (Walser 1980). That reabsorption would be puzzling and counterproductive if NH_4^+ were merely a toxic endproduct to be eliminated, as has conventionally been believed. It can be understood, however, when NH_4^+ is recognized to be an essential metabolite as the source of protons needed for maintenance of pH homeostasis. To the extent that the ammonium is reabsorbed in excess of any HCO_3^- absorption that may occur, the result is that part of the NH_4^+ is recycled and can participate in the titration of additional HCO_3^-. Ammonium is required also for other metabolic processes such as synthesis of dietarily non-essential amino acids if the dietary supply is inadequate.

The relative rates of reabsorption of NH_4^+ and HCO_3^- have not been quantitated, but indirect evidence argues against significant absorption of HCO_3^-. Surgical diversion of the ureters into the colon in treatment of certain states of severe bladder disfunction may cause striking acidosis. This is a consequence of the increased colonic load of ammonium, which is derived from both the NH_4^+ contained in urine and the much larger amount of ammonium bicarbonate that results from hydrolysis of urinary urea by urease of colonic bacteria. The NH_4^+ is absorbed into the portal circulation (McConnell et al. 1979). There is also enhanced mucosal Cl^-/HCO_3^- exchange (Davis et al. 1983). The net result is increased entry of NH_4^+ and Cl^- into the portal vein. As in the analogous case of an oral load of NH_4Cl, the consequence is metabolic acidosis.

These observations appear to reflect the fate of the smaller amount of urea that moves into the gut under normal physiological conditions, which we suggest is an essential part of the pH homeostatic mechanism. It would be interesting to determine whether the movement of urea into the gut is regulated and increases in alkalosis. If the proposed mechanism is valid, the result is that part of the metabolic HCO_3^- is excreted in feces or titrated by acidic products of metabolism of colonic microorganisms.

The lack of effect of a sodium bicarbonate load on the observed rate of urea synthesis in the rat (Oliver and Bourke 1975) or man (Gougeon-Reyburn and Marliss 1989; Bjerrum et al. 1990), which has been said to argue against the importance of urea synthesis as a defense against alkalosis, is scarcely surprising. Urinary urea excretion cannot increase in response to administration of an excess of bicarbonate if that administration is not accompanied by an additional ammonium source. Once ammonium ion is recognized to be an essential metabolite (in its role of provider of H^+ for titration of HCO_3^-), it is evident that an acute NH_4^+ deficiency is as likely as a deficiency of any other metabolite, and that such a deficiency will limit reactions or sequences for which the deficient metabolite is needed. The only apparent source of additional NH_4^+ is secretion of urea to the gut, movement to the colon, hydrolysis, and absorption of NH_4^+ (Sect. 4.1.4). This sequence of events cannot be instantaneous and might be of limited effect in the 2-hour period used by Bjerrum et al. (1990). More importantly, the rate of urea synthesis was estimated from urinary urea excretion, with the implicit assumption that the fraction of urea that was lost to the gut did not change. Thus even a large increase in the rate of urea synthesis would have been invisible to the authors if it involved use of an increased amount of NH_4^+ recycled from the gut.

5 Summary

The metabolic component of acid–base homeostasis has traditionally been considered to involve only one organ, the kidney. Recent conceptual developments, however, point to an involvement of the liver as a major pH homeostatic organ. The catabolism of proteins generates large amounts of bicarbonate and thus a potential threat of alkalosis. Different strategies for its disposal have evolved in different organisms. In mammals there is a liver-specific pathway for irreversible removal of metabolically generated bicarbonate, namely titration by protons liberated in urea synthesis. Modulation of the rate of this synthesis by blood pH provides a sensitive and complex control of bicarbonate disposal and contributes to pH homeostasis.

Nitrogen is incorporated in the liver into urea and glutamine. The enzymes for those processes are spatially separated along the liver sinusoid so that ureagenesis occurs before the blood reaches cells that contain glutamine synthetase. This sequential ordering of the two processes that consume nitrogen allows both to be regulated by appropriate signals without mutual interference, and provides regulatory priority to urea synthesis.

In acidosis part of the metabolic bicarbonate is removed by reaction with the administered or metabolically-generated acid, leaving less that must be titrated by protons liberated in urea synthesis. The resulting reduction in consumption of NH_4^+ does not lead to serious hyperammonemia, because perivenous scavenger hepatocytes downstream from the site of ureagenesis regulate the blood level of ammonium by removing it via an alternate route, incorporation into glutamine. Renal glutamine hydrolysis and subsequent urinary ammonium excretion provide a final sink for this residual ammonium. Both hepatic ureagenesis and renal ammoniagenesis are subject to complex and sensitive short- and long-term regulation. Hepatic glutamine synthesis and renal ammoniagenesis, acting in concert, can be seen as a finely tuned spillover for the elimination of excess ammonium without concomitant bicarbonate consumption. Renal ammoniagenesis may therefore function primarily in ammonium homeostasis rather than in pH homeostasis. The pH homeostatic response precedes the glutamine-synthesis response by virtue of the sequential arrangement of enzyme capacities in the hepatic acinus. Systemic acid–base status is thus a major determinant of the amount of NH_4^+ that is available for glutamine synthesis.

The role of urea synthesis in generating protons to combat the metabolic alkalosis that would otherwise result from catabolism of amino acids seems well established. However, the sites at which ureagenesis is regulated in response to pH have not been uniquely identified. In two papers that appeared after this manuscript was submitted, Boon and colleagues report that the uptake of amino acids by the liver was lower during acute acidosis induced by jugular infusion of HCl (Boon et al. 1994a) and during mild chronic acidosis induced by addition of HCl to the diet (Boon et al. 1994b) than during alkalosis induced by infusion or dietary supply of HCO_3^-. They propose that reduced availability of amino acids, which provide HCO_3^- and NH_4^+, substrates of urea synthesis, is the cause of the reduction in the rate of urea synthesis that is seen in acidosis. According to that view, amino acid transport into the hepatocyte would be the primary site of regulation of the rate of urea synthesis in response to variation in pH.

The relevant chemical relationships suggest, however, that regulation of the rate of ureagenesis by varying the availability of amino acids could not contribute to stabilization of pH. Catabolism of amino acids produces HCO_3^-, and thus tends to cause an increase in pH. Synthesis of urea liberates H^+ (since H^+ reacts with HCO_3^-, this is equivalent to the consumption of HCO_3^-) and tends to cause a decrease in pH. If an increase in pH caused an increase in the rate of amino acid uptake and catabolism, the result would be conterproductive, since the generation of HCO_3^- would exacerbate the alkalosis. An increase in the rate of urea synthesis, using the NH_4^+ that is also liberated in amino acid catabolism, could dispose of the additional HCO_3^- generated from the accelerated amino acid catabolism, but could contribute nothing towards reversal of the preexisting alkalosis. Thus, an increase or decrease in the rate of amino acid entry in response to alkalosis or acidosis could not modulate urea synthesis in a way that would contribute to pH homeostasis.

Further, the suggestion that the rate of urea synthesis is controlled by the availability of amino acids as substrates for catabolism is not supported experimentally. Boon and colleagues report no difference in the intracellular concentrations of amino acids between acidosis and alkalosis. If amino acid availability were a major factor regulating ureagenesis, it would be expected that amino acid concentrations would be markedly higher in alkalosis, when urea is made much more rapidly. The results are instead consistent with the nearly universal generalization that metabolic regulatory systems operate through changes in the properties of key enzymes, which cause fluxes to vary appropriately to meet metabolic needs and hold substrate concentrations nearly constant.

An alternate explanation for the results of Boon and colleagues can be proposed: The synthesis of urea is regulated by pH. The uptake of amino acids is regulated by intracellular concentrations. The rate of proteolysis has been shown to increase at low pH, at least in muscle (May et al. 1986). Thus, it can be proposed that as pH falls the rate of proteolysis increases. Perhaps energy-generating pathways shift in acidosis from carbohydrate oxidation (a pH-neutral pathway) to a greater reliance on amino acid oxidation (which generates HCO_3^-). If intracellular generation of amino acids by proteolysis increases, less uptake is needed to maintain intracellular concentrations in the normal physiological range, and the rate of uptake of amino acids decreases. Urea synthesis decreases simultaneously because of its modulation by pH, but there is no direct or causal connection between the rates of amino acid uptake and of ureagenesis.

An increase in the rate of generation of HCO_3^- (resulting from accelerated proteolysis and amino acid catabolism) and a decreased rate of utilization of HCO_3^- (resulting from a lower rate of ureagenesis) would contribute to correction of acidosis. Because less urea is made than at normal or alkalotic values of pH, more NH_4^+ will remain to be secreted by the kidney.

Boon and colleagues find no increase in glutamine leaving the liver in acidosis, in contrast to earlier reports from several laboratories, including those of Lueck and Miller (1970) and Häussinger (1983). Almond and colleagues (Almond et al. 1992) report that glutamine efflux increases with decrease in pH but, on the basis of experiments with an inhibitor of glutamine synthase, conclude that the difference does not result from a decrease in the rate of glutamine synthesis. Genetically programmed strict localization of glutamine synthase to the last rank of cells around the hepatic venules could have arisen and been maintained only if that arrangement confers significant functional advantage, however. The localization suggests that glutamine synthesis must be related inversely, at least under some conditions, to urea synthesis, which occurs in cells that are encountered earlier by blood flowing from portal to hepatic venules. However, the findings of Almond and colleagues suggest the existence of additional interrelationship that are not presently understood.

References

Almdal T, Vilstrup H, Bjerrum K, Østergaard Kristensen L (1989) Decrease in ureagenesis by partial hepatectomy does not influence acid–base balance. Am J Physiol 257: F696–F699

Almond MK, Iles RA, Cohen RD (1992) Hepatic glutamine metabolism and acid–base regulation. Miner Electrolyte Metab 18: 237–240

Almond MK, Smith A, Cohen RD, Iles RA, Flynn G (1991) Substrate and pH effects on glutamine synthesis in rat liver. Biochem J 278: 709–714

Andreoli TE (1992) Disorders of fluid volume, electrolytes and acid–base balance. In: Wyngaarden JB, Smith LH, Bennett JC (eds) Cecil textbook of medicine, 19th edn WB Saunders, Philadelphia, pp 499–528

Atkinson DE (1992) Functional roles of urea synthesis in vertebrates. Physiol Zool 65: 243–267

Atkinson DE, Bourke E (1984) The role of ureagenesis in pH homeostasis. Trends Biochem Sci 9: 297–300

Atkinson DE, Bourke E (1987b) Metabolic aspects of the regulation of systemic pH. Am J Physiol 252: F947–F959
Atkinson DE, Bourke E (1987b) Ammonium, urea, and systemic pH regulation (Letter to the editor.) Am J Physiol 253: F200–F202
Atkinson DE, Camien MN (1982) The role of urea synthesis in the removal of metabolic bicarbonate and the regulation of blood pH. Curr Top Cell Regul 21: 261–302
Bean ES, Atkinson DE (1984) Regulation of the rate of urea synthesis in liver by extracellular pH. A major factor in pH homeostasis in mammals. J Biol Chem 259: 1552–1559
Bjerrum K, Vilstrup H, Almdal TP, Østergaard Kristensen L (1990) No effect of bicarbonate-induced alkalosis on urea synthesis in normal man. Scand J Clin Lab Invest 50: 137–141
Boon L, Blommaart PJE, Meijer AJ, Lamers WH, Schoolwerth AC (1994a) Acute acidosis inhibits hepatic amino acid uptake: implications for regulation of acid–base balance. In: Tizianello A, Baverel G, Endou H, Schoolwerth AC, O'Donovan DJ (eds) Renal ammoniagenesis and interorgan cooperation in acid–base homeostasis. Contrib Nephrol, vol. 110. Karger, Basel, pp 133–137
Boon L, Blommaart PJE, Meijer AJ, Lamers WH, Schoolwerth AC (1994b) Effect of chronic acidosis on hepatic amino acid uptake and gene regulation: implications for control of acid–base balance. In: Tizianello A, Baverel G, Endou H, Schoolwerth AC, O'Donovan DJ (eds)Renal ammoniagenesis and interorgan cooperation in acid–base homeostasis. Contrib Nephrol, vol 110. Karger, Basel, pp 138–143
Bourke E (1977) Studies on metabolic aspects of acid–base regulation. Irish J Med Sci 146: 119–129
Bourke E, Atkinson DE (1988) The development of acid–base concepts and their application to mammalian pH homeostasis. In: Häussinger, D (ed) pH homeostasis: mechanisms and control. Acad Press, London, pp 163–179
Bourke E, Frindt G, Rubio-Paez D, Schreiner GE (1971) Effects of chronic alkalosis and acidosis on the glutaminase II path in the dog kidney in vivo. Am J Physiol 220: 1033–1036
Carroll HJ, Oh MS (1989) Principles of renal physiology. In: Carroll HJ, Oh MS (eds) Water, electrolyte and acid–base metabolism: diagnosis and management, 2nd edn. JB Lippincott Co, Philadelphia, pp 64–65
Cohen JJ, Kamm DE (1981) Renal metabolism: Relation to renal function. In: Brenner BM, Rector FC Jr (eds) The kidney, 2nd edn. PA Saunders, Philadelphia, pp 144–248
Cohen RD, Woods HF (1988) Disturbances of acid–base homeostasis. In: Weatherall DJ, Ledingham JGG, Warrell DA (eds) The Oxford textbook of medicine, 2nd edn. Oxford Univ Press, Oxford, pp 9-164– 9-175
Davis GR, Morawski SG, Santa Ana CA, Fordtran JS (1983) Evaluation of chloride/bicarbonate exchange in the human colon in vivo. J Clin Invest 71: 201–207
DuBose TD Jr, Good DW, Hamm LL, Wall SM (1991) Ammonium transport in the kidney: new physiological concepts and their clinical implications. J Am Soc Nephrol 1: 1193–1203
Fine A, Carlyle JE, Bourke E (1977) The effects of administrations of HCl, NH_4Cl and NH_4 HCO_3 on the excretion of urea and ammonium in man. Eur J Clin Invest 7: 587–589
Foley WJ, Hume ID (1987) Nitrogen requirements and urea metabolism in two arboreal marsupials, the greater glider *(Petauroides volans)* and the brushtail possum (*Trichosurus volpecula*) fed eucalyptus foliage. Physiol Zool 60: 241–250
Gougeon-Reyburn R, Marliss EB (1989) Effects of sodium bicarbonate on nitrogen metabolism and ketone bodies during very low energy protein diets in obese subjects. Metabolism 38: 1222–1230
Guppy M (1986) The hibernating bear: why is it so hot and why does it cycle urea through its gut? Trends Biochem Sci 11: 247–276
Halperin ML, Jungas RL (1983) Metabolic production and renal disposal of hydrogen ions. Kidney Int 24: 709–713
Halperin ML, Chen C, Cheema-Dhadli S, West RL, Jungas RL (1986) Is urea formation regulated primarily by acid–base balance in vivo? Am J Physiol 250: F605–F612
Häussinger D (1983) Hepatocyte heterogeneity in glutamine and ammonia metabolism and the role of an intercellular glutamine cycle during ureagenesis in perfused rat liver. Eur J Biochem 133: 269–275

Häussinger D (1990) Nitrogen metabolism in liver: structural and functional organization and physiological relevance. Biochem J 267: 281–290
Häussinger D, Gerok W (1983) Hepatocyte heterogeneity in glutamate uptake by isolated perfused rat liver. Eur J Biochem 136: 421–425
Häussinger D, Gerok W, Seis H (1984) Hepatic role in pH regulation: role of the intercellular glutamine cycle. Trends Biochem Sci 9: 300–302
Häussinger D, Gerok W, Sies H (1986) The effect of urea synthesis on extracellular pH in isolated perfused rat liver. Biochem J 236: 261–265
Häussinger D, Steeb R, Gerok W (1990) Ammonium and bicarbonate homeostasis in chronic liver disease. Klin Wochenschr 68: 175–182
Häussinger D, Meijer AJ, Gerok W, Sies H (1988) Hepatic nitrogen metabolism and acid–base homeostasis. In: Häussinger D (ed) pH homeostasis: mechanisms and control. Academic Press, London, pp 337–377
Henderson LJ (1911) A critical study of the process of acid excretion. J Biol Chem 9: 403–424
Kamemoto E, Atkinson DE (1985) Modulation of the activity of rat liver acetylglutamate synthase by pH and arginine concentration. Arch Biochem Biophys 243: 100–107
Knepper MA, Packer R, Good DW (1989) Ammonium transport in the kidney. Physiol Rev 69: 179–249
Lamers WH, Hilberts A, Furt E, Smith J, Jonges GN, van Noorden CJF, Gaasbeek Janzen JW, Charles R, Moormam AFM (1989) Hepatic enzymic zonation: a reevaluation of the concept of the liver acinus. Hepatology 10: 72–76
Leal-Pinto E, Park HC, King F, MacLeod M, Pitts RF (1973) Metabolism of lactate by the intact functioning kidney of the dog. Am J Physiol 224: 1463–1467
Lueck JD, Miller LL (1970) The effect of perfusate pH on glutamine metabolism in the isolated perfused rat liver. J Biol Chem 245: 5491–5497
May RE, Kelly RA, Mitch WE (1986) Metabolic acidosis stimulates protein degradation in rat muscle by a glucocorticoid-dependent mechanism. J Clin Invest 77: 614–621
McConnell JB, Murison J, Stewart WK (1979) The role of the colon in the pathogenesis of hyperchloremic acidosis in ureterosigmoid anastomosis. Clin Sci 57: 305–312
Meijer AJ, Lamers WH, Chamuleau RAFM (1990) Nitrogen metabolism and ornithine cycle function. Physiol Rev 70: 701–748
Oliver J, Bourke E (1975) Adaptations in urea ammonium excretion in metabolic acidosis in the rat: a reinterpretation. Clin Sci Mol Med 48: 515–520
Oliver J, Koelz AM, Costello J, Bourke E (1977) Acid–base induced alternations in glutamine metabolism and ureogenesis in perfused muscle and liver of the rat. Eur J Clin Invest 7: 445–449
Pitts RF (1948) Renal excretion of acid. Fed Proc 7: 418–426
Pitts RF (1975) Production of CO_2 by the intact functioning kidney of the dog. Med Clin N Am 59: 507–518
Pörtner HO (1989) The importance of metabolism in acid–base regulation and acid–base methodology. Can J Zool 67: 3005–3017
Preuss HG, Roxe DM, Bourke E (1987) Acidotic alterations in oxidative metabolism influencing rat renal slice ammoniagenesis. Life Sci 41: 1695–1702
Randall DJ, Wood CM, Perry SF, Bergman H, Maloiy GMO, Mommesen TP, Wright PA (1989) Urea excretion as strategy for survival in a very alkaline environment. Nature 337: 165–166
Schoolwerth AC (1991) Regulation of renal ammoniagenesis in metabolic acidosis. Kidney Int 40: 961–973
Szweda L and Atkinson DE (1990) Response of rat liver glutaminase to pH, ammonium, and citrate. Possible regulatory role of glutaminase in ureagenesis. J Biol Chem 265: 20869–20873
Toto RD (1990) Metabolic acid–base disorders: In: Kokko JP, Tannen RL (eds) Fluids and electrolytes, 2nd edn. WB Saunders, Philadelphia, pp 301–390
Van Slyke DD, Phillips RA, Hamilton PB, Archibald RM, Fructor PH, Hiller A (1943) Glutamine as source material of urinary ammonia. J Biol Chem 150: 481–482
Walser M (1980) Determinants of ureagenesis, with particular reference to renal failure. Kidney Int 17: 709–721
Walser M (1986) Roles of urea production, ammonium excretion, and amino acid oxidation in acid–base balance. Am J Physiol 250: F181–F188

Watford M, Smith EM (1990) Distribution of hepatic glutaminase activity and mRNA in perivenous and periportal rat hepatocytes. Biochem J 267: 265–267
Welbourne TC (1988) Interorgan glutamine fluxes in acid–base disorders. In: Häussinger D (ed) pH homeostasis: mechanisms and control. Academic Press, London, pp 379–401
Wood CM, Perry SF, Wright PA, Bergman HL, Randall DJ (1989) Ammonia and urea dynamics in the Lake Migadi tilapia, a ureotelic teleost fish adapted to an extremely alkaline environment. Respir Physiol 77: 1–20

Chapter 2

Renal Transport of Organic Acids and Bases in Nonmammalian Vertebrates

W.H. Dantzler

Contents

1 Introduction

Organic acids (or anions) and organic bases (or cations) are transported by vertebrate renal tubules. In terms of renal transport, the organic acids are considered to encompass a diverse group of exogenous compounds – for example, para-aminohippurate (PAH) and phenolsulfonphthalein (PSP, phenol red) – that exist as anions at physiological pH and that apparently share a common transport system for net secretion. Because many of these compounds are dyes or are easily measured chemically, they were among the first substances used to demonstrate net tubular secretion, primarily in nonmammalion vertebrates (Dantzler 1987). More recently, studies with nonmammalian vertebrates have

Department of Physiology, College of Medicine, University of Arizona, Tucson, Arizona 85724, USA

Advances in Comparative and Environmental Physiology, Vol. 22

helped to define the mechanism involved in tubular transport of these compounds (Pritchard and Miller, 1991).

The endogenous organic acid urate is often considered a member of this common organic acid group. However, its secretion by the renal tubules of nonmammalian vertebrates may or may not involve the same system that transports other organic acids. Moreover, the urate transport by the renal tubules of some nonmammalian vertebrates may be related to the regulation of acid/base balance. In the current discussion, I consider the transport of both urate and this general group of organic anions.

In a strict sense, the transport of organic acids by the renal tubules includes the transport of amino acids and specific endogenous monocarboxylic acids, such as lactate, or dicarboxylic acid, such as α-ketoglutarate. However, the primary transport systems for these compounds are quite specific and separate from the general secretory transport system for most exogenous organic acids. Moreover, tubular transport of most of these compounds has been poorly studied in nonmammalian vertebrates. Therefore, I do not consider them in this discussion.

Renal tubular transport of organic bases (or cations), both endogenous ones such as choline and N^1-methylnicotinamide (NMN) and exogenous ones such as tetraethylammonium (TEA), has only recently begun to attract significant attention. In part, this lack of attention stems from the potential toxicity of many of these compounds that makes their study in vivo difficult. In fact, most in vivo studies have taken advantage of the renal portal system of nonmammalian vertebrates to bypass the systemic circulation and bring relatively high concentrations of these compounds directly to the renal tubules. However, in vitro techniques have now made study of tubular transport more feasible and the importance of tubular transport for the regulation of the systemic concentrations of potentially toxic compounds has made study more attractive. Moreover, the transport of these compounds may be intimately related to tubular regulation of proton transport and thus acid/base balance. For these reasons, I consider our limited present knowledge of organic cation transport in this discussion.

2 Organic Acids (or Anions)

2.1 Direction and Tubule Sites of Net Transport (General Group of Compounds)

Net transport of organic acids from the peritubular blood to the lumen of the renal tubules (net secretion) occurs in all vertebrate species studied except a species of urodele amphibian, *Necturus maculosus*, and a species of hagfish, *Myxine glutinosa* (Weiner 1973; Dantzler 1992). Such net secretion or net transport in the opposite direction (net absorption) of PAH or iodopyracet can occur in *N. maculosus*, but net absorption is generally observed (Tanner,

1967; Tanner and Kinter 1966). However, no net transepithelial transport of phenol red is found in this species (Tanner et al. 1979), and no net transport of phenol red or, it is assumed, any other organic anion that shares this system is found in hagfish (Fange and Krog 1963; Rall and Burger 1967).

Net transport has been localized to the proximal tubule in *N. maculosus* by micropuncture and in chickens (*Gallus gallus domesticus*) by preliminary studies with isolated, perfused tubules (Dantzler et al. 1991). Perfusions of isolated renal tubules have demonstrated that net secretion against a concentration gradient from peritubular fluid to lumen occurs in all segments of the proximal tubule of flounders (*Pseudopleuronectes americanus*) (Burg and Waller 1969), only in the proximal and intermediate segments of the proximal tubule of frogs (*Rana catesbeiana*) (Irish and Dantzler 1976), and only in the distal portion of the proximal tubule of snakes (*Thamnophis* spp.) (Dantzler 1974a).

2.2 Magnitude, Direction, and Sites of Net Transport (Urate)

Birds, most reptiles (except chelonians from aquatic, semi-aquatic, and mesic habitats), and some amphibians (African tree frogs of the genus *Chiromantis* and South American tree frogs of the genus *Phyllomedusa*) eliminate urate as their major excretory end product of nitrogen metabolism (Dantzler 1978). Urate appears to be freely filtered by the glomeruli in amphibian, reptilian, and, most likely, avian kidneys (Dantzler 1978). However, there may be a small amount of binding to plasma proteins in birds (Dantzler 1978).

In all those uricotelic nonmammalian vertebrates studied, clearance measurements reveal only net tubular secretion (Dantzler 1978). There is no evidence for net tubular absorption. However, this does not preclude the presence of some unidirectional absorption (*vide infra*). Indeed, infusions of the diuretics ethacrynic acid and furosemide into the renal portal system of chickens can increase net urate secretion, suggesting that these agents may be inhibiting some mediated absorptive flux (Shideman et al. 1981). In addition, of the urate excreted, about 3% in fasted chickens, about 20% in normally fed chickens, about 50% in chickens infused with hypoxanthine, and about 17% in normally fed alligators is synthesized by the renal tubule cells (Chin and Quebbemann 1978; Lemieux et al. 1985). No data are available concerning the synthesis of urate by the renal tubule cells of any other nonmammalian species.

The tubule site or sites of net secretion have been explored in European starlings (*Sturnus vulgaris*) by micropuncture and in chickens and garter snakes (*Thamnophis* spp.) by perfusion of isolated renal tubules. Net secretion is found along the proximal tubules of superficial reptilian-type nephrons in starlings (Laverty and Dantzler 1983) and along the proximal tubules of both superficial and deep reptilian-type nephrons in chickens (Dantzler et al. 1991). However, it is not yet known whether net secretion can occur along more distal segments of reptilian-type nephrons or along any segments of mammalian-type nephrons.

Although no net absorption is observed, preliminary perfusions of chicken proximal tubules in vitro do show an apparently passive paracellular absorptive flux (*vide infra*) (Dantzler et al. 1991). This observation appears to conflict with the less direct in vivo observation of possible inhibition of a medicated absorptive flux (*vide supra*) (Shideman et al. 1981) and requires more detailed study.

In vitro perfusions of garter snake nephrons have demonstrated net secretion from bath to lumen against a concentration gradient along the entire proximal tubule but not along the distal tubule (Dantzler 1973, 1976). These studies have also demonstrated a passive, apparently paracellular absorptive flux from lumen to bath (*vide infra*), but have not shown any evidence of net absorption.

2.3 Mechanism of Transport (General Group of Compounds)

2.3.1 Net Tubular Absorption

As pointed out above, net absorption of these compounds against a concentration gradient has only been shown to occur along the proximal tubules of *N. maculosus* (Tanner 1967). Such net absorption can be inhibited by arterial injections of fatty acids, such as octanoate (Tanner 1967), but the mechanism of this effect is unknown and nothing further is known about the actual transport process. In most other vertebrate species there is only a small, apparently passive absorptive flux that has little effect on net transport (Burg and Weller 1969; Dantzler 1974a; Irish and Dantzler 1976). Moreover, in isolated, perfused frog and snake renal tubules, this measured lumen to bath absorptive flux can be accurately predicted from the independently measured luminal and peritubular membrane permeabilities (Fig. 1), suggesting that the entire passive flux crosses the cells (Dantzler 1976; Irish and Dantzler 1976). However, recent preliminary data from studies on isolated perfused chicken proximal tubules suggest that the absorptive flux, although dominated by the much larger secretory flux, involves transport of these anions into the cells against an electrochemical gradient at the luminal membrane (Dantzler et al. 1991)

2.3.2 Net Tubular Secretion

Transepithelial Transport Kinetics. Perfusions of isolated proximal tubules from flounders, frogs, and snakes (Burg and Weller 1969; Dantzler 1974a; Irish and Dantzler 1976) as well as clearance studies in numerous species (Weiner 1973) have demonstrated that net transepithelial secretion saturates. Moreover, saturation occurs at about the same peritubular concentration with isolated, perfused tubules from all three species studied. In frogs, this peritubular concentration at which saturation occurs in vitro actually corresponds closely to the plasma concentration at which saturation occurs in vivo (Schmidt-Nielsen and Forster 1954; Irish and Dantzler 1976). The apparent K_m for net PAH secretion is about the same in isolated, perfused frog tubules as in snake tubules (about 10–15 μM) whereas the apparent J_{max} is about twice as great in frog tubules as in snake tubules (Dantzler 1974a, 1976; Irish and Dantzler 1976). Thus, although the affinity of the transepithelial transport system for organic anions appears to be about equal in the two species, the capacity is much greater in frog than in snake tubules.

Transport at Peritubular Membrane. Recent studies have concentrated on the steps in transepithelial transport occurring at the individual cell membranes, particularly the peritubular membrane. PAH, phenol red, and other members of this group of organic anions are taken up across the peritubular membrane of proximal tubules teased from or in slices from kidneys of fish, amphibians, reptiles, and birds (Dantzler 1969, 1974a; Hoshi and Hayashi 1970; Weiner 1970, 1985; Irish and Dantzler 1976). The steady-state concentration in the cells is always substantially greater than the concentration in the incubation medium. Moreover, during net secretion of PAH or iodopyracet against a transepithelial concentration difference by isolated, perfused proximal tubules of flounders, frogs, snakes, and chickens, the concentration in the tubule cells is greater than that in either the bath or the lumen (Burg and Weller 1969; Dantzler 1974a; Irish and Dantzler 1976, Dantzler et al. 1991). Because these substances exist as anions at physiological pH, because the inside of the cells is negative compared to the bathing medium, and because there is no evidence of significant binding of the organic acids within the cells, these data are compatible with transport into the cells against an electrochemical gradient at the peritubular membrane (Fig. 1).

The way in which such transport against an electrochemical gradient – the primary energy requiring step in the net secretory process – is coupled to an energy source is only beginning to be understood and that understanding is by no means complete. There is no evidence that transport is coupled directly to the hydrolysis of ATP. Therefore, it appears to be a secondary active process, probably involving coupling to gradients for other ions (Fig. 1). In part, this has long appeared to involve gradients for or transport of inorganic cations. First, it has been repeatedly demonstrated that the removal of potassium from the medium bathing teased flounder tubules, snake and chicken kidney slices, and isolated, perfused snake renal tubules inhibits transport of organic anions into the cells against an electrochemical gradient without influencing the passive permeability for these ions (Fig.1; Dantzler 1969, 1974b; Weiner 1973, 1985). However, the mechanism involved in this potassium requirement is unknown and may be related, in part, to the effect of potassium removal on sodium transport (*vide infra*).

Second, sodium has also long been shown to be essential for transport into the cells at the basolateral membrane. For example, double perfusions of frog kidneys by Vogel and his colleagues (Vogel et al. 1965; Vogel and Kroger 1966; Vogel and Stoeckert 1966) suggested that both sodium absorption and the

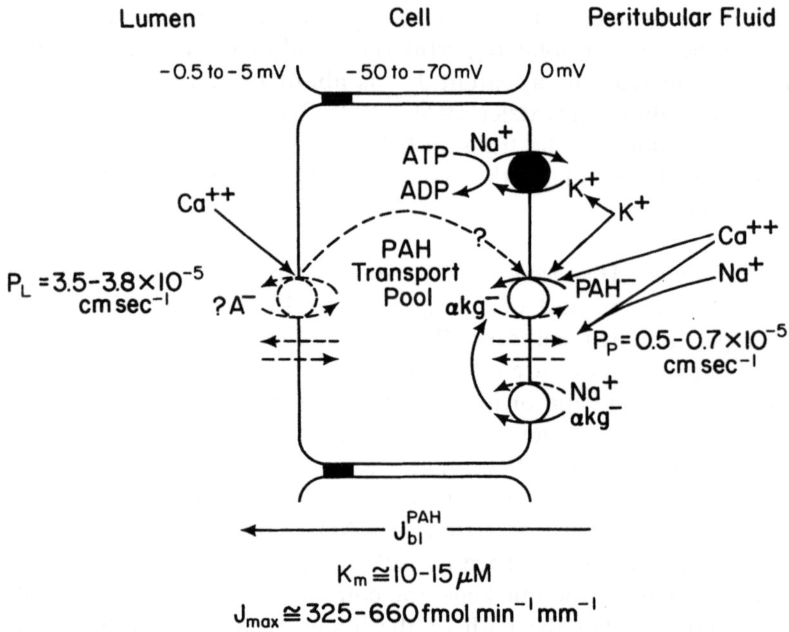

Fig. 1. Model for net tubular secretion of organic anions (with PAH as the prototype anion) based on studies with fish, amphibians, reptiles, and birds. *Solid circles with solid arrows* indicate either primary or secondary active transport against electrochemical gradient. *Broken circles with broken arrows* indicate mediated transport down electrochemical gradient. *Broken arrows alone* across membranes indicate simple passive movement. *Question marks* indicate tentative suggestions. *A^-* indicates anion of unspecified nature. *Arrows* from K^+, Na^+, and Ca^{2+} indicate sites of effects of alterations of concentrations of these inorganic cations. *Broken arrow with question mark* leading from transport step at luminal membrane to transport step at peritubular membrane indicates possible feedback coupling between transport steps. Apparent permeabilities to PAH of luminal (P_L) and peritubular (P_p) membranes for control conditions in snake and frog renal tubules are shown. Values for K_m and J_{max} based on studies with snake and frog renal tubules are also shown at the bottom of the figure. Electrical potentials shown at the top of the figure are in reference to the peritubular fluid at 0 mV

intracellular sodium concentration might be important for net PAH secretion. Early studies with isolated, perfused snake renal tubules also demonstrated that the removal of sodium from the bathing medium reversibly depressed net PAH secretion, this depression apparently involving both inhibition of the concentrative transport step across the peritubular membrane and an increased passive permeability of that membrane (Fig. 1) (Dantzler and Bentley 1976). These effects were not attributable to the increased concentration of calcium in the cytosol that is secondary to the removal of extracellular sodium (Dantzler and Brokl 1984a). Moreover, except for one series of studies on newts (*Triturus pyrrhogaster*) in which measurements of PAH uptake and potential differences across the peritubular membrane suggested that organic anions might enter the cells in a cotransport step with sodium driven by the sodium electrochemical potential (Kikuto et al. 1979; Kikuto and Hoshi 1979), studies on nonmammalian vertebrates provided no evidence for direct cotransport of organic anions with sodium. However, the inhibitory effect of 4-acetamido-4'-isothiocyano-2,2'-disulfonic stilbene (SITS) on the peritubular transport step for PAH in isolated, perfused snake renal tubules suggested that this step might involve countertransport for other anions whose transport might, in turn, be dependent on or coupled to sodium transport (Fig. 1; Dantzler and Bentley 1980). Indeed, studies with mammalian renal basolateral membrane vesicles suggested that PAH transport into the cells against an electrochemical gradient could involve exchange for an intracellular dicarboxylate (e.g. α-ketoglutarate) that is itself transported into the cells by direct coupling to the inwardly directed sodium gradient (Pritchard 1987, 1988; Shimada et al. 1987). The energy for this transport process is derived from the Na-K-ATPase at the basolateral membrane that maintains the inwardly directed sodium gradient. The effect of potassium removal on organic anion uptake and transepithelial transport could then result largely from its inhibition of Na-K-ATPase. Recent studies with southern flounder (*Paralicthys lethostigma*) and garter snake (*Thamnophis* spp.) renal tubules have demonstrated that this process definitely can account, at least in part, for transport into the cells across the peritubular membrane and thus for net transepithelial secretion in nonmammalian vertebrates (Fig. 1; Chatsudthipong and Dantzler 1991; Miller and Pritchard 1991).

The uptake of PAH by teased masses of flounder tubules is stimulated by the presence of micromolar concentrations of glutarate in the bathing medium and this stimulation is prevented by the presence of lithium, an inhibitor of the sodium–dicarboxylate coupled entry step (Miller and Pritchard 1991). Moreover, in these tubules in which the broken tubule ends seal off so that the fluid-filled lumen only communicates with the bath across the epithelium, the concentration of fluorescein, study by epifluorescence, increases in both the cells and the lumen in the presence of glutarate (Miller and Pritchard 1991). This observation suggests that transepithelial secretion as well as uptake across the peritubular membrane is stimulated. Evidence for indirect coupling of PAH uptake and net secretion to a sodium gradient via dicarboxylate/PAH countertransport has been even more rigorously demonstrated with isolated perfused and nonperfused snake renal tubules (Chatsudthipong and Dantzler 1991). Preloading the tubules with α-ketoglutarate markedly stimulated both intracellular accumulation and net PAH secretion in these isolated, perfused tubules (Fig. 2). Stimulation of PAH accumulation and net secretion were inhibited when the sodium-coupled uptake of α-ketoglutarate was prevented by the substitution of N-methyl-D-glucamine for sodium in the bathing medium or the addition of lithium to the bathing medium. The addition of α-ketoglutarate to the bathing medium also stimulated the efflux of PAH across the peritubular membrane of tubules preloaded with PAH, providing additional evidence for PAH/α-ketoglutarate countertransport. Finally, preliminary data indicate that

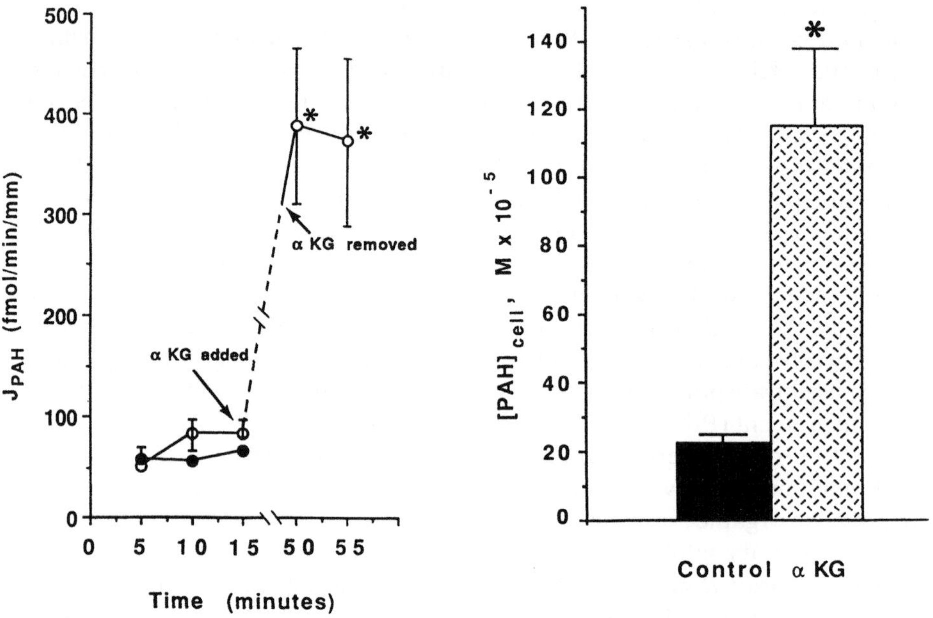

Fig. 2. Comparison of net PAH secretion (J_{PAH}) and cell concentration of PAH between control and α-ketoglutarate (αKG)-loaded tubules from the same kidney studied in parallel. Results are means + SE. Group means that were significantly different from control values are indicated by*. (Chatsudthipong and Dantzler 1991)

preloading reptilian-type chicken proximal tubules with α-ketoglutarate can stimulate PAH uptake in this species (Brokl and Dantzler, unpubl. obser.)

Although PAH/dicarboxylate countertransport can explain part of the sodium dependence of uphill organic anion transport across the peritubular membrane, it does not explain all of it. Nor does it explain all of the uphill transport. In the absence of any dicarboxylate in the bathing medium, there is still substantial transport of PAH into the renal tubule cells against an electrochemical gradient and substantial net transepithelial secretion (Chatsudthipong and Dantzler, 1991; Miller and Pritchard 1991). A portion, but not all, of this transport can be inhibited by SITS, suggesting that countertransport of PAH for α-ketoglutarate (or perhaps some other anion) already present in the cells may account for part of the baseline transport (Chatsudthipong and Dantzler 1991). Moreover, the removal of sodium from the bathing medium can inhibit PAH transport in the absence of any added dicarboxylate in the bathing medium (Dantzler and Bentley 1976). Of course, this effect may result from a failure to reaccumulate dicarboxylate that has entered the bathing medium in exchange for PAH, but this does not appear to account for the whole effect. As noted above, sodium removal can increase the passive permeability of snake peritubular membranes to PAH (Fig. 1; Dantzler and Bentley 1976). In addition, kinetic analyses of phenol red uptake by teased goldfish (*Carassius auratus*) renal tubules and PAH uptake by newt (*T. pyrrhogaster*) kidneys suggest that the sodium concentration may influence the relationship between the organic anion and the carrier at the peritubular membrane (Fig. 1; Hoshi and Hayashi 1970; Kikutu et al. 1979; Kikutu and Hoshi 1979).

Studies involving the effects of a low calcium concentration in the medium and the effects of lanthanum and verapamil on PAH transport by isolated, perfused snake tubules suggest that the entry of calcium into the cells, but probably not the actual cytosolic concentration of calcium, may be important for the transport step into the cells at the peritubular membrane and for maintenance of the normal passive permeability of this membrane to PAH (Fig. 1; Dantzler and Brokl 1984a,b). In addition, the mercaptide-forming reagent r-chloromercuribenzoate inhibits the peritubular transport step in isolated, perfused snake tubules, suggesting that this step also involves sulfhydral groups (Dantzler and Bentley 1983).

Transport at Luminal Membrane. Transport across this membrane, although less well studied than transport across the peritubular membrane, is of considerable interest, because it may play a regulatory role in the secretory process. As noted above, during net secretion of organic anions by isolated, perfused flounder, frog, snake, and chicken tubules the concentration of the transported anion in the cell water is greater than that in the lumen, indicating that transport into the lumen across this membrane is down an electrochemical gradient (Burg and Weller 1969; Dantzler 1974a; Irish and Dantzler 1976; Dantzler et al. 1991). Although this movement could occur by simple diffusion, the initial studies with isolated, perfused frog and snake tubules revealed an apparent permeability of the luminal membrane to PAH 6–7 times that of the peritubular membrane (Fig. 1), thereby suggesting that movement across the luminal membrane was mediated in some fashion (Dantzler 1974a,b; Irish and Dantzler 1976). This concept is supported by additional studies with isolated, perfused snake renal tubules (Dantzler and Bentley 1979, 1980, 1981) and by studies with brush border membrane vesicles (BBMV) from flounder and snake kidneys (Eveloff et al. 1979; Benyajati and Dantzler 1988). With either intact tubules or membrane vesicles, movement of radioactively labeled PAH across this membrane is inhibited by the presence of unlabeled PAH, phenol red, probenecid, and SITS (Dantzler and Bently 1979, 1980; Eveloff et al. 1979; Benyajati and Dantzler 1988). The effect of SITS strongly indicates that the mediated step at this membrane involves some form of anion exchange (Fig. 1). This is also suggested by studies with isolated, perfused snake tubules indicating that, although the movement of PAH from cells to lumen is not dependent on the presence of chloride in the lumen, it is probably dependent on the presence of an equivalent quantity of some other anion to which the membrane is highly permeable (Dantzler and Bentley 1981). However, the possibility of anion exchange at this membrane, although extremely likely, has yet to be examined rigorously.

The studies with flounder BBMV also indicate that movement across this membrane may be electrogenic (Eveloff et al. 1979). If this is so, it would explain an observed anion concentration in the lumen greater than that in the cells during stopped flow, steady-state conditions in isolated flounder tubules (Kinter 1966, 1975; Miller and Pritchard 1991). Even if transport across this membrane during secretion appears to be uphill with respect to concentration under these circumstances, it is actually downhill with respect to electrochemical potential. On the other hand, as noted above, preliminary studies with chicken tubules, indicate that any absorptive flux that occurs involves movement into the cells against a small electrochemical gradient across the luminal membrane (Dantzler et al. 1991). The mechanism involved in this movement and the manner in which it relates to movement in the opposite direction have yet to be explored.

In addition, in intact snake tubules, the movement of PAH across the luminal membrane during net secretion is not influenced by the presence or absence of sodium or potassium in the lumen (Dantzler 1974b; Dantzler and Bentley 1976), but movement into snake BBMV is enchanced by an inwardly directed sodium gradient (Benyajati and Dantzler 1988). These conflicting observations have yet to be resolved. Finally, transport of PAH from cells to lumen, at least in snake tubules, may be dependent on the entry of calcium into the cells (Fig. 1; Dantzler and Brokl 1984a,b) and may involve sulfhydral groups (Dantzler and Bentley 1983).

Coupling Between Transport at the Two Membranes. Although direct coupling between the transport at the two membranes during net anion secretion has yet to be examined rigorously, the transport of PAH into snake tubule cells across the basolateral membrane appears to be partially inhibited when movement from the cells to the lumen is inhibited (Dantzler and Bentley 1979, 1980, 1981; Dantzler and Brokl 1984a,b). This observation suggests that there may be some form of feedback coupling between transport processes at the luminal and peritubular membranes (Fig. 1).

2.4 Mechanism of Transport (Urate)

2.4.1 Net Tubular Secretion

As noted above, only net secretion of urate by the renal tubules of nonmammalian vertebrates has been demonstrated. A few studies with snake and chicken kidney slices have contributed to our understanding of this process, but most insights into the mechanism of secretory transport have come from studies with isolated perfused and nonperfused snake and chicken renal tubules and snake BBMV (Dantzler 1978; Benyajati and Dantzler 1988; Dantzler et al. 1991). Net secretion of urate, like net secretion of other organic anions, apparently involves transport into the cells against an electrochemical gradient at the peritubular side followed by movement from the cells into the lumen down an electrochemical

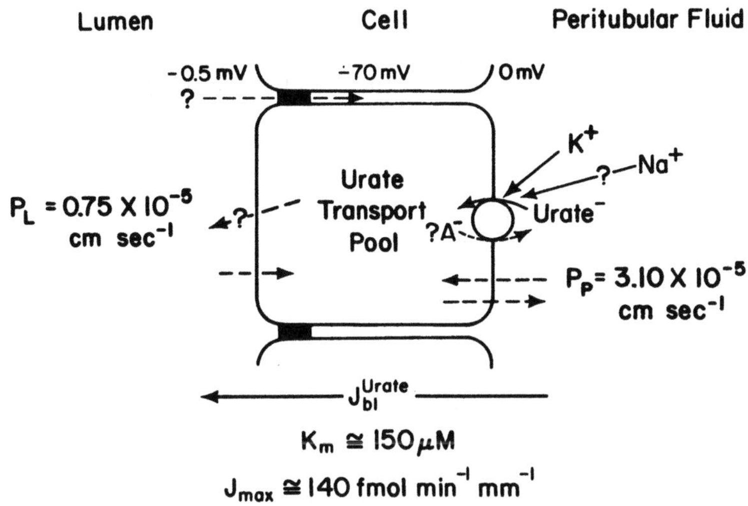

Fig. 3. Model for net tubular secretion of urate based on studies with reptiles and birds. Apparent permeabilities to urate of luminal (P_L) and peritubular (P_P) membranes for control conditions in snake tubules are shown. K_m and J_{max} for net urate secretion based on studies with snake tubules are also shown. Other *symbols* have the same meanings as in Fig. 1

gradient (Fig. 3). However, the process for urate differs from that for PAH and other organic anions in a number of important ways.

Transepithelial Transport Fluxes and Kinetics. The kinetics of net transepithelial secretion of urate by isolated perfused snake renal tubules are quite different from those for the net transepithelial secretion of PAH. The concentration of substrate required to saturate the urate secretory system is much higher than that required to saturate the PAH secretory system (Dantzler 1973, 1974a). The K_m for net urate secretion (about 150 μM), determined from these saturation data, is about 10–15 times that for net PAH secretion, but the V_{max} for net urate secretion is only about one-half that for net PAH secretion (Dantzler 1973, 1974a, 1982; Figs. 1 and 3). Thus, both the affinity and the capacity of the urate transport system are lower than those of the PAH transport system. This K_m for urate secretion is still well below the normal plasma urate concentrations (400–500 μM) in garter snakes, suggesting that the secretory process is normally saturated and that changes in plasma urate concentration can have little effect on net secretion.

Evidence related to the saturation of the transport systems in reptiles and birds supports the concept of separate transport systems for urate and PAH (and other organic anions that share this system). In snakes, net urate secretion by the renal tubules is not inhibited by high concentrations of PAH in vivo or in vitro and net PAH secretion is not inhibited by high concentrations of urate in vitro (Dantzler 1978). In normal chickens, in vivo, plasma PAH levels sufficient to completely saturate the PAH renal tubular secretory system only partially inhibit net urate secretion (Cacini and Quebbemann 1978). Moreover, the administration of adenine at this time further inhibits net urate secretion without affecting net PAH secretion (Cacini and Quebbemann 1978). In a strain of chickens with gout, net urate secretion is impaired (apparently because of a defect in the transport step at the peritubular membrane), but net PAH secretion shows evidence of impairment only when the plasma PAH concentrations are near the level for maximum transport (Austic and Cole 1972; Zmuda and Quebbemann 1975). These data support the concept originally put forward by Weiner and Tinker (1972) that in avian renal tubules there are two pathways for organic anion secretion, both of which transport PAH but only one of which transports urate.

Net urate secretion by isolated, perfused snake tubules, in contrast to net PAH secretion, varies directly with perfusion rate, suggesting that there is a significant backleak from the lumen to the bathing medium (Dantzler 1973). In fact, an absorptive flux from lumen to bath that appears to be passive and that varies with perfusion rate has been demonstrated in these tubules and a similar flux appears to exist in chicken proximal tubules (Dantzler 1973; Dantzler et al. 1991). Of particular interest is the transepithelial permeability of the snake tubules determined from the lumen-to-bath flux is about four times the permeability calculated from the independently measured permeabilities to the luminal and peritubular membranes (Dantzler 1976), suggesting that most of this absorptive flux occurs via a paracellular route (Fig. 3). Although the membrane permeabilities for urate have not been determined in chicken tubules, preliminary data indicating the absence of any detectable urate in the tubule cells during the lumen-to-bath flux (Dantzler et al. 1991) suggest that the absorptive flux is paracellular in this species as well. If, as indicated above, changes in plasma urate concentrations cannot greatly influence net secretion, then the rate of flow along the lumen and the corresponding degree of paracellular diffusion from lumen to bath may be particularly important in determining net secretion and final net excretion (Dantzler 1988).

Transport at Peritubular Membrane. The mechanism involved in transport into the cells against an electrochemical gradient at this membrane is completely unknown. As in the case of the transport of PAH (and other organic anions that share that pathway), there is no evidence of direct coupling to ATP hydrolysis. Moreover, the inhibitory effect of SITS on this urate transport step in isolated, perfused snake renal tubules suggests that the process involves anion exchange (Fig. 3; Mukherjee and Dantzler 1985). However, the possible anion that could be exchanged for urate is much less clear than in the case of the general organic anion secretory system. In the case of snake tubules, there is absolutely no requirement for sodium for urate transport (Randle and Dantzler 1973). This not only indicates that there is no direct or indirect coupling of urate transport to sodium, but also makes countertransport for dicarboxylates unlikely. Indeed, a few preliminary experiments show no involvement of dicarboxylates with urate transport in this species (Dantzler, unpubl. observ.). On the other hand, early studies on the uptake of urate by chicken kidney slices suggest that sodium may play some role in the process in avian kidneys (Dantzler 1969, 1978). This has yet to be examined with more refined techniques. However, the ambiguity concerning the possible requirement for sodium by this transport step is indicated by the arrow with the question mark in Fig. 3.

In contrast to the possible requirement for sodium, the requirement for potassium is unambiguous in all species studied. Removal of potassium from the bathing medium suppresses urate uptake by lizard, snake, and chicken kidney slices (Dantzler 1969, 1978) and reversibly inhibits net transepithelial urate secretion by isolated, perfused snake proximal renal tubules (Randle and Dantzler 1973). During the time of maximum inhibition of urate secretion, the urate concentration in the tubule cells is below that in the bathing medium but greater than that in the lumen, indicating that uptake into the cells against an electrochemical gradient at the peritubular side is completely inhibited and that urate moves from bath to lumen by a passive process (Randle and Dantzler 1973). This potassium requirement is like that of the peritubular transporter for other organic anions, but, in view of the questionable sodium requirement for urate transport, the mechanism by which potassium influences urate transport is even more obscure. Finally, in contrast to the peritubular transport step for other organic anions, neither calcium entry into the cells nor the apparent cytosolic concentration of calcium has any effect on the peritubular transport step for urate (Dantzler and Brokl 1984c).

In the snake tubules, in which the apparent permeabilities of the individual cell membranes to urate have been measured, the permeability of the peritubular membrane is about four times that of the luminal membrane (Fig. 3; Dantzler 1976). This is just the opposite of the situation for other organic anions such as PAH (Fig. 1; Dantzler 1974a,b), and suggests a very inefficient system, if urate transported into the cells is to move into the lumen and not back into the peritubular fluid. Finally, this transport step for urate into the cells appears to be stimulated by the presence of glomerular filtrate (or an equivalent artificial perfusate) moving along the tubule lumen.

Transport at Luminal Membrane. As in the case of the other organic acids, urate that is transported into the cells against an electrochemical gradient at the basolateral membrane apparently moves from the cells into the lumen down an electrochemical gradient in the process of net secretion (Fig. 3). However, as noted above, the apparent permeability of the luminal membrane to urate is much lower than that of the peritubular membrane (Fig. 3; Dantzler 1976). Indeed, this apparent permeability is only about one-fourth of that for PAH (Figs. 1 and 3).

In accord with the low apparent permeability of the luminal membrane to urate is the lack of evidence for any form of mediated urate transport across this membrane. Neither unlabeled urate, probenecid, nor SITS in the lumen of isolated, perfused snake tubules has any effect on the flux of radioactively labeled urate from the cells to the lumen or on the apparent permeability of the luminal membrane to urate (Dantzler and Bentley 1979; Mukherjee and Dantzler 1985). Similarly, studies with BBMV from these same snake tubules reveal no sodium dependence or effects of probenecid or SITS on the urate transport process (Benyajati and Dantzler 1988). Moreover, the preliminary perfusions of isolated chicken tubules indicating that the reabsorptive flux of urate does not even enter the cells (*vide supra*; Dantzler et al. 1991) suggests a lack of mediated transport across the luminal membrane in this species as well. All of these observations are compatible with the movement of urate from the cells to the lumen by simple passive diffusion, although they do not prove it. Moreover, because a simple diffusive flux seems unlikely in view of the substantial net transepithelial secretory flux, the interaction of urate transport with this luminal membrane warrants further detailed study.

Interactions Between Membranes : Physiological Significance and Control. In most nonmammalian vertebrates, nephrons normally filter intermittently. For uricotelic nonmammalian vertebrates such as reptiles, it appears worth considering that the apparent dependence of the peritubular transport step on perfusion, the relatively high apparent passive permeability of the peritubular membrane, the relatively low apparent passive permeability of the luminal membrane, the lack of evidence for mediated transport across the luminal membrane, and the apparent large passive backleak between the cells may function to reduce the accumulation of urate in the lumens or cells of nephrons that are not filtering.

Net secretion of urate by the renal tubules of uricotelic vertebrates, and thus the transport steps at one or both of the cell membranes, may be related in part to the acid/base status of the animals. Although the phosphate buffer system is probably important for the secretion of acid (via the formation of titratable acid) and the generation of new bicarbonate, the uric acid/urate system (pK about 5.7) could be equally or more important in uricotelic nonmammalian vertebrates, because the excretion of urate is much greater than the excretion of phosphate. Moreover, the regulation of net tubular secretion of urate could be related, in part, to the acid/base status of the animals. Although this possibility has not been well studied, it is suggested by one study on water snakes (*Nerodia* spp.; Dantzler 1968). In this study, induction of an acute alkalosis by an intravenous infusion of bicarbonate produced a marked increase in net tubular secretion of urate. This increased secretion was apparently related to the increase in pH of the blood, because the administration of a carbonic anhydrase inhibitor, which made the urine alkaline with minimal effect on the pH of the blood, had no effect on net urate secretion (Dantzler 1968). Because urine is highly alkaline during alkalosis and at least a portion of the increased urate secreted should be excreted as a salt, this process will help eliminate base. The possible relationship of net urate secretion to acid/base status in uricotelic vertebrates certainly merits further study.

3 Organic Bases (or Cations)

3.1 Direction and Tubule Sites of Net Transport

Although studies of renal organic cation transport have been relatively few, in vivo renal portal infusions in chickens have provided evidence for net tubular secretion of a number of endogenous organic cations (e.g. NMN and catecholoamines) and exogenous organic cations (e.g. TEA and morphine) (Rennick 1981). In vivo clearance studies on teleost fish (winter flounder, *P. americanus*) also demonstrate significant net tubular secretion of TEA and NMN (Miller and Holohan 1987). On the other hand, the in vivo studies with chickens indicate that a few endogenous organic cations (most obviously, choline) undergo net tubular absorption at normal or low plasma levels and net tubular secretion at elevated plasma levels (Rennick 1981). Similarly, studies with isolated, perfused snake (*Thamnophis* spp.) renal tubules indicate that either net secretion or net absorption can occur, depending on the cation (Hawk and Dantzler 1984; Dantzler and Brokl 1986).

Studies with masses of teased flounder tubules and with isolated nonperfused and perfused snake tubules indicate that net transepithelial organic cation transport occurs throughout the proximal tubule (Dantzler, unpubl. observ.; Hawk and Dantzler 1984; Dantzler and Brokl 1986; Miller and Holohan 1987). It seems most likely that the proximal tubule is also the site of net transepithelial transport in birds, although this has not been studied directly.

3.2 Mechanism of Transport

3.2.1 Net Tubular Absorption

Transepithelial Fluxes and Kinetics. The only studies on net tubular absorption at the cellular or membrane level involve the transport of NMN by isolated perfused and nonperfused snake renal tubules and brush border membrane vesicles from these tubules (Dantzler and Brokl 1986, 1987; Dantzler et al. 1991). The transport of NMN appears to be completely separate from the transport of TEA, which undergoes net secretion (*vide infra*). Net absorption of NMN results from a small difference between a relatively large saturable unidirectional flux from lumen to bath and an almost equally large saturable unidirectional flux from bath to lumen (Fig. 4; Dantzler and Brokl 1986). Direct measurements of net transport indicate that a small reabsorptive flux always occurs (Dantzler and Brokl 1986). Indeed, this net flux is essentially equal to the difference between the mean values for the two unidirectional fluxes when the net and unidirectional fluxes are measured at the same NMN concentrations (Dantzler and Brokl 1986). The K_m for the unidirectional absorptive flux (about 60 μM) is slightly lower than that for the unidirectional secretory flux (about 85μM), indicating a greater affinity for transepithelial absorption. However,

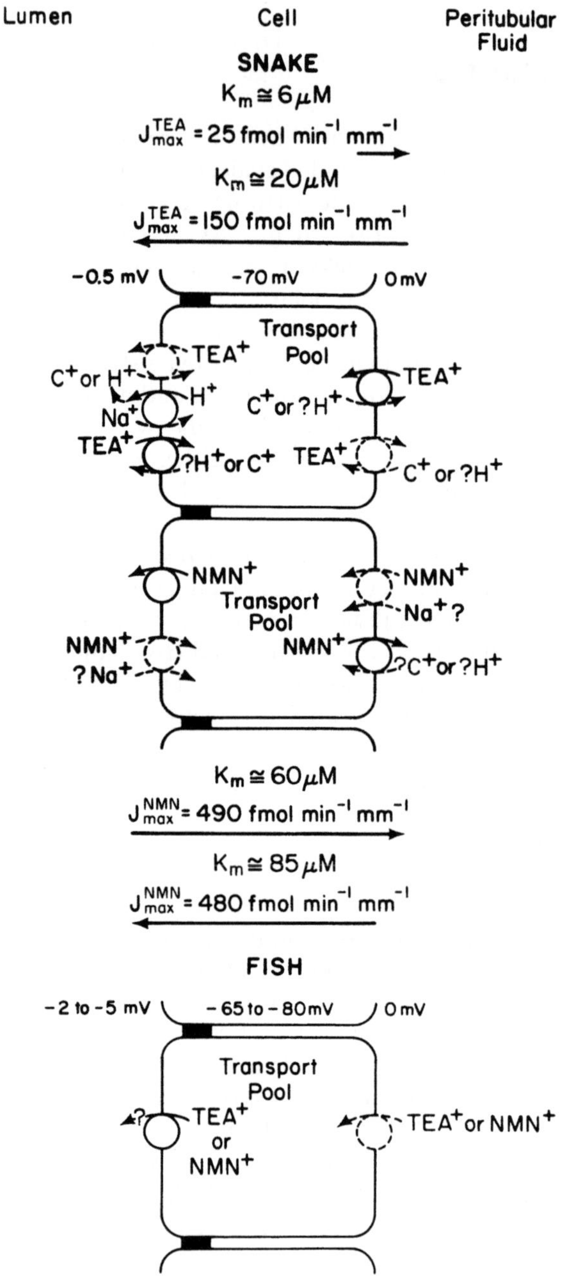

Fig. 4. Models for net tubular transport of tetraethylammonium (TEA^+) and N^1-methylnicotinamide (NMN^+). The *upper part* of the figure shows models based on studies with snake proximal renal tubules or membrane vesicles. These models indicate differences in directions of net transport and mechanisms of net transport for TEA^+ and NMN^+ in snake tubules. C^+ indicates unspecified organic cation. Other *symbols* are the same as in Fig. 1. K_m and J_{max} values for each unidirectional flux are shown J_{max} values are also illustrated by the *length* of the *arrows*. The *bottom part* of the figure shows a model for common net secretory transport of TEA^+ and NMN^+ based on studies with fish renal tubules. All *symbols* have the same meanings as in Fig. 1

neither of these affinities for NMN are as great as those for TEA (*vide infra*). Moreover, even millimolar concentrations of TEA have absolutely no effect on the unidirectional fluxes of NMN (Dantzler and Brokl 1987).

Transport at Peritubular Membrane. At steady-state during the unidirectional absorptive and secretory fluxes and during the net absorptive flux, the concentration of NMN within the cells is below that expected at electrochemical equilibrium and falls even lower when the entry step from either side is blocked (Dantzler and Brokl 1986). This indicates that NMN moves into the cells down an electrochemical gradient and is transported out of the cells against an electrochemical gradient (Fig. 4). At the peritubular membrane, movement into the cells is carrier-mediated (because inhibitable by mepiperphenidol and several structural analogues of NMN) and sodium dependent (Fig. 4; Dantzler and Brokl, 1986). The inhibitor studies with structural analogues indicate that a ring configuration containing the quaternary ammonium is essential for transport (Dantzler and Brokl 1986).

Studies of the efflux of radioactively labeled NMN across the peritubular membrane of tubules with oil-filled lumens indicate that transport out of the cells at this membrane (apparently the dominant transport step in the net absorptive process) can involve countertransport of organic cations and possibly activation by protons or proton-NMN exchange (Fig. 4; Dantzler and Brokl 1987). However, these observations do not explain how such NMN transport against an electrochemical gradient actually occurs physiologically. Any proton gradient, regardless of acid/base status, that exists across this membrane should be in the opposite direction from that needed to drive NMN exit from the cells. Moreover, the endogenous organic cation choline does not share this transport process (Dantzler and Brokl 1987), and it is difficult to conceive of another organic cation that could be exchanged for NMN in a physiological transport process.

Transport at Luminal Membrane. As noted above, the low steady-state NMN concentration in the cells even during the bath-to-lumen flux indicates that NMN enters the cells down an electrochemical gradient and is transported out against an electrochemical gradient at the luminal membrane (Fig. 4; Dantzler and Brokl 1986). Like the entry step at the peritubular membrane, the entry step down an electrochemical gradient at the luminal membrane is carrier-mediated (because inhibitable by mepiperphenidol and several structural analogues of NMN) and sodium-dependent (Fig. 4; Dantzler and Brokl 1986). The inhibitory effects of the structural analogues at this membrane again indicate that transport requires a ring configuration containing the quaternary ammonium (Dantzler and Brokl 1986). Studies of the efflux of radiolabeled NMN across the luminal membrane of tubules covered with oil give no indication that NMN transport out of the cells across this membrane can involve countertransport for either organic cations or protons (Fig. 4; Dantzler and Brokl 1987). Preliminary studies with brush border membrane vesicles from these tubules also fail to provide definitive evidence of a countertransport step for NMN at this membrane (Dantzler et al. 1991). Although this failure may have resulted from the lability of the transporter and its loss during the preparation of the vesicles, the absence of any evidence for countertransport or even self-exchange in intact tubules is much more difficult to explain. Thus, the mechanism involved in the apparent transport step for NMN out of the cells against an electrochemical gradient at the luminal membrane remains obscure.

3.2.2 Net Tubular Secretion

Transepithelial Fluxes and Kinetics. Studies on the possible mechanisms involved in net tubular secretion at the cellular and membrane levels come from work with fish (winter flounder, *P. americanus*; southern flounder, *P. lethostigma*, and killifish, *Fundulus heteroclitus*) and snake (*Thamnophis* spp.) renal tubules and brush border membrane vesicles (Hawk and Dantzler 1984; Miller and Holohan 1987; Dantzler and Brokl 1988; Smith et al. 1988). In fish the same mechanism apparently functions for both prototype organic cations, TEA and NMN, whereas in snakes the net tubular secretory system involves only TEA (Fig. 4).

Transepithelial flux data are only available for isolated, perfused snake renal tubules (Hawk and Dantzler 1984). In these tubules, net secretion of TEA results from a substantial difference between a small, saturable unidirectional flux from lumen to bath and a much larger, saturable flux from bath to lumen (Fig. 4). Kinetic analyses of these unidirectional fluxes demonstrate that the K_m for the bath-to-lumen flux (about 20 μM) is three times the K_m for the lumen-to-bath flux (about 6 μM), indicating a greater affinity for TEA in the lumen-to-bath direction than in the bath-to-lumen direction. However, the affinities for TEA transport in both directions are much higher than those for NMN (*vide supra*). Moreover, the J_{max} for the bath-to-lumen flux of TEA (about 150 fmol min^{-1} mm^{-1}) is about six times the J_{max} for the lumen-to-bath flux (about 25 fmol min^{-1} mm^{-1}), resulting in net secretion (Fig. 4).

Transport at Peritubular Membrane. In nonperfused tubules teased from teleost kidneys, TEA and NMN enter the cells across the peritubular membrane by a saturable process that can be inhibited by other organic cations but not organic anions (Miller and Holohan 1987). However, the affinity of the transporter for TEA appears to be greater than that for NMN (Miller and Holohan 1987). Hyperpolarization of the potential across the peritubular membrane increases carrier-mediated TEA transport into the cells whereas depolarization decreases it (Smith et al. 1988). Moreover, the addition of TEA or mepiperphenidol to the medium bathing the tubules depolarizes the peritubular membrane potential (Smith et al. 1988). These data are compatible with the concept that carrier-mediated uptake of organic cations at the peritubular membrane in these and, perhaps, all teleost species is an eletrogenic process driven by the membrane potential (Fig. 4; Smith et al. 1988). In other words, the organic cations enter the cells across the peritubular membrane down an electrochemical gradient during the process of net secretion. This is the same pattern described above for the movement of NMN into snake tubule cells across the peritubular membrane.

During the bath-to-lumen flux of TEA in isolated, perfused snake renal tubules, the steady-state concentration of TEA in the cells is far above that expected for passive distribution at electrochemical equilibrium (Hawk and Dantzler 1984). This does not appear to be the result of intracellular binding or trapping. At least, the accumulation above electrochemical equilibrium is eliminated by cyanide (Hawk and Dantzler 1984). If this concentration does reflect the activity of TEA in the cell water, then transport into the cells across the peritubular membrane during the lumen-to-bath flux must be against an electrochemical gradient (Fig. 4). This is just the opposite of the situation for NMN in these tubules or for both TEA and NMN in teleost tubules (*vide supra*). Studies of TEA efflux across the peritubular membrane similar to those described for NMN above indicate that the transport step can involve countertransport for organic cations (including choline) and may involve proton-TEA exchange or proton activation (Fig. 4; Dantzler and Brokl 1988). Under physiological circumstances, countertransport for choline that is being absorbed may be important in the movement of TEA into the cells.

Transport at Luminal Membrane. In teleost renal tubules, TEA enters the cells down an electrochemical gradient across the peritubular membrane (*vide supra*) and the steady-state concentration is maintained below that expected at electrochemical equilibrium (Smith et al. 1988). These findings indicate that, during net secretion, TEA must be transported out of the cells against an electrochemical gradient by some energy-requiring process at the luminal membrane (Fig. 4). Indeed, preliminary work on brush border membrane vesicles from flounder renal tubules suggests that there is some form of mediated transport step at the luminal membrane (Miller and Holohan 1987). Unfortunately, no further information is available on the nature of this process from studies on either intact tubules or isolated cell membranes.

If the activity of TEA in snake tubule cells during net bath-to-lumen transport is indeed above that expected at electrochemical equilibrium (*vide supra*), then, in contrast to net secretion in teleost tubules, TEA moves from cells to lumen down an electrochemical gradient (Fig. 4). However, this could still be a carrier-mediated process (Fig. 4). Indeed, the intial studies with isolated, perfused snake tubules indicated that during the small absorptive (lumen-to-bath) fluxes, the TEA concentration is above that expected at electrochemical equilibrium (Hawk and Dantzler 1984). Again, if this is a true reflection of the free activity in the cells, it indicates that transport into the cells at this membrane during the lumen-to-bath flux is against an electrochemical gradient. Moreover, these same initial studies of TEA transport with intact snake tubules revealed that transport across the luminal membrane is at least partly sodium dependent.

The transport step for TEA at the luminal membrane of snake tubules has been explored in studies of the efflux of radioactively labeled TEA across this membrane in isolated, perfused tubules and in studies of TEA transport by luminal (brush border) membrane vesicles from these tubules (Dantzler and Brokl; 1988; Dantzler et al. 1991). Efflux from intact tubules is stimulated by inwardly directed gradients of protons, unlabeled TEA, or other organic cations (Dantzler and Brokl 1988). These data indicate that transport across this membrane can involve countertransport of organic cations or proton-TEA exchange (Fig. 4). These concepts are strongly supported by the studies with isolated vesicles from these membranes (Dantzler et al. 1991). In these studies, an outwardly directed proton gradient stimulated uptake of TEA into the BBMV and supported concentrative accumulation. Uptake of radioactively labeled TEA was also stimulated by outwardly directed gradients of unlabeled TEA and choline. Moreover, TEA uptake was not influenced by changes in the electrical potential differences across the membranes. These data, in agreement with those from intact tubules, indicate that TEA transport across the luminal membrane involves electroneutral countertransport for protons or organic cations. Under physiological conditions, countertransport of TEA for protons appears most likely. In this regard, data from the vesicle studies clearly indicate that the sodium dependency observed in intact tubules is related to establishment of a proton gradient via the sodium-proton exchanger also located in this membrane (Dantzler et al. 1991). Studies with intact tubules (Hawk and Dantzler 1984) suggested that net secretion of TEA involves transport into the cells against an electrochemical gradient at the peritubular side and movement from the cells down an electrochemical gradient at the luminal side. If this is true, then coupling the efflux across the luminal membrane to proton countertransport would tap the energy in the inwardly directed proton gradient (generated, at least in part, by the sodium-proton exchanger also located in this membrane) and make the transport of the organic cation into the lumen an electroneutral process. Both of these effects would provide thermodynamic enhancement to the secretion of TEA and make the process more efficient. If the measured intracellular TEA concentrations during the transport process are much higher than the true intracellular activities and transport into the lumen is against an electrochemical gradient, then such transport could actually be driven by the inwardly directed proton gradient. This is the process that appears most likely for mammalian renal tubules (Holohan and Ross 1981). In any case, generation of an appropriate proton gradient, which may in turn relate to acid/base status, is of great importance for transport across this membrane.

References

Austic RE, Cole RK (1972) Impaired renal clearance of uric acid in chickens having hyperuricemia and articular gout. Am J Physiol 223: 525–530

Benyajati S, Dantzler WH (1988) Enzymatic and transport characteristics of isolated snake renal brush-border membranes. Am J Physiol 255: R52–R60

Burg MB, Weller PF (1969) Iodopyracet transport by isolated perfused flounder proximal renal tubules. Am J Physiol 217: 1053–1056
Cacini W, Quebbemann AJ (1978) The metabolism and active excretion of hypoxanthine by the renal tubules in the chicken. J Pharmacol Exp Ther 207: 574–583
Chatsudthipong V, Dantzler WH (1991) PAH/a-KG countertransport stimulates PAH uptake and net secretion in isolated snake renal tubules. Am J Physiol 261: F858–F867
Chin TY, Quebbemann AJ (1978) Quantification of renal uric acid synthesis in the chicken. Am J Physiol 234: F446–451
Dantzler WH (1968) Effect of metabolic alkalosis and acidosis on tubular urate secretion in water snakes. Am J Physiol 215: 747–751
Dantzler WH (1969) Effect of K, Na, and ouabain on urate and PAH uptake by snake and chicken kidney slices. Am J Physiol 217: 1510–1519
Dantzler WH (1973) Characteristics of urate transport by isolated perfused snake proximal renal tubules. Am J Physiol 224: 445–453
Dantzler WH (1974a) PAH transport by snake proximal renal tubules: differences from urate transport. Am J Physiol 226: 634–641
Dantzler WH (1974b) K^+ effects of PAH transport and membrane permeabilities in isolated snake renal tubules. Am J Physiol 227: 1361–1370
Dantzler WH (1976) Comparison of uric acid and PAH transport by isolated, perfused snake renal tubules. In: Silbernagl S, Lang F, Greger R (eds) Amino acid and uric acid transport. Thieme, Stuttgart, pp 169–180
Dantzler WH (1978) Urate excretion in nonmammalian vertebrates. In: Kelley WN, Weiner IM (eds) Handbook of experimental pharmacology, vol 51. Uric acid, Springer, Berlin Heidelberg New York, pp 185–210
Dantzler WH (1982) Studies on nonmammalian nephrons. Kidney Int 22: 560–570
Dantzler WH (1987) Comparative renal physiology. In: Gottschalk CW, Berliner RW, Giebisch G (eds) Renal physiology: people and ideas. Am Physiol Soc, Bethesda, pp 437–481
Dantzler WH (1988) Comparative physiology of the vertebrate kidney. Springer, Berlin Heidelberg New York
Dantzler WH (1992) Comparative aspects of renal function. In: Seldin DW, Giebisch G (eds) The Kidney: physiology and pathophysiology, 2nd ed, vol 1. Raven Press, New York
Dantzler WH, Bentley SK (1976) Low Na^+ effects on PAH transport and permeabilities in isolated snake renal tubules. Am J Physiol 230: 256–262
Dantzler WH, Bentley SK (1979) Effects of inhibitors in lumen on PAH and urate transport by isolated renal tubules. Am J Physiol 236: F379–F386
Dantzler WH, Bentley SK (1980) Bath and lumen effects of SITS on PAH transport by isolated perfused renal tubules. Am J Physiol 238: F16–F25
Dantzler WH, Bentley SK (1981) Effects of chloride substitutes on PAH transport by isolated perfused renal tubules. Am J Physiol 241: F632–F644
Dantzler WH, Bentley SK (1983) Effects of sulfhydral reagent, p-chloromercuribenzoate, on p-aminohippurate transport by isolated, perfused snake renal tubules. Renal Physiol 6: 209–217
Dantzler WH, Brokl OH (1984a) Effects of low [Ca^{2+}] and La^{3+} on PAH transport by isolated perfused renal tubules. Am J Physiol 236: F175–F187
Dantzler WH, Brokl OH (1984b) Verapamil and quinidine effects on PAH transport by isolated perfused renal tubules. Am J Physiol 246: F188–F200
Dantzler WH, Brokl OH (1984c) Lack of effect of low [Ca^{2+}], La^{3+}, pyrazinoate on urate transport by isolated, perfused snake renal tubules. Pflügers Arch Eur J Physiol 401: 262–265
Dantzler WH, Brokl OH (1986) N^1-methylnicotinamide transport by isolated perfused snake proximal renal tubules. Am J Physiol 250: F407–F418
Dantzler WH, Brokl OH (1987) NMN transport by snake renal tubules: choline effects, countertransport, H^+-NMN exchange. Am J Physiol 255: F656–F663
Dantzler WH, Brokl OH (1988) TEA transport by snake renal tubules: choline effects, countertransport, H^+-TEA exchange. Am J Physiol 255: F167–F176

Dantzler WH, Braun EJ, Brokl OH (1991) Urate and PAH transport by isolated perfused avian proximal renal tubules. J Am Soc Nephrol 2: 759 (Abstr)
Eveloff J, Kinne R, Kinter WB (1979) p-Aminohippuric acidtransport into brush border vesicles isolated from flounder kidneys. Am J Physiol 237: F291–F298
Fange R, Krog J (1963) Inability of the kidney of the hagfish to secrete phenol red. Nature 199: 713
Hawk CT, Dantzler WH (1984) Tetraethylammonium transport by isolated perfused snake renal tubules. Am J Physiol 246: F476–F487
Holohan PD, Ross CR (1981) Mechanisms of organic cation transport in kidney plasma membrane vesicles. 2D pH studies. J Pharmacol Exp Ther 216: 294–298
Hoshi T, Hayashi H (1970) Role of sodium ions in phenol red transport by renal tubules of the goldfish. Jpn J Physiol 20: 683–696
Irish JM III, Dantzler WH (1976) PAH transport and fluid absorption by isolated perfused frog proximal renal tubules. Am J Physiol 230: 1509–1516
Kikuta Y, Hoshi T (1979) Role of sodium ions in p-aminophippurate transport by newt kidney. Biochim Biophys Acta 553: 404–416
Kikuta Y, Hayashi H, Saito Y (1979) Effect of changes in sodium electrochemical potential gradient on p-aminohippurate transport in newt kidney. Biochim Biophys Acta 556: 354–365
Kinter WB (1966) Chlorophenol red influx and efflux: microspectrophotometry of flounder kidney tubules. Am J Physiol 211: 1152–1164
Kinter WB (1975) Structure and function of renal tubules isolated from fish kidneys. Fortschr Zool 23: 233–231
Laverty G, Dantzler WH (1983) Micropuncture study of urate transport by superficial nephrons in avian (*Sturnus vulgaris*) kidney. Pflügers Arch Eur J Physiol 397: 232–236
Lemieux G, Berkofsky J, Quenneville A, Lemieux C (1985) Net tubular secretion of bicarbonate by the alligator kidney, antimammalian response to acetazolamide. Kidney Int 28: 760–766
Miller DS, Holohan PD (1987) Organic cation secretion in flounder renal tissue. Am J Physiol 253: R861–R867
Miller DS, Pritchard JB (1991) Indirect coupling of organic anion secretion to sodium in teleost (*Paralicthys lethostigma*) renal tubules. Am J Physiol 261: R1470–R1477
Mukherjee SK, Dantzler WH (1985) Effects of SITS on urate transport by isolated, perfused snake renal tubules. Pflügers Arch Am J Physiol 403: 35–40
Pritchard JB (1987) Luminal and peritubular steps in renal transport of p-aminohippurate. Biochim Biophys Acta 906: 295–308
Pritchard JB (1988) Coupled transport of p-aminohippurate by rat kidney basolateral membrane vesicles. Am J Physiol 255: F597–F604
Pritchard JB, Millar DS (1991) Comparative insights into the mechanisms of renal organic anion and cation secretion. Am J Physiol 261: R1329–R1340
Rall DP, Burger JW (1967) Some aspects of hepatic and renal excretion in *Myxine*. Am J Physiol 212: 354–356
Randle HW, Dantzler WH (1973) Effects of K^+ and Na^+ on urate transport by isolated perfused snake renal tubules. Am J Physiol 225: 1206–1214
Rennick BR (1981) Renal tubule transport of organic cations. Am J Physiol 240: F83–F89
Schmidt-Nielsen B, Forster RP (1954) The effect of dehydration and low temperature on renal function in the bullfrog. J Cell Comp Physiol 44: 233–246
Shideman JR, Zmuda MJ, Quebbemann AJ (1981) The acute effects of furosemide, ethacrynic acid and chlorothiazide on the renal tubular handling of uric acid in chicken. J Pharmacol Exp Ther 216: 441–446
Shimada H, Moewes B, Burckhardt G (1987) Indirect coupling to Na^+ of p-aminohippuric acid uptake into rate renal basolateral membrane vesicles. Am J Physiol 253: F795–F801
Smith PM, Pritchard JB, Miller DS (1988) Membrane potential drives organic cation transport into teleost renal proximal tubules. Am J Physiol 255: R492–R499
Tanner GA (1967) Micropuncture study of PAH and diodrast transport in *Necturus* kidney. Am J Physiol 212: 1341–1346
Tanner GA, Kinter WB (1966) Reabsorption and secretion of p-aminohippurate and diodrast in *Necturus* kidney. Am J Physiol 210: 221–231

Tanner GA, Carmines PK, Kinter WB (1979) Excretion of phenol red by the *Necturus kidney.* Am J Physiol 236: F442–F447

Vogel G, Kroger W (1966) Die Bedeutung des Transporters, der Konzentration and der Darbietungsrichtung von Na^+ für den tubulären Glucose- und PAH-Transport. Pflügers Arch Eur J Physiol 288: 342–358

Vogel G, Stoeckert I (1966) Die Bedeutung des Anions für den renal tubulären Transport von Na^+ und die Transporte von Glucose und PAH. Plügers Arch Eur J Physiol 292: 309–315

Vogel G, Lauterbach F, Kroger W (1965) Die Bedeutung des Natriums für die renalen Transporte von Glucose und para-Aminohippursäure. Pflügers Arch Eur J Physiol 283: 151–159

Weiner IM (1973) Transport of weak acids and bases. In: Orloff J, Berliner RW (eds) Handbook of physiology. Renal physiology, Sect 8. Am Physiol Sci, Washington, DC, pp 521–554

Weiner IM (1985) Organic acids and bases and uric acid. In: Seldin DW, Giebisch G (eds) The kidney: physiology and pathophysiology, vol 2. Raven Press, New York, pp 1703–1724

Weiner IM, Tinker JP (1972) Pharmacology of pyrazinamide: metabolic and renal function studies related to the mechanism of drug-induced urate retention. J Pharmacol Exp Ther 180: 411–434

Zmuda MJ, Quebbemann AJ (1975) Localization of renal tubular uric acid transport defect in gouty chickens. Am J Physiol 229: 820–825

Chapter 3

pH Homeostasis in Terrestrial Vertebrates: A Comparison of Traditional and New Concepts

H.O. Pörtner

Contents

1 Introduction

The view of how metabolism influences acid–base status has been changed and questioned several times during the past few years. For aerobic steady state metabolism a major stimulus arose from a paper by Atkinson and Camien (1982), which by the application of stoichiometric principles led to a reconsideration of the importance of protein/amino acid catabolism and of urea synthesis in acid–base regulation. At the same time using stoichiometric principles also led to a quantification of the influences of non-steady state (anaerobic, exercise) metabolism on the acid–base status (Pörtner et al. 1984; Pörtner 1987, 1989). All of these considerations provide a basis for a general and quantitative understanding of the relationships between metabolic and acid–base regulation.

Much experimental work has been stimulated by the view that urea synthesis contributes to acid–base regulation (reviewed by Atkinson and Bourke, this vol.) and much of this evidence is in favour of a central role of the urea cycle in terrestrial vertebrate acid–base regulation. Since traditional concepts of

Alfred-Wegener-Institut für Polar- und Meeresforschung, Abt. BiologieI/Ökophysiologie, D-27568 Bremerhaven, FRG

Advances in Comparative and Environmental Physiology, Vol. 22

acid–base regulation in vertebrates did not recognize that production of both base and ammonium (not just ammonia) occurs during protein and amino acid catabolism, it was not anticipated that a fine control of urea synthesis could occur via acid–base parameters (pH, HCO_3^-, CO_2). Thus, urea synthesis may also be a metabolic means to regulate the systemic acid–base status.

The present chapter is not intended to question this modified view of the role of urea synthesis but rather to examine whether the traditional concepts of acid–base regulation can be modified accordingly. A second major focus of this contribution is to ask whether these modifications must lead to any revision of the established methods of acid–base physiology. Such a revision, focussing on the physico-chemical characteristics of the metabolites involved in acid–base regulation (especially of ammonium, NH_4^+), has been requested by Atkinson and his colleagues (see Atkinson and Camien 1982; Atkinson 1992; Atkinson and Bourke, this Vol.). If such a revision is accepted traditional and new concepts would yield different results for the analysis of net proton equivalent ion exchange between the terrestrial animal and its environment (i.e. net acid excretion). However, if a different perspective is adopted taking the generation and fate of all ammonium, acid and base in metabolism and their transfer across epithelia into account (Pörtner 1989), both the traditional and the new approach will lead to identical conclusions in quantitative terms. The methodological modifications proposed by Atkinson and Bourke would lead to fatal errors in the analysis of metabolic influences on acid–base status and, thus, net acid excretion. Therefore, this brief chapter is meant to address these discrepancies and to present a counter-argument to the conclusions drawn by Atkinson and Bourke.

2 Nutrients and Acid–Base Status

The composition of the diet determines the net amount of protons formed or consumed in steady state metabolism. The major components are usually fat, carbohydrates, protein and, to a minor extent, nucleic acid, nucleotides, phospholipids, and other organic phosphates, and possibly free organic acids. For an estimate of the quantitative influence on acid–base regulation it must be considered that metabolism of these substrates is complex and occurs in different body compartments. Acid–base relevant ion (and gas) exchange between the body and the environment occurs across various epithelia: For a clear understanding, Pörtner (1989) proposed that the metabolic processes occurring in the lumen of the digestive system can be considered as taking place in the environment of the animals. Substances will then become effective as they cross the gut epithelium and enter the body. However, net ion exchange processes by which the organism influences or regulates the acid–base status in the digestive system have to be taken into account in an overall balance of acid–base events.

Table 1. Proton balance of aerobic catabolism in the post-absorptive organism. (Modified after Pörtner 1989)

Substrate	Chemical reaction	Net stoichiometric turnover (H^+, NH_4^+)
Carbohydrates	$C_6H_{12}O_6 + 6O_2 \rightarrow 6CO_2 + 6H_2O$	(0,0)
Fatty acids	$CH_3(CH_2)_{n-2}COO^- + [(3n-2]/2]O_2$	
	$+ H^+ \rightarrow nCO_2 + nH_2O$	(–1, 0)
Fat[a]	$R\text{–}COO\text{–}R\text{–}H_2O \rightarrow R\text{–}COO^- + ROH + H^+$	(+1,0)
Glycerol	$C_3H_8O_3 + 3.5O_2\text{–}3CO_2 + 4H_2O$	(0,0)
Ketobodies	β-hydroxybutyrate$^-$ + H^+	
	$+ 4.5O_2 \rightarrow 4CO_2 + 4H_2O$	(–1,0)
Ketogenesis	Palmitate$^-$ + H^+	
	$+ 5O_2 \rightarrow$ 4 β-hydroxybutyrate$^-$ + $4H^+$	(+3,0)
Amino acids:		
General	$R\text{–}CH(NH_3^+)\text{–}COO^- + O_2 + H^+$	
	$\rightarrow R\text{–}H + 2CO_2 + NH_4^+$	(–1,+1)
Neutral	Alanine + $3O_2 + H^+ \rightarrow 3CO_2 + 2H_2O + NH_4^+$	
Acidifying	Methionine + $7.5O_2 + H^+$	
	$\rightarrow 5\,CO_2 + 3H_2O + NH_4^+ + SO_4^{2-} + 2H^+$	(–1, + 1)
Alkalizing	Glutamate – + $6O^2 + 2H^+$	
	$\rightarrow 5\,CO_2 + 3H_2O + NH_4^+$	(–2, +1)
Organic phosphates:		
Monoesters	$R\text{–}PO_4^{2-} + H_2O \rightarrow R\text{–}OH + HPO_4^{2-}$	
Phosphagens	$R\text{–}NH\text{–}PO_3^{2-} + H_2O \rightarrow R\text{–}NH_2 + HPO_4^{2-}$	
Oligoesters[b]	$R_1\text{–}PO_4\text{–}R_2^- + 2H_2O \rightarrow R_1OH + R_2OH + HPO_4^{2-} + H^+$	
All[c]	$HPO_4^{2-} + xH^+ \rightarrow xH_2PO_4^- + (1 - X)\,HPO_4^{2-}$	

Note that phosphate monoesters (and phosphagens, i.e. guanidino-phosphates like phosphocreatine) cause net H^+ consumption, whereas phosphate oligoesters like ATP cause H^+ release.
[a] Fatty acid-glycerol erster.
[b] E.g. R_1, $R_2 = -PO_3^{2-}$
[c] pK around 6.8 (cf. Pörtner 1990 for exact values), $x = 1/(10^{pH-pK} + 1)$.

These processes and the metabolism of absorbed nutrients inside the body determine the net contribution of the kidney to acid–base regulation.

2.1 Sources of Acid or Base in Metabolism

Table 1 summarizes the proton balance of the oxidative catabolism of nutrients and body stores in the resting, fully aerobic post-absorptive organism and reveals to what extent acid–base regulation may be influenced by the different classes of substances. Homeostatic maintenance of body stores (e.g. gluconeogenesis in the liver, cf. Jungas et al. 1992) is neglected in this approach since by catabolism of the synthesized substrate any net effect of this process on systemic acid–base status will be eliminated.

The equations focus on the actual dissociation equilibria of substrates and products and, therefore, the actual generation or consumption of molecules directly and indirectly relevant for the acid–base status (in this case: H^+, NH_4^+). Generally, the oxidative catabolism of amino and other organic acids involves the oxidative decarboxylation of carboxylic acid anions. Net proton consumption during this process will lead, via a contribution of carbonic anhydrase, to the net formation of bicarbonate, which equals the amount of protons removed during oxidative decarboxylation (Pörtner 1987, 1989). Thus, in a more simplified view it may be stated that protein and amino acid catabolism yields ammonium and bicarbonate ions (Atkinson and Camien 1982; protein hydrolysis by itself has no effect on the acid–base status, it can be seen to replenish the pool of free amino acids).

Since most diets include a mixture of proteins and/or amino acids, ammonium ions and bicarbonate may be seen to form the bulk of acid–base relevant substances. However, only "neutral" amino acids like alanine would yield an equimolar quantity of ammonium and bicarbonate. This equimolarity originates from the fundamental chemical structure of any amino acid which is $R{-}CH(NH_3^+){-}COO^-$ at physiological pH (in principle valid for proline, too, cf. Pörtner 1989). During catabolism of –R, the ratio between ammonium and bicarbonate quantities is varied according to the chemical nature of the respective residue. For example, catabolism of dicarboxylic-monoamino acids like glutamate would yield 2 mol bicarbonate per mole of ammonium, whereas lysine yields 2 mol ammonium per mole of bicarbonate. In addition, the oxidation of functional groups like the SH-group in methionine causes net proton formation (Table 1). These protons may be seen to titrate and reduce the amount of bicarbonate. Overall, catabolism of a typical mixture of protein and amino acids yields less bicarbonate than ammonium.

Any additional net consumption or generation of protons would also form or titrate bicarbonate (CO_2 being retained or released by respiratory gas exchange, such that Pco_2 remains constant and respiratory changes in the acid–base status do not occur; Pörtner, 1987). With a fixed amount and composition of the amino acid mixture such processes would not affect the generated quantity of ammonium but rather vary the quantity of bicarbonate. In a simplified view metabolic effects on acid–base status in addition to those exerted by the respective mixture of amino acids modify the overall ratio between bicarbonate and ammonium quantities even further although direct compensation of such disturbances via Na^+/H^+ exchange or bicarbonate excretion is certainly possible. Since the discrepancy between traditional and new concepts arises only when it comes to a discussion of the importance of amino acid catabolism in acid–base balance, such a simplified view, which does not change the quantitative conclusions, provides a clearer basis for the following sections of this chapter.

2.2 Ureagenesis and Acid–Base Turnover

Urea formation removes both the excess base and ammonium formed during amino acid catabolism (Atkinson and Camien 1982; Atkinson 1992; Atkinson and Bourke, this Vol.) This strategy mostly used by terrestrial animals is certainly more costly than the one water-breathing animals can use where diffusive release of ammonia (NH_3) via the body surface or gills (Cameron and Heisler 1983; Heisler 1989) leaves a proton behind which neutralizes accumulating bicarbonate. In water breathers this mechanism fulfils two goals which are tightly coupled: removal of a toxic waste product and elimination of a base formed in amino acid catabolism (cf. Pörtner 1989). An imbalance between ammonium and bicarbonate quantities as caused by additional metabolic disturbances of the acid–base status (see above) is compensated by net H^+ equivalent ion exchange between animals and water (including the potential removal of excess ammonium; for review of the respective mechanisms, see Heisler 1986, 1989).

In air-breathing vertebrates and invertebrates these two coupled functions are taken over by the urea cycle or by urate and guanine synthesis (cf. Pörtner 1989). Evidence is strong that removal of base is an important function of urea synthesis in air breathers, and in water breathers when the removal of an incoming base load may take precedence over the diffusive elimination of ammonia (e.g. in alkaline lakes; Randall et al. 1989). However, the large quantities of base removed by urea formation should not mislead in that, in accordance with the traditional view, the detoxification of excess ammonium is tightly coupled to this process and may have been one additional driving force for the use of the urea cycle by terrestrial vertebrates (for a review of ammonium toxicity, see Meijer et al. 1990). The need to eliminate the ammonium is also reflected by the use of glutamine (and glycine) as an ammonium shuttle to the kidney which allows ammonium levels to be kept low when urea synthesis is impaired by acidosis (see below).

If the biochemical reality is considered, there is not just a net release of H^+ in urea synthesis but a net consumption of bicarbonate and of NH_3 withdrawn from the dissociation equilibrium of $NH_4{}^+$ (at high mitochondrial pH, reviewed by Pörtner 1989).

Thus :

$$NH_4^+ \longrightarrow NH_3 + H^+; \quad (1)$$

$$HCO_3^- + NH_3 + \text{aspartate}^- \longrightarrow \text{urea} + \text{fumarate}^{2-}. \quad (2)$$

These details might become important when the molecular mechanisms regulating urea synthesis are analysed. Aspartate is formed during transamination from another amino acid. When the consumption of one additional proton during oxidative decarboxylation of the respective amino acid carboxyl group after transamination is considered, it is clear that urea synthesis removes base and ammonium in equal quantities (cf. Pörtner 1989; Atkinson and Bourke, this Vol.)

The focus of this chapter is to understand the effects of urea synthesis on acid–base regulation from a quantitative point of view and to see whether the quantitative conclusions must differ between traditional and new concepts as requested by Atkinson and Bourke. First, the question arises what an amino

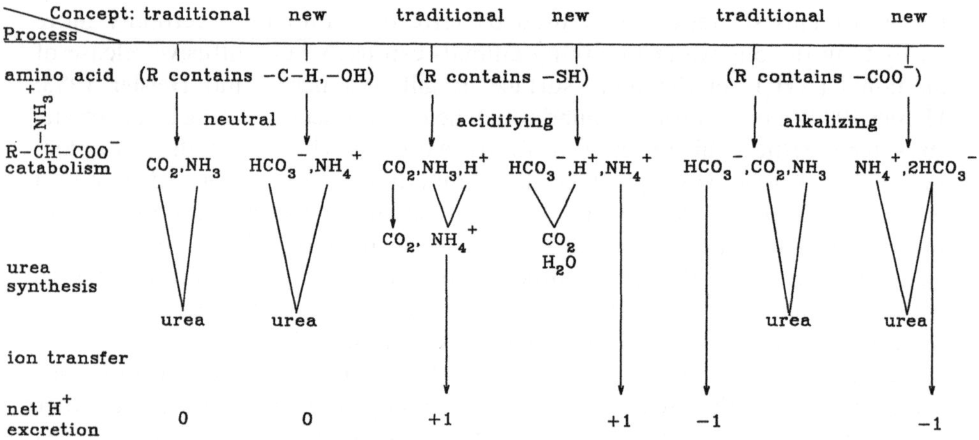

Fig. 1. Simplified, schematic presentation of the quantitative equivalence of traditional and new concepts of how the catabolism of various amino acids influences acid–base regulation in terrestrial vertebrates. In quantitative terms the graph focusses on the equimolar generation of bicarbonate and ammonium from the fundamental amino acid skeleton (Table 1) and the modulation of this ratio by the metabolism of the residues R considering an equimolar consumption of base and ammonium in urea synthesis. Values of net acid excretion are identical for traditional and new concepts. For further explanations, see text and Table 1

acid metabolism neutral for the acid–base status would mean in the two concepts (Fig. 1). In the traditional view, the formation of neutral urea from neutral amino acids like alanine would have been neglected; no net acid excretion would occur. It turns out that the same quantitative conclusions hold true for the new concept as well: Neutrality requires that all base and ammonium are taken up by urea synthesis without influencing the acid–base status. This condition is only fulfilled when ammonium and bicarbonate are generated in equal quantities, for example from "neutral" amino acids like alanine, and are removed in equal quantities during urea formation. In principle, the 1:1 generation of bicarbonate and ammonium from the characteristic carboxyl and α-amino groups of any amino acid is mirrored by the 1:1 stoichiometry of urea synthesis.

However, urea synthesis in vivo occurs based on a ratio of ammonium and bicarbonate quantities which has specifically been modified during oxidation of amino acid residues and has also been modified by other metabolic processes (see Table 1 and Fig.1). Generation of excess bicarbonate from organic acids and amino acids like glutamate would, therefore, cause an alkalosis with the result that both urea formation and ionic base release occur. This result is identical in both concepts. On the other hand, acidifying sulphate generation in methionine catabolism would lead to the titration of bicarbonate yielding excess ammonium. Complete removal of ammonium by urea formation would then lead to the development of metabolic (non-respiratory) acidosis, since more base would be removed than is being generated. Therefore, excess ammonium is eliminated by ionic exchange. Again, this view is quantitatively equivalent to the traditional conclusion that aerobic metabolism of protein is linked to net proton production and causes net acid excretion to avoid systemic acidosis. However, this conclusion is obviously not shared by Atkinson and Bourke.

2.3 The Role of Glutamine

Based on the regulatory mechanisms reviewed by Atkinson and Bourke an acidosis would lead to a reduced formation of urea, with the effect of protecting the bicarbonate pool of the body. Since this reduction reflects an acidotic threat imposed by metabolism the question arises, how can it adequately be quantified? It is important in this context that compensation of an acidotic trend also means an increased export of ammonium to the kidney proportional to the reduced rate of urea synthesis (Fig. 1). In the whole system it is predominantly glutamine (and to a minor extent glycine) which serves as a vehicle for the net transport

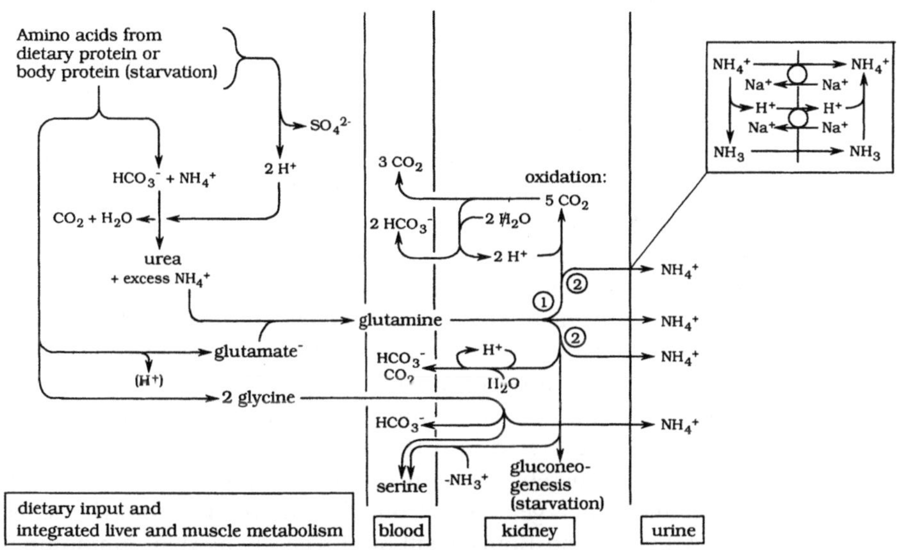

Fig. 2. Schematic presentation of the role of glutamine (and glycine) as an ammonium shuttle between muscle, liver and kidney. In acidosis, an increase in glutamine catabolism in the kidney is balanced by an increased export from muscle and liver. Ammonium is excreted and the remaining base is used to avoid and compensate for metabolic acidification (modified after Pörtner 1989). The basic conclusions remain unchanged when glycine is used as a shuttle metabolite and when serine is formed from glutamine and glycine and released into the circulation (Jungas et al. 1992). Net synthesis of non-dietary glutamate (as a monoamino-dicarboxylate anion) in liver or muscle involes proton production (H^+) when it is synthesized from monoamino-monocarboxylate precursors (like alanine). ① Glutaminase; ② glutamate dehydrogenase; $-NH_3^+$ indicates occurrence of transamination

of ammonium (Pörtner 1989, for a more recent analysis see Jungas et al. 1992). Glutamine synthesis and export from the liver (cf. Häussinger et al. 1988; Häussinger 1990; Almond et al. 1991) and from muscle (see Pörtner 1989) increase in acidosis. pH and hormonal changes stimulate glutamine uptake and oxidation in the kidney (Bogusky and Dietrich 1989; Welbourne 1989; Sastrasinh and Sastrasinh 1990), emphasizing the crucial role of glutamine and the kidney in acid–base homeostasis. Glutamine is deaminated to yield α-ketoglutarate and part of it is completely oxidized or used in gluconeogenesis (Fig. 2). Oxidation of this dicarboxylic acid involves a local net generation of base which has been seen to provide new base for systemic acid–base regulation when ammonium is released into the urine. However, as outlined in more detail by Pörtner (1989) and in accordance with the more recent treatment of Jungas et al. (1992), non-dietary glutamate/glutamine is resynthesized in other organs (e.g. muscle, implying release of protons or consumption of base, Fig. 2) such that, from a whole organism perspective, new or net base generation can be excluded as a special function of glutamine metabolism in the kidney. This view is more to the point than focussing on that the catabolism of glutamine may either occur in the kidney or in other organs (Atkinson and Bourke, this Vol.).

In summary, glutamine synthesis in muscle and liver and its export to the kidney are equivalent to an ammonium shuttle system to the kidney (Pörtner 1989). Excess ammonium not used by urea synthesis is excreted into the urine. The base generated from glutamine (or glycine) catabolism in the kidney represents base originating from overall amino acid catabolism which is protected from being consumed in urea synthesis and can, thus, contribute to maintain systemic acid–base homeostastis. That the kidney plays an active role in the metabolic control of acid–base parameters is emphasized by the observation that acidosis may cause an increase in kidney glutamine catabolism in some species like humans, rats and sheep to an extent that plasma glutamine levels decrease (Welbourne 1987). Thus, the rate of glutamine catabolism in the kidney does not just depend upon the supply by the liver and peripheral tissues but may influence the release of glutamine from other tissues by modulating the plasma concentration. Obviously, liver and kidney cooperate in the fine control of acid–base regulation.

Traditional and new concepts come together when it is appreciated that any excess ammonium not excreted into the urine would have to be used in urea synthesis and would thereby cause metabolic acidification (see below). Only when excess ammonium is generated from diamino-monocarboxylic acids, the contention of Atkinson and Bourke will hold true saying that ammonium production and release will not reflect a direct acidifying influence (titration of base) by metabolism. It would, nonetheless, cause acidosis when NH_4^+ was consumed in ureagenesis instead. Contrary to the emphasis of Atkinson and Bourke, such a threat of acidosis is also imposed when NH_4Cl is added to the system although, by itself, this salt does not represent an acid. Glutamine synthesis and its export to the kidney are thus part of a regulatory fine tuning for the export of excess ammonium to the kidney and prevent both acidosis and also large fluctuations in plasma ammonium levels (Pörtner 1989).

3 Renal Ammonium and Net Acid Excretion

For an evaluation of net acid excretion by the kidney both titratable acidity of the urine and ammonium excretion have traditionally been measured. Thus,

$$\dot{V}u\ ([NH_4^+] + [TA]–[HCO_3^-]) = \text{net acid excretion (mmol kg}^{-1}\ \text{h}^{-1}), \qquad (3)$$

where $\dot{V}u$ is the urinary flow rate, $[NH_4^+]$ and $[HCO_3^-]$ are the ammonium and bicarbonate concentrations in the urine and [TA] is the titratable acidity determined by titrating urine samples back to plasma pH. Titratable acidity measurements evaluate the proton quantities transferred by the sodium/proton exchanger (cf. Fig. 2) and take into account to what extent phosphate and other buffers determine the final urine pH. Urinary acidification causes ammonium trapping and supports ammonium excretion (which otherwise occurs by Na^+/NH_4^+ exchange) as well as "bicarbonate reabsorption" (bicarbonate "reabsorption" means the titration of bicarbonate in the acidified urine. Accumulating CO_2 diffuses back into the plasma and is rehydrated forming bicarbonate when the sodium/proton exchanger removes the protons from the respective dissociation equilibrium of carbonic acid; Valtin and Gennari 1987). Atkinson and Bourke have concluded from their point of view that ammonium excretion cannot represent acid excretion, because ammonium is not an acid in a chemical sense. It should, thus, not be considered or included in the term for net acid excretion. How important it is to come to a valid conclusion is reflected by the fact that, according to traditional analyses, changes in net acid excretion during acid or base loads are predominantly due to the respective changes in ammonium excretion (cf. Knepper et al. 1989).

Although Atkinson and Bourke's statement reflects the physicochemical properties of the ammonium ion it is not important in this context whether ammonium functions as an acid but rather what ammonium excretion into the urine means for the acid–base status of the animal: since neutrality requires a strict 1:1 removal of base and ammonium in urea synthesis to match a catabolic 1:1 generation of base and ammonium in metabolism (Fig. 1), ammonium release does certainly not cause an acidosis in the urine by itself, but quantifies bicarbonate which is not used in urea synthesis. It has been stated above that this bicarbonate is (a) consumed by other, acidifying metabolic processes or (b) is not available for the removal of excess ammonium generated from diamino-monocarboxylic acid. In addition (c), it might be bicarbonate which had been lost as base from the body (released into the urine and considered by Eq. (3), or released into the intestinal lumen, see below). In an overall balance of metabolic proton production and the responses of the mechanisms of acid–base regulation this summed amount of bicarbonate must be considered. It is most adequately quantified by analyzing urinary ammonium excretion since, based on the smaller background levels of ammonia in the urine, changes in ammonia excretion are much more easily monitored than small reductions relative to the large background of urea excretion (via kidney and gut). As a corollary, quantities of excreted ammonium must be taken into account even when accepting the modified concept of urea formation. For a quantification of metabolic influences on acid–base regulation they must be included in the numbers evaluated as net acid excretion. Such a procedure is also in line with the established procedures of measuring net acid excretion in aquatic animals (cf. Heisler 1986, for review).

These conclusions are not changed under conditions when the diet causes an excess of base over the amount of available ammonium (cf. Table 1; Fig. 1). As outlined by Atkinson (1992) and Atkinson and Bourke (this Vol.), urea may be recycled through the gut and ammonium be reabsorbed. It should be emphasized that bicarbonate originally consumed in urea synthesis is then regenerated in the gut and may be disposed of that way, the reabsorbed ammonium being available for amino acid and protein synthesis, and also for urea synthesis, if the additional removal of systemic bicarbonate is required. Should the reabsorbed ammonium be excreted via the kidney instead, it will again reflect the fate of one bicarbonate in the body somewhere else (released via the gut) than in urea synthesis.

4 Conclusions and Summary

In accordance with traditional concepts of pH homeostasis in terrestrial (and aquatic) vertebrates the reabsorbed components of the diet and their metabolism determine the net proton outcome of metabolism. For the analysis of the resulting net acid excretion via the kidney Atkinson and Bourke concluded, based on a new view of the importance of urea formation in acid–base regulation, that ammonium excretion should no longer be considered as acid excretion. A different conclusion arises from a previous and the present analysis (Pörtner 1989). In a comparison and synthesis of traditional and new concepts of pH homeostasis in terrestrial vertebrates the interpretation of the role of renal ammonium excretion needs to be modified: The traditional concept looked at NH_4^+ as a proton vehicle being formed from ammonia in the urine and removing protons from the body. The new, more realistic concept should rather consider the ratio between ammonium and bicarbonate quantities formed in metabolism. Based on the 1:1 generation of base and ammonium from the α-amino and carboxyl groups characteristic for any amino acid and on the 1:1 consumption of base and ammonium in urea synthesis, any ammonium not being used by urea synthesis and excreted into the urine quantifies a loss or deficit of base in excess of what is measured as $[TA]–[HCO_3^-]$ in the urine. Metabolic influences on aerobic acid–base parameters can be understood as causing a deviation from the 1:1 generation of base and ammonium in "neutral" amino acid catabolism. Such a deviation may also be caused by net H^+ equivalent ion exchange across the gut epithelia. Accepting these preconditions, both traditional and new concepts will come to the conclusion, that a protein diet causes net acid excretion, which is reflected by ammonium formation and excretion being in excess of the amount of bicarbonate removed during urea synthesis. To quantify the base deficit, the analysis of "net acid excretion" must still include a measurement of ammonium excretion. Although the modified view of the importance of urea synthesis in acid–base regulation is the more realistic concept, the new and the traditional concepts of acid–base regulation can adequately be brought together in their identical quantitative conclusions. The methodology for investigating acid–base regulation is still accurate and does not need to be modified (Pörtner 1989).

Acknowledgement. Supported by the Deutsche Forschungsgemeinschaft, Alfred-Wegener-Institute Publication No. 909.

References

Almond MK, Smith A, Cohen RD, Iles RA, Flynn G (1991) Substrate and pH effects on glutamine synthesis in rat liver. Consequences for acid–base regulation. Biochem J 278: 709–714

Atkinson DE (1992) Functional roles of urea synthesis in vertebrates. Physiol Zool 65: 243–267

Atkinson DE, Camien MN (1982) The role of urea synthesis in the removal of metabolic bicarbonate and the regulation of blood pH. Curr Top Cell Regul 21: 261–302

Bogusky RT, Dietrich RL (1989) Effect of acute metabolic acidosis on ammonia metabolism in kidney. Am J Physiol 256: F321–F328

Cameron JN, Heisler N (1983) Studies of ammonia in the rainbow trout: physico-chemical parameters, acid–base behaviour and respiratory clearance. J Exp Biol 105: 107–125

Häussinger D (1990) Nitrogen metabolism in liver: structural and functional organization and physiological relevance. Biochem J 267: 281–290

Häussinger D, Meijer AJ, Gerok W, Sies H (1988) Hepatic nitrogen metabolism and acid–base homeostasis. In: Häussinger D (ed) pH homeostasis: mechanisms and control. Academic Press, New York, pp 337–377

Heisler N (1986) Buffering and transmembrane ion transfer processes. In: Heisler N (ed) Acid–base regulation in animals. Elsevier, Amsterdam, pp 309–356

Heisler N (1989) Interactions between gas exchange, metabolism, and ion transport in animals: an overview. Can J Zool 67: 2923–2935

Jungas RL, Halperin ML, Brosnan JT (1992) Quantitative analysis of amino acid oxidation and related gluconeogenesis in humans. Physiol Rev 72: 419–448

Knepper MA, Packer R, Good DW (1989) Ammonium transport in the kidney. Physiol Rev 69: 179–249

Meijer AJ, Lamers WH, Chamuleau RAFM (1990) Nitrogen metabolism and ornithine cycle function. Physiol Rev 70: 701–748

Pörtner HO (1987) Contributions of anaerobic metabolism to pH regulation in animal tissues: theory. J Exp Biol 131: 69–87

Pörtner HO (1989) The importance of metabolism in acid–base regulation and acid–base methodology. Can J Zool 67: 3005–3017

Pörtner HO (1990) Determination of intracellular buffer values after metabolic inhibition by fluoride and nitrilotriacetic acid. Respir Physiol 81: 275–288

Pörtner HO, Heisler N, Grieshaber MK (1984) Anaerobiosis and acid–base status in marine invertebrates: a theoretical analysis of proton generation by anaerobic metabolism. J Comp Physiol 155(B): 1–12

Randall DJ, Wood CM, Perry SF, Bergman H, Maloiy GMO, Mommsen TP, Wright PA (1989) Urea excretion as a strategy for survival in a very alkaline environment. Nature 337: 165–166

Sastrasinh M, Sastrasinh S (1990) Effect of acute pH change on mitochondrial glutamine transport. Am J Physiol 259: F863–F866
Valtin H, Gennari FJ (1987) Acid–base disorder. Basic concepts and clinical management. Little, Brown and Company, Boston
Welbourne TC (1987) Interorgan glutamine flow in acidosis. Am J Physiol 253: F1069–F1076
Welbourne TC (1989) Glucocorticoid and acid–base homeostasis: effects on glutamine metabolism and transport. Am J Kidney Dis 14: 293–297

Chapter 4

Ammonia vs Ammonium: Elimination Pathways of Nitrogenous Wastes in Ammoniotelic Fishes

N. Heisler

Contents

1 Introduction

Nitrogenous waste from protein catabolism is produced and eliminated by animals in three different forms: ammonia, urea and uric acid. Production of the primary endproduct ammonia is energetically much more efficient than synthesis and subsequent elimination of urea or production of uric acid. The pronounced toxicity of nonionized ammonia, however, limits utilization of ammonia as a

Department of Animal Physiology, Humboldt University at Berlin, D-10115 Berlin, and Max-Planck-Institut für experimentelle Medizin, D-37075 Göttingen, FRG

Advances in Comparative and Environmental Physiology, Vol. 22

nitrogenous waste to animals which can readily dump the substance at low body fluid concentrations into the environment. Similar to CO_2, ammonia can be voided as a respiratory gas, or eliminated in the form of its ionization product, the ammonium ion. Elimination as a respiratory gas as well as sizable elimination of ammonium ions can be performed at the required small partial pressure or concentration gradients, respectively, by species capable of exploiting the favorable effective capacitance coefficient of aqueous solutions, provided that large volumes of environmental water of low ammonia and ammonium concentrations are available. These prerequisites are generally fulfilled by aquatic animals, whereas terrestrial species are restricted by the small volumes of eliminated solutes, mainly urine, in their capacity to excrete ammonia. Thus, the energetic advantage of ammonia as the main nitrogenous metabolic end product is generally utilized only by teleost fish and aquatic amphibia.

The similarity in many physicochemical properties of gaseous ammonia and CO_2 has often led to a comparative description of these main gaseous metabolic end products of ammoniotelic vertebrates (e.g. Heisler 1990b). CO_2 is the endproduct of the carbonic fractions of substrates, whereas ammonia as nitrogenous end product is produced in ammoniotelic animals at a rate of up to 15% of the oxygen consumption (Heisler 1984, 1986b, 1988a). CO_2 as well as NH_3 exist in the body fluids mainly ionized and their ionization/deionization processes challenge the organismic acid–base regulation. While CO_2 represents the anhydride of a weak acid, NH_3 is a weak base of approximately the same numeric strength (pH_{pl}–pK'_1 ~ 1.6–2.0). Thus, immediately after production the largest fractions of both of these metabolic end products produce or consume, respectively, H^+ ions, the process being reversed upon elimination from the body fluids in non-ionized form. In contrast, elimination of the ionization products will affect acid–base status. It has been well known for many years that CO_2 is transferred in vertebrates through membranes and epithelia mainly by non-ionic diffusion, with hardly any contribution of ionic mechanisms. The similarity in diffusivity between the two gases has led to the view that the same mechanism operates for NH_3, but the pattern is not quite as clear. Recently, ammonia has been postulated to be transferred mainly by ionic diffusion or active carrier-mediated ion exchange processes by at least some authors, challenging the traditional concept.

Rather than providing a complete review, this chapter is mainly focused on characterizing the contributions of different transfer mechanisms for ammonia at aqueous, cellular and epithelial sites, and it re-evaluates the role of ammonia as a respiratory gas, mainly on the basis of relevant recent data.

2 Production, Ionization and Partial Pressure

The rate of ammonia production in ammoniotelic animals (e.g. teleost fish) is quite variable and directly related to the relative contribution of protein catabolism to overall energy production, but it rarely exceeds 0.15 of molar oxygen consumption (cf. Heisler 1984, 1986b, 1988a). Ammonia production represents about 60–90% of the total nitrogenous waste of teleost fish (Smith 1929b; Wood 1958; Fromm 1963). In aquatic and semi-aquatic animals net production of ammonia is generally larger than net production of nonvolatile acid–base relevant endproducts (Heisler 1988b; re: amphibia: Heisler et al. 1982; Boutilier and Heisler 1988; re: fishes: Heisler 1984, 1986b). On a smaller scale this

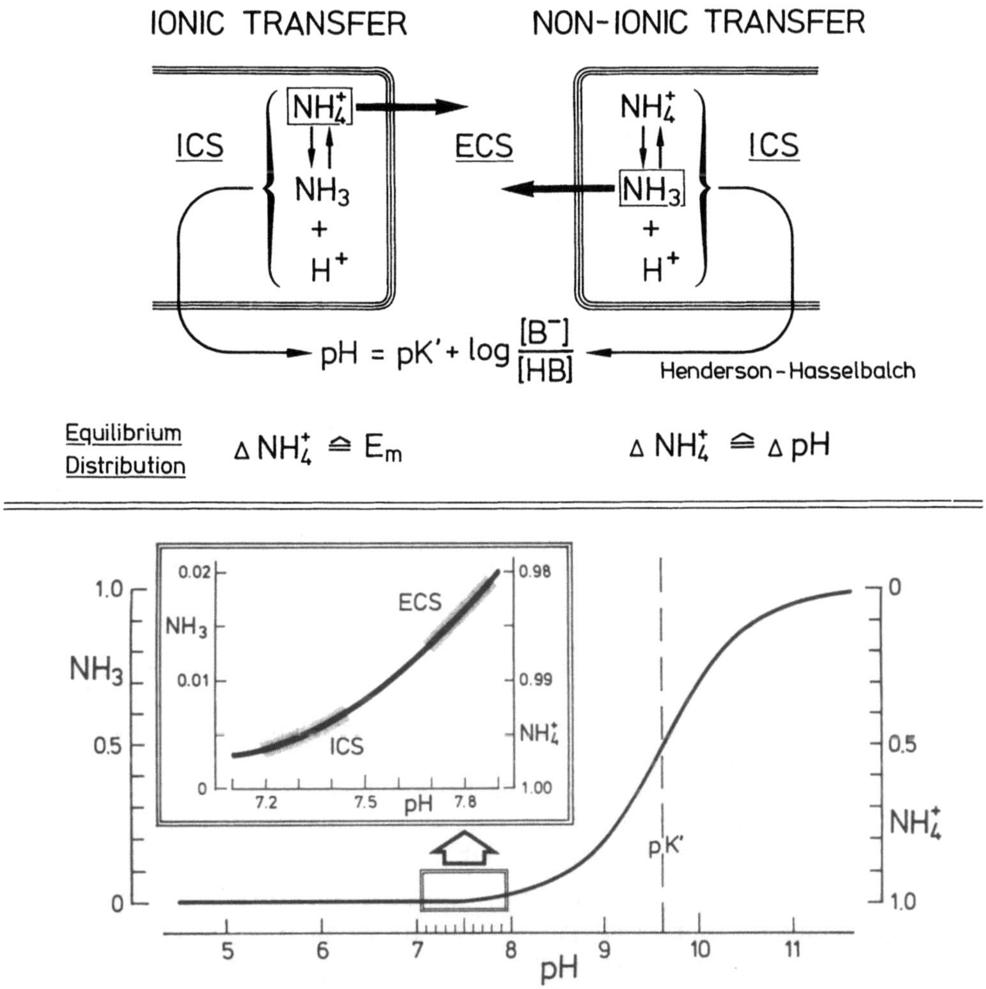

Fig. 1. Physicochemical relationships of the ammonia/ammonium buffer pair. *Upper panel* Depending on the form of main transfer across cell membranes total ammonia is distributed between intracellular and extracellular space at equilibrium according to the membrane potential (mainly ionic transfer) or to the intra-/extracellular pH difference. If both forms are transferred to a significant extent, intermediate distribution ratios can be obtained. *Lower panel* As a function of the high pK of the buffer system (for constants: Cameron and Heisler 1983), total ammonia is present at physiological pH mainly in ionized form (NH_4^+). See also text

also holds for ureotelic species like the elasmobranchs (Heisler 1984, 1986b, 1988a), whereas the reverse applies for terrestrial animals (cf. Heisler 1988b).

After production by deamination of a-amino acids (mainly glutamine) at various sites like liver, kidneys and gills (Goldstein et al. 1964; Pequin 1962; Pequin and Serfaty 1963; Aster et al. 1982; Cameron and Heisler 1983), ammonia is ionized by uptake of H^+ ions from the body fluids according to the Henderson-Hasselbalch equation:

$$pH = pK' + \log \frac{[NH_3]}{[NH_4^+]} = pK' + \log \frac{\alpha_{NH_3} \cdot P_{NH_3}}{[NH_4^+]}$$

The pK′ value for this reaction is in the range of 9.0 to 9.8, depending on temperature, salinity and protein content (cf. Cameron and Heisler 1983). Accordingly, in the case of teleost fish, > 97% of intracorporeal ammonia is present in form of NH_4^+ (Fig. 1). The physical solubility of ammonia, αNH_3, is larger than that for CO_2 by a factor of about 1000 in aqueous fluids (Cameron and Heisler 1983). According to high pK and solubility, plasma P_{NH_3} in teleost fish is in the range of only about 50×10^{-6} mmHg at physiological pH and ammonium concentrations (Cameron and Heisler 1983).

3 Aqueous Diffusion

Diffusive translocation of ammonia and ammonium in aqueous solutions such those found in cytosolic or interstitial pathways follows Fick's law of diffusion:

$$\dot{M} = D \cdot F \frac{\Delta C}{\Delta d} = D \cdot F \frac{\alpha_x \cdot \Delta P_x}{\Delta d}$$

where: $\dot{M}$ = the amount of substance transferred along the distance 'Δd'
D = diffusion coefficient
F = the surface area available for diffusion,
ΔC = the concentration difference along the diffusion path,
α_x and P_x represent solubility and partial pressure of gas 'x'.

The aqueous diffusion coefficient 'D' for ammonia (1.77×10^{-5} cm²/s; Radford 1964) is similar to that for CO_2 ($D_{CO_2} = 1.77 \times 10^{-5}$ cm²/s, Washburn 1929; 1.72 – 1.75×10^{-5} cm²/s, Gros and Moll 1974) and slightly larger than that of NH_4^+ (1.4×10^{-5} cm²/s; Arruda et al. 1984). The concentration gradients for NH_4^+, however, will always be considerably larger at physiological pH values than those for NH_3, because of 30–100 times larger relative concentrations of NH_4^+ (high pK′ value of the ammonia system; cf. Fig. 1). Passive aqueous diffusion is therefore expected to take place mainly in form of NH_4^+, given simultaneous transfer of associated H^+ ions back to the origin of the diffusion pathway.

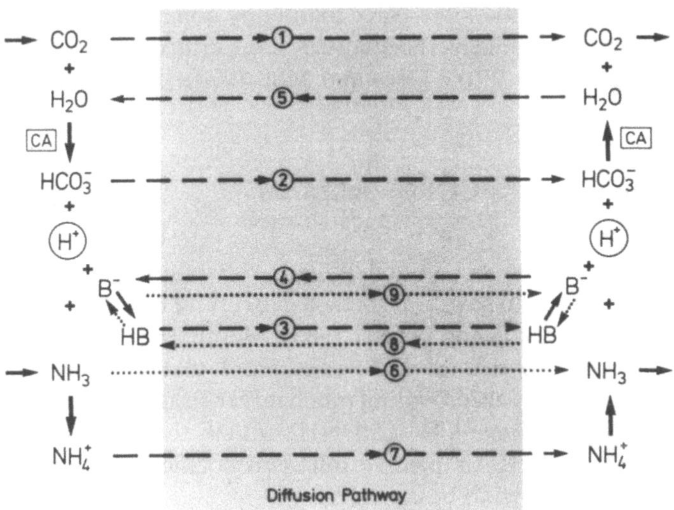

Fig. 2. Facilitated diffusion of ammonia and CO_2 along aqueous pathways. Concurrent ionic movement of NH_4^+ and HCO_3^- requires no additional mechanism such as a shuttling buffer system (pathways 8 and 9 for ammonia, and 3 and 4 for CO_2) for recolation of H^+ ions net transferred by the isolated diffusion processes, as long as rapid hydration of CO_2 is facilitated by sufficient activity of carbonic anhydrase. Facilitated diffusion is advantageous for the elimination of both metabolic endproducts because of the much larger ionic than non-ionic diffusion gradients with similar ionic and non-ionic diffusion coefficients (for details, see text and Heisler 1989)

Relocation of H^+ ions in aqueous solutions, however, cannot sufficiently be provided by simple diffusion: although the diffusion coefficient for H^+ is favorable and even larger than for NH_4^+ (9.2×10^{-5} cm²/s, Kortüm 1972), the diffusion gradient for free H^+ ions in the physiological pH range is definitely too small for significant counter-diffusion. Accordingly, for diffusive transfer of NH_4^+ rather than NH_3, relocation of the associated H^+ ions can only be facilitated by other mechanisms, such as shuttling buffer systems with relatively high diffusivity (Fig. 2, pathways 8 and 9), e.g. inorganic phosphate ($D = 0.95 \times 10^{-5}$ cm²/s). In vivo, however, such an additional mechanism is not required for translocation of NH_4^+. The elimination of the weak base ammonia is generally accompanied by transfer of the weak-acid-equivalent CO_2. Facilitated diffusion of CO_2 requires movement of H^+ ions opposite (Fig. 2, pathways 3 and 4) to that for facilitated diffusion of ammonia (pathways 8 and 9), so that the simultaneous equimolar ionization of these two waste products eliminates the need for an additional cycling buffer system, as long as sufficient local carbonic anhydrase activity allows for rapid hydration and dehydration of CO_2 (cf. Heisler 1990b). Although ammonia production is quite heterogeneous among tissues (taking place mainly in the liver, kidneys and gills; cf. Cameron and Heisler 1983) the rate of local CO_2 elimination will usually exceed the rate of ammonia production, thus providing a vehicle for ionic diffusion of NH_4^+ in free aqueous phases.

Aqueous ammonia diffusion will thus take place mainly by concurrent passive diffusion of NH_4^+ and HCO_3^-, facilitated by the relatively large diffusion coefficient 'D' for bicarbonate (e.g. 0.38–0.40 × 10^{-5}, Gros and Moll 1974).

4 Transfer of Ammonia Across Cell Membranes

Although the lipid solubility of ammonia is often quoted to be superior compared to that of NH_4^+, its water/chloroform partition coefficient of about 8 does actually not support this notion (cf. Cameron and Heisler 1983). Membrane diffusion of ammonia is therefore, similar to that of CO_2, considered to take place via aqueous, charged membrane channels providing much larger resistance for ions than for uncharged molecules. Since CO_2 and NH_3^+ utilize the same transmembrane pathway, the ratio of their passive diffusion coefficients through membranes and epithelia is expected to be constant.

Conclusive evidence as to the mechanisms mainly responsible for membrane transfer of ammonia cannot easily be provided. Direct measurement of ammonia and ammonium concentrations, as well as of interface conditions with respect to local pH, electrical charges and other factors is presently impossible for single cells. For a conclusive analysis, however, single cell approaches are definitely required. In complex preparations interstitial diffusion pathways may provide a specific diffusion resistence many orders of magnitude lower than the cell membrane matrix and the small relative cross-sectional area of water channels, thus largely biasing determination of the apparent membrane diffusion coefficient. These complications for direct measurement of cell membrane transfer capacity have led to alternative, indirect approaches.

4.1 Direct Measurement of Ammonia Transfer in Complex Structures

Although turtle bladder epithelium falls short of an ideal preparation for the analysis of ammonia transfer mechanisms across cell membranes the preparation has nonetheless been used for this purpose. Measurement of translocation through membrane-like tissue sheets in an Ussing-type chamber is generally impaired by tissue damage at the circumferential seal of the chamber, and by the large unaccounted relative contribution of paracellular pathways with low diffusional resistance (Arruda et al. 1984). Also, active, preferentially directed transfer by carrier-mediated mechanisms provides considerable problems for the interpretation of results.

The wide range of reported relative ammonia/ammonium diffusion coefficients in turtle bladder epithelium is reflected by two rather contrasting studies: D_{NH_3} was reported to be 24 times, and $D_{NH_4^+}$ to be more than 310 times smaller than the aqueous diffusion coefficients, presenting a ratio of $D_{NH_3}/D_{NH_4^+}$ of about 13 (Arruda et al. 1984), or, in another case, the diffusivity of $D_{NH_4^+}$ was about 50 times smaller than that for NH_3 (Schwartz and Tripolone 1983). Despite the variability of data on turtle bladder, prevalent concentration differences along the diffusion pathway of about 30–100 times larger for NH_4^+ (cf. Fig. 1), together with ratios for $D_{NH_3}/D_{NH_4^+}$ in the range of 10 to 50 suggest that a large if not main fraction of membrane transfer takes place by NH_4^+ diffusion, at least in this preparation.

However, the inadequacy of the turtle bladder preparation with its large paracellular diffusion pathways for estimation of membrane transfer becomes apparent by comparison with results obtained from other systems. Ratios of $D_{NH_3}/D_{NH_4^+}$ of ~ 100 in seawater-acclimated in contrast to ~ 1200 in 5%-seawater-acclimated toadfish gills (Evans et al. 1989), of ~ 1000 in shark pub gills (Evans and More 1988), and of ~ 350 in the rabbit proximal straight tubule (e.g. Garvin et al. 1987) support the notion that relatively high tissue diffusivity of NH_4^+ (D-ratios below 350) is associated with 'leaky' paracellular pathways.

4.2 Estimation of Relative Diffusivities on the Basis of Ammonia Distribution

The distribution of a buffer system, like ammonia/ammonium, between intra- and extracellular space is, at least theoretically, a function of the diffusivities of the two components of the buffer (charged and uncharged). Predominant transfer of the non-ionized form will result in distribution of total ammonia according to the pH difference, whereas mainly ionic transfer is expected to result in distribution according to the membrane potential (Fig. 1). With sizable diffusion of both forms, intermediate distribution ratios have to be expected (cf. Heisler 1990a,b).

Quite dissimilar patterns have resulted from a number of investigations on intracellular/extracellular ammonia distribution performed during the last few years. In a few studies employing determination of plasma and tissue total ammonia levels after mechanical homogenization, the obtained intracellular/extracellular concentration ratios were found to be much higher than expected on the basis of the pH difference. They were close to values predicted from the membrane potential, suggesting predominant transfer of ammonium (Wright et al. 1988b; Wright and Wood 1988; Wood et al. 1989). $D_{NH_3}/D_{NH_4^+}$ was estimated in the range of 75 to 1, with values close to 1 in fishes and rising values with decreasing degree of ammoniotelism (Wood et al. 1989). In contrast, similar other studies employing tissue homogenization under liquid nitrogen have recently indicated much lower concentration ratios in tissues of *Bufo marinus* (H.O. Pörtner and D.P. Toews, unpubl. data), Illex illecebrosus (H.O. Pörtner, D.M. Webber, R.G. Boutilier, and R.K. O'Dor, unpubl. data), *Cyprinus carpio* (N. Heisler and C. Schorer, unpubl. data) and *Rattus* (N. Heisler and C. Schorer, unpubl. data), approaching values of $D_{NH_3}/D_{NH_4^+}$ close to distribution according to the pH difference.

High tissue total ammonia concentrations as reported in some of the above studies (Wright et al. 1988b; Wright and Wood 1988; Wood et al. 1989) imply

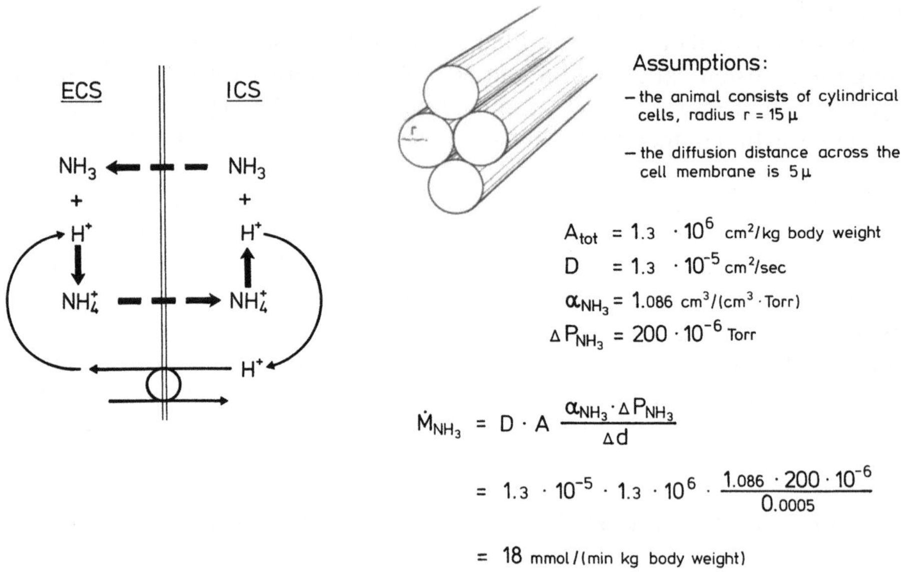

Fig. 3. *Left* Predominantly ionic translocation of NH_4^+ across cell membranes will effect non-ionic diffusion of NH_3 in opposite directions according to the implicit P_{NH_3} gradient. An ammonia shuttle (forward NH_4^+, backwards NH_3) will result in equimolar net transfer of H^+ from extracellular (*ECS*) to intracellular space (*ICS*), to be corrected for by active acid–base homoiostatic mechanisms. *Right* Model calculations on the basis of Fick's law of diffusion, literature data on the distribution of ammonia across cell membranes and the indicated parameters (see also text). This probability model results in an NH_3 flux (equivalent to the rate of H^+ extrusion), which is orders of magnitude too large to be expected in vivo. See text

considerable P_{NH_3}-disequilibrium at the cell membrane, with partial pressure differences in the range of up to 1000 μtorr (Wright et al. 1988b). The similarity of the diffusion characteristics of CO_2 and NH_3 in combination with the reported partial pressure gradients strongly imply simultaneous, oppositely directed, diffusive transfer of NH_3 and NH_4^+ of considerable magnitude across the cell membrane. The associated cycling of ammonia and unidirectional translocation of H^+ ions towards the intracellular compartment represents a considerable acid–base relevant load to be corrected for by appropriate mechanisms (Fig. 3, left).

The magnitude of ammonia flux to be expected on the basis of the above partial pressure gradients can roughly be estimated from application of Fick's law of diffusion (see above), based on conservative estimates of the parameters: the diffusion coefficient for NH_3 determined for gill epithelium (Cameron and Heisler 1983); the assumption that the cell surface area is represented by the circumferential area of cylinders with a diameter of 30μ; an average diffusion distance across the cell membrane of 5μ; and an average partial pressure gradient of 200 μtorr (low range value from Wright et al. 1988b). The rate of active H^+ extrusion from the cells during steady state, equivalent to the rate of ammonia shuttling across the cell membrane, calculates to as high as 18 mmol/min × kg body weight) (Fig. 3, right), even with the applied parameters leading towards a low estimate. This estimate of transmembrane transfer is almost 2 orders of magnitude higher than the highest rate ever reported for lower vertebrates in vivo during most extreme acid–base disturbances (e.g. exhausting exercise, *Scyliorhinus stellaris:* < 250 μmol/(min × kg), Holeton and Heisler (1983), and 3 orders of magnitude larger than the maximal H^+ excretion rates of organs dedicated to organismic acid–base regulation (mammalian kidney: 5 mmol/(min · kg); fish gills: 15–25 mmol/(min × kg), Heisler 1986c). The obvious disparity clearly suggests that an equivalent capacity of active H^+ transfer is not available in vivo. Furthermore, the energetic implications render predominant ionic transfer as a most unlikely mechanism of transmembrane ammonia transfer.

Surprisingly, ammonia distribution of fish blood during hypercapnia is also reported to deviate considerably from that expected on the basis of the intra-/extracellular pH difference (Wright et al. 1988a). The implied P_{NH_3} gradients are not quite as pronounced as those for systemic tissues, but a surface area per volume for erythrocytes larger by 1 to 2 orders of magnitude in comparison to the above model leads to very similar conclusions. The required active H^+ extrusion could be handled by erythrocytes with their very limited scope of metabolism even less than by the systemic tissues.

Active transfer of NH_4^+ by carrier-mediated mechanisms of cell membranes certainly invalidates any conclusions as to the ratio of diffusivities of NH_3 and NH_4^+ from intracellular/extracellular concentration ratios. Although Na^+/K^+-ATPase-related transfer does not seem to exist in fish erythrocytes, as tested by application of ouabain (Wright et al. 1988a), it cannot be ruled out for systemic tissues. Evidence has been provided that this mechanism may be involved in the transfer of ammonium in fish gill epithelium (Evans et al. 1989). Also, active translocation of NH_4^+ may be related to other, ouabain-insensitive mechanisms, such as a furosemide/bumetanide sensitive NH_4^+-$2Cl^-$$Na^+$ co-transporter (e.g. Evans and More 1988; cf. Fig. 7). The capacity of such mechanisms will, however, be much smaller than the above estimated rates of ammonia shuttling. However, nature and capacity of mechanisms responsible for NH_4^+ accumulation in the intracellular space will leave the conclusions from the above plausibility calculations unaffected.

These considerations underline the evident discrepancy between the two groups of reports, conforming or not conforming with ammonia distribution according to the pH difference (see above), suggesting the desirability of looking for possible methodological artefacts. One of the possibilities is related to the thermolability of terminal protein amino groups, which may result in liberation of NH_3 during homogenization at temperatures above 0 °C, causing falsely high tissue ammonia concentrations. Considering the large amount of energy introduced by mechanical homogenizers into small volumes of tissue homogenates, considerable local elevation of sample temperature, even with external cooling, must be envisioned. Literature data support this notion: all studies utilizing mechanical homogenizers (Omnimix, Ultraturrax) indicate high tissue ammonia levels, whereas homogenization tissues deep frozen under liquid

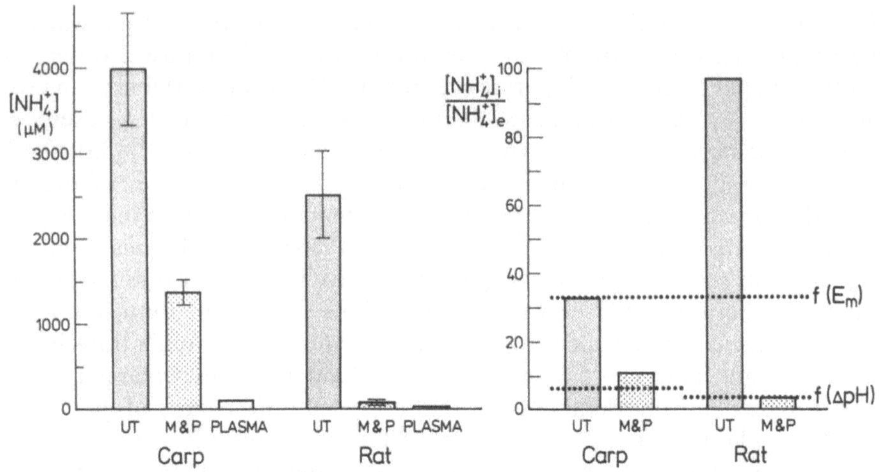

Fig. 4. Total ammonia concentration in intracellular muscle tissues of carp and rat, determined by application of mechanical homogenizers (*UT* Ultraturrax, Omnimix etc.) and by the conventional method of homogenization with mortar and pestle under liquid nitrogen (*M&P*), compared to the respective plasma concentrations. The intracellular/extracellular distribution ratio is close to that expected from the pH difference with M&P homogenization, whereas UT homogenization results in values close to or much higher than those expected on the basis of distribution according to the membrane potential (E_m). See also text

nitrogen generally resulted in distribution ratios close to those expected on the basis of the pH difference (see above). Recent re-evaluation in tissues of rat and carp resulted in distribution ratios more or less according to the membrane potential with use of mechanical homogenizers (Ultraturrax), whereas analysis of the same tissues after homogenization with mortar and pestle under liquid nitrogen yielded values only slightly higher than with pH-equivalent distribution in the same tissues (Fig. 4; C. Schorer, G. Forcht, S. Glage and N. Heisler, unpubl. data). When the deep frozen tissue homogenates were treated with a mechanical homogenizer (Ultraturrax) subsequent to the mortar and pestle procedure, the total tissue ammonia concentrations were higher by factors of 6–25. In this context, mechanical homogenization is accordingly to be viewed with a certain amount of caution. During homogenization under liquid nitrogen the degree of NH_3 liberation from terminal amino groups is evidently smaller, but may still bias the obtained ratio of $D_{NH_3}/D_{NH_4^+}$ to a certain extent.

4.3 Distribution of Infused Ammonia

Determination of the tissue ammonia content is evidently a critical step in the analysis of the intracellular/extracellular distribution ratios. Even when homogenization can be carried out at temperatures well below 0 °C, the

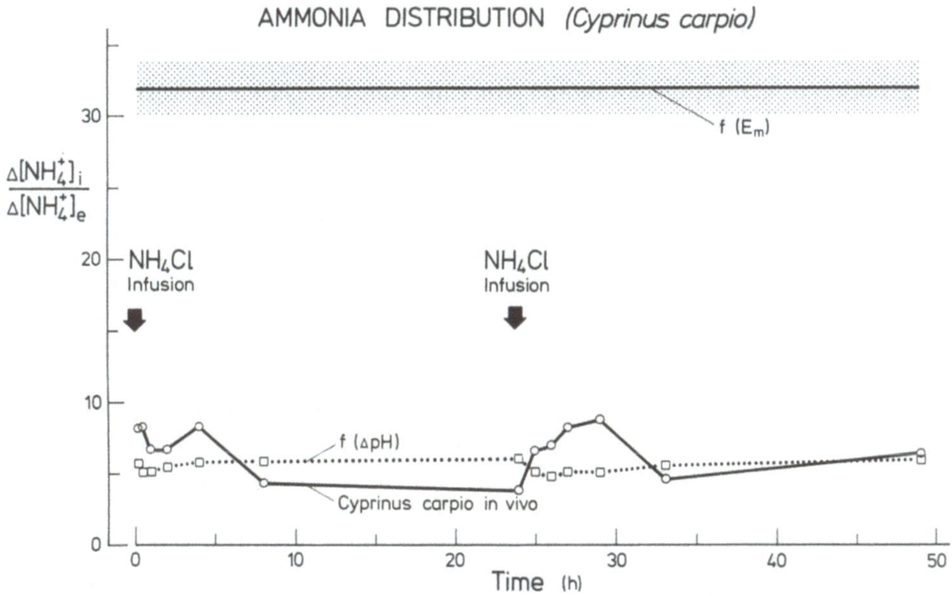

Fig. 5. Distribution of ammonia infused into the bloodstream of carp (*Cyprinus carpio*) between intracellular and extracellular body fluid compartments, compared to distribution according to pH difference [f(ΔpH)] and membrane potential [f(E_m)]. See text

temperature will always rise for short periods during transfer and deproteinization. Unknown is the extent to which deproteinization itself adds to enhancing the homogenate ammonia concentration. On the other hand, although deep frozen, the tissue powder of the homogenate is vigorously bubbled with nitrogen and the amount of ammonia evaporating during the process is difficult to estimate.

In order to approach the problem from a completely different side, ammonia was infused into the blood stream of carp, and the distribution of additional ammonia was monitored for 24 h after the infusion. The additional ammonia is expected to distribute between intra- and extracellular space as a function of the ratio of $D_{NH_3}/D_{NH_4^-}$ and relative fluid volumes. Complete analysis on the basis of changes in plasma pH and ammonia concentration, the ammonia elimination to the environment, and separately determined data on intracellular pH and fluid volumes (Heisler 1984, 1986b; Neumann and Heisler, unpublished data) by application of mass conservation models (Heisler 1984, 1986a) indicated a ratio of $\Delta[NH_4^+]_i/\Delta[NH_4^+]_e$ of about 3–8. This is slightly lower than ratios determined by deep frozen tissue homogenate analysis, and is close to that predicted from the pH difference (~ 6). It is significantly different from that expected from distribution according to the membrane potential (~ 32) (Fig. 5; Heisler, F. Forcht, R. Neumann and N.A. Andersen, unpubl. data).

4.4 Indirect Evidence

The notion of main transmembrane ammonia transfer by non-ionic diffusion of NH_3 across cell membranes is further supported by indirect evidence. Exposure of isolated cell preparations generally results in an initial very fast alkalinization, attributable to non-ionic diffusion with subsequent intracellular ionization of NH_3, and recovery due to slow diffusive entry of NH_4^+ (or simultaneous regulatory transfer of H^+ ion-equivalents, e.g. Boron and De Weer 1976). Also, infusion of NH_4Cl results in a pronounced extracellular acidosis, induced by non-volatile H^+ ions left behind during preferential diffusion of NH_3 to intracellular and extracorporeal fluid compartments (Cameron and Heisler 1983; J.B. Claiborne, N.A. Andersen and N. Heisler, unpubl.).

If transfer of ammonia was mainly by non-ionic diffusion, then alkalinization of the body fluids should enhance its elimination by increasing the P_{NH_3} diffusion gradient, whereas the relative concentration of ammonium and thus ionic elimination would hardly be affected. Testing this hypothesis by infusion of bicarbonate into trout, inducing an extracellular non-respiratory alkalosis, did actually result in a decrease in total ammonia elimination, providing another piece of indirect evidence towards non-ionic elimination of ammonia (Tang et al. 1992).

5 Elimination of Ammonia from the Body Fluids

Although ammonia can be voided to a certain small extent also at aerial interfaces, such as lung alveolar epithelium or the skin, these pathways are quantitatively negligible due to the physicochemical properties of ammonia, namely its low capacitance coefficient in air. In contrast, large aqueous solubility and high pK both result in small partial pressure activity of the highly toxic NH_3 in ammoniotelic waterbreathing fishes, with partial pressure differences between blood and water in the range of $\sim 50 \times 10^{-6}$ mmHg (e.g. Cameron and Heisler 1983). For elimination at a gaseous interface partial pressure and, at constant pH, total ammonia concentration had to rise by a factor of $\sim 10^5$, on the basis of the much smaller capacitance for ammonia of air as compared to water ($\sim 10^{-3}$) and a rate of aerial ventilation smaller by the capacitance ratio of CO_2 between water and air (~ 0.03). The resulting range of partial pressure gradients for aerial elimination would be prohibitive in face of the pronounced toxicity of NH_3 and may actually have provided evolutionary pressure towards the energetically less efficient ureo- or uricotelism of air-breathing vertebrates.

In contrast to aerial elimination, ammonia may be voided to the aquatic environment also in the form of NH_4^+, either by ionic diffusion or by active carrier-mediated mechanisms. As at the cell membrane level, low total ammonia concentrations and in particular, P_{NH_3} values are required at the environmental level in order not to introduce complicating factors, such as back-diffusion of

NH_3. Thus, ammonia elimination is limited by pH and the amount of available volume. Due to the rising fraction of highly diffusible NH_3, high water pH is definitely a major problem in this context. While some fish species may have solved this problem by utilising an unstirred layer at the water interface in combination with the acidifying effect of CO_2 elimination (see below), others have resorted to urea as the main nitrogenous endproduct (e.g. Wood et al. 1994).

Numerous investigations and intensive discussion with respect to the form in which ammonia is transferred across the gill epithelium of fishes (for review: Evans 1980a,b, 1984, 1986; Heisler 1984, 1986b; Evans and Cameron, 1986; Wood 1993) has generally confirmed that all three possible modes, non-ionic diffusion, ionic diffusion and ionic carrier-mediated active transfer associated with movement of Na^+ may actually be utilized under certain conditions. Experimental conditions, however, have often forced utilization of mechanisms not used under physiological conditions.

5.1 Elimination of Ammonia and Homoiostasis

Metabolic production and ionization of NH_3 has an alkalinizing effect, which is reversed when NH_3 is eliminated by non-ionic diffusion. The effect of ionic elimination of NH_4^+ depends on the associated counter- or co-ions. Simultaneous facilitated diffusion of NH_4^+ and HCO_3^-, utilized in intracorporeal aqueous pathways, is electro- as well as acid–base neutral, but will considerably affect osmoregulation. Any other combination of fluxes will affect acid–base regulation, except exchange of NH_4^+ against H^+. This is, however, most unlikely to occur on a passive basis, because of minute diffusion gradients for H^+ ions as compared to NH_4^+.

Ionic ammonium elimination on a sizeable scale will result in a tremendous alkalinizing load for the acid–base regulatory system of up to 15% of the oxygen consumption (cf. Heisler 1984b, 1986b, 1988a). This disturbance of homoiostasis is to be corrected for by appropriate epithelial acid-base regulatory mechanisms. This implication renders ionic transfer an alternative of energetic disadvantage during physiological conditions.

5.2 Ionic Diffusion of Ammonium

Passive diffusive elimination of NH_4^+ takes place mainly through relatively 'leaky' paracellular junctions of the gill epithelium in marine fish (e.g. Goldstein et al. 1982; Evans and More 1988; Evans et al. 1989; cf. Evans and Cameron 1986). In freshwater fish this pathway is less important, because the ion conductivity of paracellular pathways is smaller by orders of magnitude than in marine fishes, as indicated by the unidirectional flux rates of N_a^+ and Cl^- (cf. Evans 1980, 1984, 1986).

5.3 Gill Diffusion Coefficient for Ammonia

Voiding of ammonia to the environmental water is facilitated by the high aqueous solubility of NH_3, and the high relative ventilation rate of water breathing animals, induced by the much lower oxygen content of the breathing medium, water (Rahn 1966; cf. Heisler 1986b). Passive elimination of ammonia by non-ionic diffusion follows Fick's law of diffusion and is accordingly a function of partial pressure difference, diffusion coefficient and effective gill, surface area. Estimation of the NH_3 diffusion coefficient of fish gills is difficult because of the possible interaction of ionic diffusion and active carrier-mediated transfer of ammonium. However, based on general conclusions, carrier-mediated mechanisms can be assumed to be inactive at low environmental and plasma ammonia concentrations, and passive ionic diffusion of ammonium is considered negligible in freshwater-acclimated teleosts (cf. Evans 1984, 1986).

Determination of the transfer rate between blood and environmental water of rainbow trout at a low environmental ammonia concentration resulted in an estimate of the 'apparent' ammonia gill diffusion coefficient (D) of 1.31×10^{-5} cm^2/s (Cameron and Heisler 1983). This value complies well with the diffusivity of gill tissue for CO_2 and is about 25% less than the aqueous diffusion coefficient (Cameron and Heisler 1983). Another approach, utilizing similar technology, but attempting to completely block NH_4^+/Na^+ ion exchange (which might affect estimation of D) by exposure to low water pH of 4 (Wright and Wood 1985) resulted in a 35% smaller value. The apparently lower diffusive conductance determined under these conditions may be effected by the considerably increased mucus production (generally observed at low environmental water pH), which results in a pronounced increase of transepithelial diffusion distance (Ultsch and Gros 1979). On the other hand, the acidifying effect of CO_2 hydration in an unstirred water layer on the epithelium may generally result in an enchanced gradient for P_{NH_3}, leading to larger diffusion coefficients.

Differentiation at normal environmental pH between active ion exchange and passive NH_3 diffusion is possible by utilizing specific characteristics of the involved mechanisms, thereby circumventing these sources of interference. Due to the nature of the mechanisms, diffusive transfer is linearly related to the partial pressure difference between plasma and water, whereas the activity of active ionic exchange is somewhat related to the absoulte level of total plasma concentration (Cameron and Heisler 1983). Analysis of total ammonia transfer between fish and water in a variety of conditions of external and internal total ammonia concentrations and effective ΔP_{NH_3} at the gill epithelium of rainbow trout and carp, resulted in highly significant linear correlations for data sets with total plasma ammonia concentrations below 0.2 mM, with the overall transfer factor for carp being about 20% smaller than in trout (Fig. 6). The diffusion coefficient determined from the slope of the regression for trout complies well with that originally estimated by Cameron and Heisler (1983). The tight linear relationship is a very good indicator for non-ionic diffusion as the dominating transfer

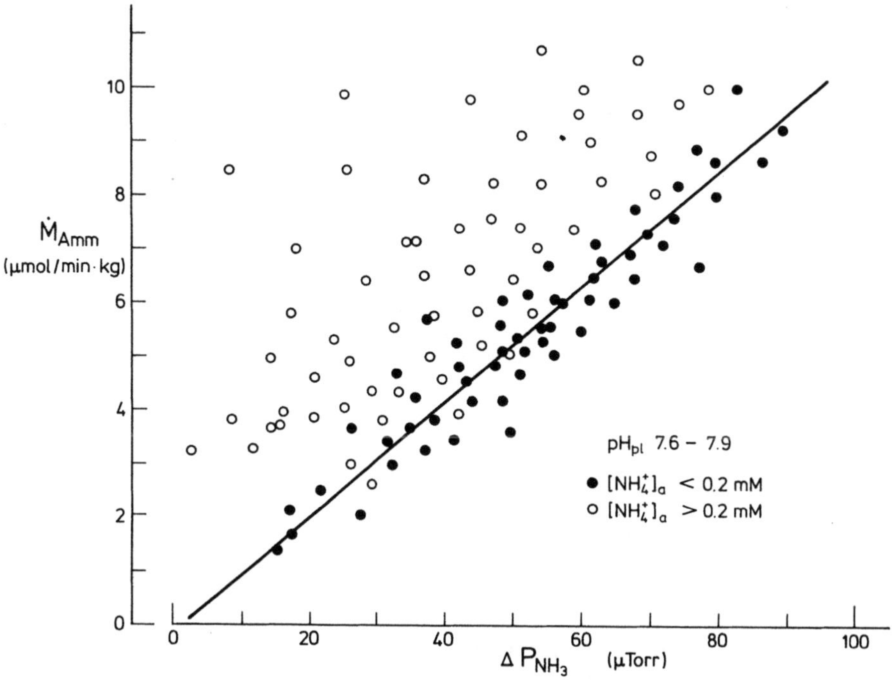

Fig. 6. Rate of ammonia elimination ($\dot{M}_{Amm}$) as a function of the effective partial pressure difference (ΔP_{NH_3}) across the gill epithelium in trout (*Oncorhynchus mykiss*). At plasma $[NH_4^+]$ below 0.2 mM the elimination rate is directly related to ΔP_{NH_3}, whereas total ammonia release generally exceeds the values predicted from this relationship at higher plasma ammonia levels. See text

mechanism. This notion is confirmed by the fact that the same data sets lack any correlation between any one of plasma or water $[NH_4^+]$ or the effective $\Delta[NH_4^+]$ between plasma and water, and the transepithelial ammonia transfer rate.

Although there is little doubt about non-ionic diffusion as the predominant elimination mechanism on the basis of the elimination characteristics, the absolute values of diffusion coefficients determined in whole animal preparations have to be viewed with caution. Too many factors may bias an absolute estimate of this parameter, such as unknown interface conditions with respect to pH, local electrical charges and finally total ammonia concentrations. This need for caution may apply in particular to the conditions in so-called unstirred water layers adherent to the gill epithelium (see below).

5.4 Active Ionic Elimination of Ammonium and Ammonia Shuttling

Ammonia elimination rates at plasma ammonia levels above 0.2 mM (associated with elevated environmental ammonia concentration), however, are generally

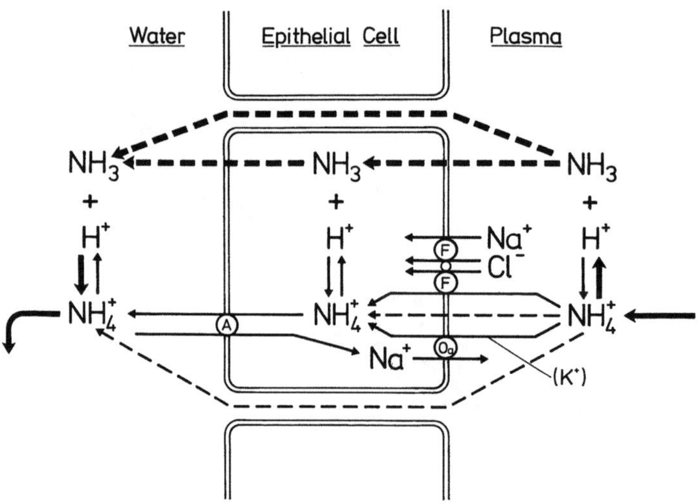

Fig. 7. Mechanisms of active ammonia extrusion. See text

higher to a significant extent than expected on the basis of the linear regression of data sets with $[NH_4^+]_{pl}$ < 0.2 mM (Fig. 6), suggesting active carrier-mediated transfer being activated after surpassing a threshold of $[NH_4^+]_{pl}$ of ~ 0.2 mM. Several ion transfer mechanisms may be responsible for active extrusion of ammonium, such as the well known apparently single step NH_4^+/Na^+ exchange (for review: Maetz 1974; Evans 1980, 1984, 1986; Evans and Cameron 1986), or an apical amiloride-sensitive NH_4^+/Na^+ exchange, supported by a basolateral furosemide/bumetanide-sensitive co-transfer of NH_4^+, 2 Cl^- and Na^+ (Evans and More 1988), or a ouabain-sensitive $NH_4^+(K^+)/Na^+$ exchange (Claiborne et al. 1982; Evans et al. 1989) (cf. Fig. 7), but their quantitative contribution under physiological conditions is still rather unclear.

The general participation of Na^+ does not lead very much further in characterizing the contribution of ionic transfer. Unidirectional Na^+ fluxes are difficult to interpret in terms of being related to carrier-mediated ammonium extrusion. Involvement of sodium in ammonia-independent exchange processes like H^+/Na^+ and the osmoregulatory Na^+ transport, compensating for diffusive Na^+ loss (freshwater fish) or entry (marine species), limits their value as an indicator for ionic NH_4^+ movement, in particular because of the acidifying effect of non-ionic NH_3 elimination possibly leading to the activation of H^+/Na^+ transfer. These interferences can be taken care of in isolated and very tightly controlled experimental approaches, for instance in isolated head preparations or skin preparations in Ussing-type chambers. In physiological experiments utilizing whole animals, however, unidirectional Na^+ flux can only depict the sum of flux of all possible counterions, representing the maximal possible contribution of a specific Na^+ related transfer process.

A further complicating factor is the possibility of ammonia shuttling across the gill epithelium. Under conditions of low environmental ammonia levels, NH_3 diffusion may well account for complete diffusive elimination at low plasma total ammonia concentration (see above), but reversal of the normal outward diffusion gradient of ΔP_{NH_3} results in ammonia uptake by non-ionic diffusion that requires active elimination of this highly toxic substance. The range of P_{NH_3} values in which this will be possible, however, may be quite small. Active transfer of NH_4^+ counteracts non-ionic entry of NH_3, but, as discussed above, continuous transfer of NH_4^+ with diffusion of NH_3 in the opposite direction is equivalent to net translocation of H^+ ions. Ammonia shuttling and associated unidirectional transfer of H^+ attains considerable rates (for nomograms: Heisler 1984, 1990a), resulting in an alkalinization of the body fluids that evidently cannot be corrected for by the limited capacity of acid–base regulatory mechanisms. (cf. Cameron and Heisler 1983). The effect of such a shuttle mechanism is an elevation of the unidirectional Na^+ influx far beyond the (net) ammonia elimination rate, complicating any comparative analysis of these parameters.

Further complicating the pattern, Na^+ may not be involved at all. On the basis of the previously described active Na^+-related mechanisms for NH_4^+ translocation, ammonia shuttling is expected to effect an equivalent elevation of unidirectional Na^+ flux. Ammonia-loading experiments in catfish, however, have not indicated any significant change in Na^+ influx, which, together with the obtained net acid–base flux data, led to the conclusion that the extrusion of NH_4^+ was performed by NH_4^+/H^+ exchange rather than by conventional NH_4^+/Na^+ translocation (Cameron 1986). These data suggest that the carriers involved may be relatively unspecific and may, under certain conditions, accept cations other than Na^+ on the inward route also. The selectivity may be regulated by ion gating mechanisms adjusted to the requirements of the specific state of homoiostasis. At least in trout and carp, however, involvement of an acid–base-neutral mechanism such as NH_4^+/H^+ exchange is unlikely, taking into account the pronounced alkalinization generally observed during ammonia loading by non-ionic diffusion from the environment after establishment of a new steady state (e.g. Cameron and Heisler 1983; Claiborne and Heisler 1984).

The ammonia shuttle mechanism results in an apparent H^+/Na^+ ion exchange, effective both for the internal fluid homoiostasis of the fish, and for the environmental water. Since both H^+ ions and NH_4^+ are considered to be competing for the same carrier mechanism (for review: Maetz 1974; Evans 1979, 1980a, 1984, 1986) the acid–base relevant ion transfer rate limitation imposed by the carrier capacity is a priori not expected to be changed by ammonia shuttling. However, affinity and K_m values of the carrier for transfer of H^+ ions and NH_4^+ may be considerably different. Also, transfer of the larger NH_4^+ ions may require more energy compared to H^+ ions. Accordingly, ammonia shuttling, although apparently conducting just the same net transfer as the H^+/Na^+ exchange, may reduce the original transfer capacity of the mechanism. Unfortunately, no data at all are available as to the above mentioned characteristics of the $NH_4^+/H^+(Na^+)$ ion exchange mechanism in vivo.

5.5 Elimination of Infused Ammonia

One way to effectively diminish the overall influence of active carrier-mediated mechanisms and assess the capacity of non-ionic diffusion is to overload the active carrier-mediated systems by infusion of ammonia into the blood stream. In carp, infusion of NH_4^+ raised plasma ammonia levels up to values of 1.5 mM, resulting in ammonia elimination rates up to 60 µmol/(min × kg) (Fig. 8), higher by a factor of > 6 than the maximal acid–base relevant transfer rates reported during severe acid–base disturbances in this species ($<$ 10 µmol/(min × kg, Claiborne and Heisler 1984, 1986). Thus, at least the major fraction of the elimination rate is expected to have occurred on the basis of passive diffusion. This view is confirmed by modeling the rate of non-ionic ammonia diffusion on the basis of the effective ΔP_{NH_3} across the gill epithelium, and previously reported gill diffusion coefficients (Fig. 8). The modeled rate of ammonia release closely resembles the experimentally determined pattern of ammonia elimination, with only slightly higher values for the modeled flux rates in a range of total ammonia output above 7–10 µmol/(min × kg). This comparison confirms the reliability of the diffusional transfer factors determined at low plasma ammonia levels (less than 0.2 mM) from the effective ΔP_{NH_3} and $\dot{M}_{NH_3}$ at neutral water pH.

Unidirectional Na^+ influx under these conditions was enhanced as compared to control conditions, generally by less than 5 mmol/(min · kg), except for the time

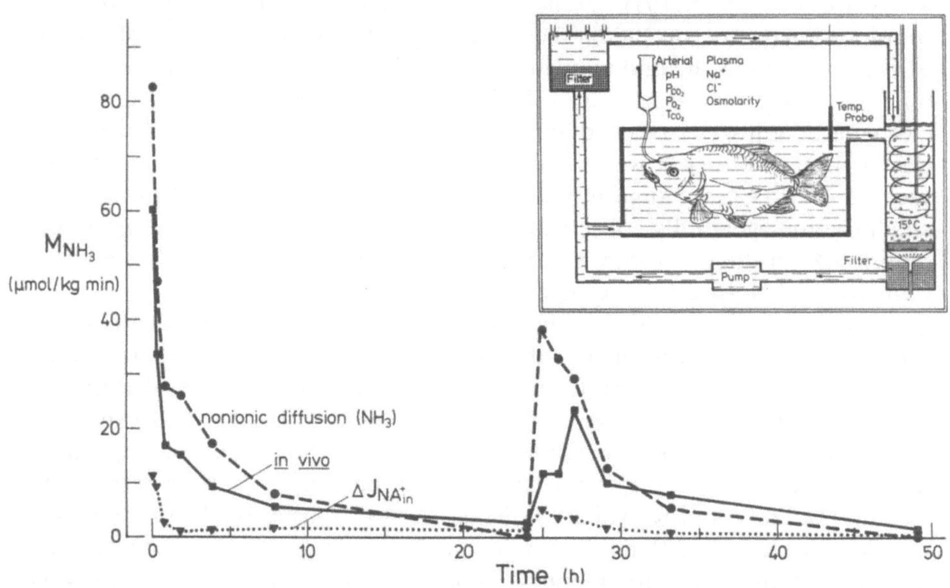

Fig. 8. Ammonia elimination in carp (*Cyprinus carpio*) after NH_4Cl infusion ('in vivo') in comparison with the transfer rate expected from non-ionic diffusion (estimated from the effective ΔP_{NH_3} and the gill diffusion coefficient), and the activity of Na^+-related ion exchange processes ($\Delta J^+_{Na_{in}}$). See text

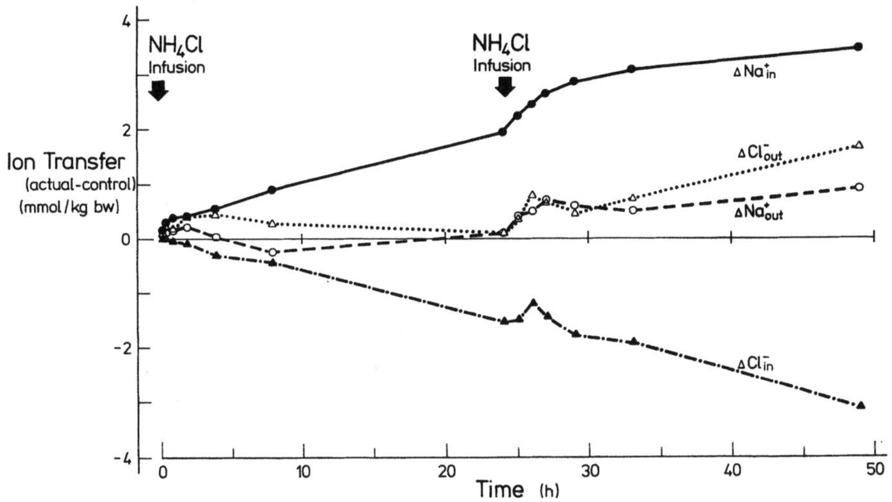

Fig. 9. Changes in unidirectional fluxes of Na^+ and Cl^- after infusion of NH_4Cl in the carp (*Cyprinus carpio*). See text

period immediately after infusion (11mmol/(min × kg)) (Fig. 9). Attributing this elevation exclusively to activation of Na^+/NH_4^+ exchange in order to correct for elevated plasma total ammonia levels, the contribution of active carrier mediated processes for ammonia elimination would generally have been less than 15%. However, the change in Na^+ flux is a poor indicator of active extrusion of NH_4^+, because of the involvement of Na^+ in other processes also. Infusion of relatively large amounts of NH_4Cl results in a considerable acidification of the body fluids, due to non-ionic elimination of NH_3. Thus, most likely, Na^+/H^+ exchange was triggered by the acidosis and was responsible for at least a major fraction of the rise in Na^+ influx.

5.6 Utilization of Mechanisms: Non-Ionic Diffusion vs Carrier-Mediated Ionic Transfer

The quantitative contribution of active carrier-mediated ion transfer mechanisms and passive diffusion is not settled. The diffusion capacities determined for the gills of fishes certainly allow complete elimination of the endogenously produced ammonia by non-ionic diffusion. NH_3 diffusion has actually been depicted as the main transfer route (generally more than 70% in freshwater fishes) by a number of experimental studies, with a fraction of 20 to 30% eliminated by ionic diffusion in marine teleost species (e.g. Holeton et al. 1983; Claiborne and Heisler 1984, 1986; Evans and More 1988; Evans et al. 1989; for review: Heisler 1984, 1986b, 1988c). Other experiments have yielded results that can only be explained on the basis of almost complete active extrusion (e.g.

Wood 1989). The general difference appears to reside in different environmental conditions. Only when the external ammonia rises to values interfering with diffusive elimination of NH_3 and plasma ammonia concentration becomes elevated, is active extrusion of NH_4^+ (cf. Fig. 7) is apparently triggered before toxic levels of ammonia are attained (Kerstetter et al. 1970; Maetz 1973; Payan and Maetz, 1973; Evans 1977, 1980b; Cameron and Heisler 1983, 1985) for review see also Maetz, 1974; Evans 1984, 1986; Evans and Cameron 1986). As long as the water P_{NH_3} (in seawater also $[NH_4^+]$) is low enough to allow establishment of the required diffusion gradients with plasma ammonia concentrations below 0.2 mM, transfer will take place by non-ionic (in seawater also ionic) diffusion. Such conditions are usually met in the natural environment, but the build-up of ammonia in experimental recirculation systems requires particular attention and may actually have led to overemphasizing the role of active ammonium extrusion at least under physiological conditions.

5.7 Unstirred Layers: Tools for Ammonia Elimination at High pH?

The amount of ammonia that can be eliminated by the energetically least costly mechanism on non-ionic diffusion is a function of ammonia production rate, rate of respiratory water flow past the epithelium and the average partial pressure gradient across the epithelium. In a typical freshwater fish at 15 °C, the addition of a resting ammonia production of ~5 μmol/(min · kg body weight) to 200 ml/min of respiratory water results in total expired ammonia water concentration of about 25 μM. At a water pH of 7.6 this concentration corresponds to a P_{NH_3} of $\sim 6 \times 10^{-6}$ mmHg, leaving a generous P_{NH_3} gradient from plasma P_{NH_3} of $\sim 95 \times 10^{-6}$ mmHg (plasma pH 7.9 and plasma total ammonia 0.2 mM) which is sufficient to rid the metabolic production rate by non-ionic diffusion without any problems (for constants of pK′ and α_{CO_2} of the Hendersen-Hasselbalch equation: Cameron and Heisler 1983).

This favorable scenario of P_{NH_3} gradients changes dramatically with relatively small rises in pH. At high water pH even very small water ammonia concentrations result in high P_{NH_3} values due to the pK' value of ammonia in the alkaline pH range (cf. Heisler 1990a). A water ammonia concentration of 25 μM at pH 9 will result in P_{NH_3} of $\sim 150 \times 10^{-6}$ mmHg, reversing the P_{NH_3} gradient and effecting inward movement of ammonia (cf. Heisler, 1990a). Although the elimination of ammonia is expected to be performed by active ionic transfer under such conditions, there is no evidence confirming a rise in movement of the potential counterion, Na^+. In contrast, data of Wright and Wood (1985) have indicated a falling tendency of Na^+ influx with rising pH, approaching zero at water pH of 11. This seems to indicate that NH_4^+/Na^+ exchange is not involved, but still leaves the possibility of elimination by NH_4^+/H^+ ion exchange, as proposed by Cameron (1986). The K_m value of the carrier for H^+ transfer, however, had to be extremely small because of the very low concentration of H^+ ions at high pH.

An unstirred water layer on the gill epithelium has been proposed as an important factor for ammonia elimination at extreme water pH values (Wright and Randall 1987) and may actually provide an alternative mechanism. Provided CO_2 diffusing the epithelial interface would be ionized on the mucosal side of the epithelium in an unstirred water layer, pH and accordingly P_{NH_3} would be reduced at the interface. The crucial point in this mechanism is the availability of carbonic anhydrase at the apical surface of the gill epithelium, facilitating rapid hydration and dissociation of CO_2 at a rate sufficient to cover the elimination rate of NH_3 (Fig. 10).

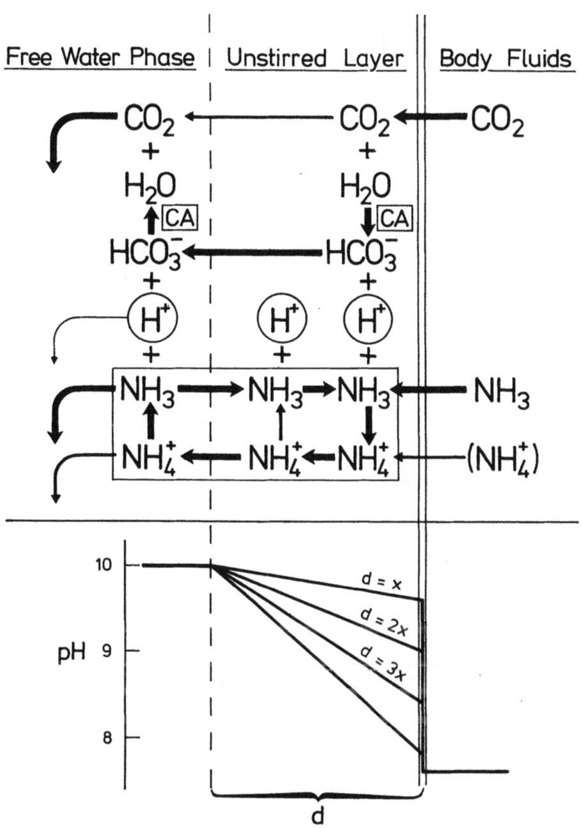

Fig. 10. Model of the effects of an unstirred layer at the gill epithelium on the elimination of ammonia at high environmental water pH. With presence of carbonic anhydrase at the apical surface of the epithelium rapid hydration and release of H^+ from almost complete dissociation will result in a considerable local reduction of pH and ionization of NH_3, thus maintaining the diffusion gradient required for further non-ionic elimination. The degree of acidification depends on various factors like $\dot{M}_{NH_3}$, the rate of CO_2 production and hydration, the rate of NH_3 back-diffusion from the bulk interface epithelium, pH of the bulk fluid and the thickness of the unstirred layer ('*d*'). Lack of carbonic anhydrase along the diffusion distance for NH_4^+ and HCO_3^- will result in an acid disequilibrium pH, further supporting elimination of NH_3 by non-ionic diffusion. See also text

The efficiency of this mechanism clearly depends on two factors: the fraction of CO_2 that is hydrated directly at the apical side of the epithelium, and the rate at which the hydration process is reversed. For basic functioning only a small fraction of the CO_2 production needs to be hydrated, providing H^+ ions for ammonia production, the rate of which is usually less than 10 (15)% of the oxygen consumption. Obviously, with a larger fraction of hydration, pH and P_{NH_3} will fall lower at the apical interface. With carbonic anhydrase being available along the entire distance of the outward diffusion way (i.e. the unstirred layer represented by a carbonic anhydrase-containing mucus layer) the process would gradually be reversed along the diffusion distance and P_{NH_3} as well as pH would be elevated to the level expected at the epithelium without any unstirred layer at the interface with the free-flowing water (Fig. 10; Heisler 1990).

The highest efficiency of this mechanism, however, would be achieved even with little activity of carbonic anhydrase if this activity were strictly attached to the apical epithelial surface. Lack of carbonic anhydrase along the diffusion pathway will not allow reconversion as the changing conditions gradually approach those of bulk environmental water, because of the slow uncatalyzed dehydration time (halftime in the range of 100 s; Kern 1960). This blockade of gradual reconversion of HCO_3^- to CO_2, and, due to relevant equilibria (cf. Heisler 1984, 1986a), also most of the reconversion of CO_3^- to HCO_3^- will leave the majority of dissociated H^+ free, producing an acid disequilibrium pH, while CO_3^-, HCO_3^- and NH_4^+ will be carried away by the bulk water flow in disequilibrium with CO_2 (Heisler 1990).

The advantage of such a mechanism is obvious: instead of energy expenditure for elimination of shuttling NH_4^+ ammonia could be eliminated by non-ionic diffusion which functions also under adverse conditions presented by high environmental pH. Unfortunately, data are not available to directly support this notion: lack of unidirectional Na^+ fluxes is not a stringent condition supporting this mechanism, for reasons lined out above. The conditions at which an unstirred layer could provide the expected function have not been worked out yet, but certainly the thickness of such a layer has to be large enough to protect the acid microenvironment against too quick diffusional escape of H^+ to the bulk water fluid. Thick unstirred water layers, on the other hand, interfere with gas exchange, a task which is performed in many fish and conditions just marginally above the requirements of the system. A fine balance between the contrasting requirements of gas exchange, ammonia elimination and ionic exchange may exist to fulfill all of these homoiostatic functions to satisfaction.

6 Conclusions

As suggested by the similarity of the aqueous diffusion coefficients of NH_3 and NH_4^+ and the much higher diffusion gradients for NH_4^+, ammonia transfer along aqueous diffusion pathways takes place mainly by 'facilitated ionic diffusion' with simultaneous transfer of NH_4^+ and HCO_3^- originating from metabolic CO_2.

Transfer across cell membranes occurs mainly by non-ionic diffusion, as indicated by intracellular/extracellular total ammonia concentration ratios. High intracellular ammonium concentrations are apparently obtained by mechanical homogenization (in contrast to mortar and pestle under liquid N_2), possibly causing liberation of terminal amino groups. This notion is strongly supported by theoretical model calculations as well as by the distribution of infused ammonia between intracellular and extracellular space.

Elimination of ammonia from the body fluids takes place mainly by non-ionic diffusion (also ionic diffusion in marine species), as long as plasma total ammonia remains low. Carrier-mediated ionic transfer processes for NH_4^+ are utilized only when the environmental conditions do not allow establishment of sufficient P_{NH_3} gradients for diffusive elimination. A microenvironment of an unstirred gill water layer acidified by carbonic-anhydrase-facilitated rapid ionization of CO_2 may provide the basis for non-ionic diffusion against an apparent P_{NH_3} gradient of bulk interface fluids.

Thus, elimination of ammonia through aqueous phases, membranes and epithelia exhibits a close similarity to the mechanisms responsible for elimination of CO_2, generally following the laws of respiratory gas exchange.

References

Aster PL, Casey CA, Vorhaben JE, Campell JW (1982) Hepatic ammoniogenesis in the catfish, *Ictalurus punctatus.* Am Zool 21: 1029

Arruda JAL, Dytko G, Withers L (1984): Ammonia transport by the turtle urinary bladder. Am J Physiol 246: F635–F647

Boron FW, De Weer P (1976) Intracellular pH transients in squid giant axons caused by CO_2, NH_3 and metabolic inhibitors. J Gen Physiol 67: 91–112

Boutilier RG, Heisler N (1988) Acid–base regulation and blood gases in the anuran amphibian, *Bufo marinus*, during environmental hypercapnia. J Exp Biol 134: 79–98

Cameron JN (1986) Responses to reversed NH_3 and NH_4^+ gradients in a teleost (*Ictalurus punctatus*), an elasmobranch (*Raja erinacea*), and a crustacean (*Callinectes sapidus*): evidence for NH_4^+/H^+ exchange in the teleost and the elasmobranch. J Exp Zool 239: 183–195

Cameron JN, Heisler N (1983) Studies of ammonia in the rainbow trout: physico-chemical parameters, acid–base behavior and respiratory clearance. J Exp Biol 105: 107–125

Cameron JN, Heisler N (1985) Ammonia transfer across fish gills: a review. In: Gilles R (ed) Circulation, respiration and metabolism. Springer, Berlin Heidelberg New York, pp 91–100

Claiborne JB, Heisler N (1984) Acid–base regulation and ion transfers in the carp (*Cyprinus carpio*) during and after exposure to environmental hypercapnia. J Exp Biol 108: 25–43

Claiborne JB, Heisler N (1986) Acid–base regulation and ion transfers in the carp (*Cyprinus carpio*): pH compensation during graded long- and short-term environmental hypercapnia and the effect of bicarbonate infusion. J Exp Biol 126: 41–61

Claiborne JB, Evans DH, Goldstein L (1982) Fish branchial Na^+/NH_4^+ exchange is via basolateral Na^+–K^+– activated ATPase. J Exp Biol 96: 431–434

Evans DH (1977) Further evidence for Na^+/NH_4^+ exchange in marine teleost fish. J Exp Biol 70: 213–220

Evans DH (1979) Fish. In: Maloiy GMO (ed) Comparative physiology of osmoregulation in animals vol I. Academic Press, New York, pp 305–390
Evans DH (1980a) Kinetic studies of ion transport by fish gill epithelium. Am J Physiol 238: R224–R230
Evans DH (1980b) Osmotic and ionic regulation by freshwater and marine fishes. In: Ali MA (ed) Environmental physiology of fishes. Plenum Press, New York, pp 93–122
Evans DH (1980c) Na^+/NH_4^+ exchange in the marine teleost, *Opsanus beta*: stoichiometry and role in Na^+ balance. In: Lahlou B (ed) Epithelial transport in the lower vertebrates. Univ Press, Chicago, pp 197–205
Evans DH (1984) The roles of gill permeability and transport mechanisms in euryhalinity. In: Hoar WS, Randall DJ (eds) Fish physiology vol XB. Academic Press, Orlando, pp 239–283
Evans DH (1986) The role of branchial and dermal epithelia in acid–base regulation in aquatic animals. In: Heisler N (ed) Acid–base regulation in animals. Elsevier, Amsterdam, pp 139–172
Evans DH, Cameron JN (1986) Gill ammonia transport. J Exp Zool 239: 17–23
Evans DH, More KJ (1988) Modes of ammonia transport across the gill epithelium of the dogfish pub (*Squalus acanthias*). J Exp Biol 138: 375–397
Evans DH, More KJ, Robbins SL (1989) Modes of ammonia transport across the gill epithelium of the marine teleost fish *Opsanus beta*. J Exp Biol 138: 339–356
Fromm PO (1963) Studies on renal and extrarenal excretion in a freshwater teleost, *Salmo gairdneri*. Comp Biochem Physiol 10: 121–128
Garvin JL, Burg MB, Knepper MA (1987) NH_3 and NH_4^+ transport by rabbit renal proximal straight tubules. Am J Physiol 252: F232–F239
Goldstein L, Forster G, Fanelli GM Jr (1964) Gill blood flow and ammonia excretion in the marine teleost, *Myoxocephalus scorpius*. Comp Biochem Physiol 12: 489–499
Goldstein L, Claiborne JB, Evans DH (1982). Ammonia excretion by two marine teleost fish: the importance of NH_4^+ permeance. J Exp Zool 219: 395–397
Gros G, Moll W (1974) Facilitated diffusion of CO_2 across albumin solutions. J Gen Physiol 64: 356–371
Heisler N (1984) Acid–base regulation in fishes. In: Hoar WS, Randall DJ (eds) Fish physiology, vol XA. Academic Press, New York, pp 315–401
Heisler N (1986a) Buffering and transmembrane ion transfer processes. In: Heisler N (ed) Acid–base regulation in animals. Elsevier, Amsterdam, pp 3–47
Heisler N (1986b) Acid–base regulation in fishes. In: Heisler N (ed) Acid–base regulation in animals. Elsevier, Amsterdam, pp 309-356
Heisler N (1986c) Mechanisms and limitations of fish acid–base regulation. In: Nilsson S, Holmgren S (eds) Fish physiology: Recent advances. Croom-Helm, London, pp 24–49
Heisler N (1988a) Acid–base regulation in elasmobranch fishes. In: Shuttleworth TJ (ed) Physiology of elasmobranch fishes. Springer, Berlin Heidelberg New York, pp 215–252
Heisler N (1988b) Ion transfer processes as mechanisms for acid–base regulation. In: Wood SC (ed) Lenfent C (exec ed) Lung biology in health and disease, vol. 39. Comparative pulmonary physiology: current concepts. Marcel Dekker, New York, pp 539–583
Heisler N (1988c) Acid–base regulation in fishes I. Mechanisms. In: Morris R et al. (eds) Acid toxicity and aquatic animals. Society of Experimental Biology Seminar Series. Univ Press. Cambridge
Heisler N (1990a) Acid–base regulation: Interrelationships between gaseous and ionic exchange. In: Boutilier RG (ed) Vertebrate gas exchange from environment to cell. Advances in environmental and comparative physiology, vol 6. Springer, Berlin Heidelberg New York, pp 211–251
Heisler N (1990b) Mechanisms of ammonia elimination in fishes. In: Mellinger J, Truchot JP, Lahlou B (eds) Animal nutrition and transport processes, II. Transport, respiration and excretion. Comparative physiology. Karger, Basel, pp 137–151
Heisler N, Forcht G, Ultsch GF, Anderson JF (1982) Acid–base regulation in response to environmental hypercapnia in two aquatic salamanders, *Siren lacertina* and *Amphiuma means*. Respir Physiol 49: 141–158
Holeton GF, Heisler N (1983) Contribution of net ion transfer mechanisms to the acid–base regulation after exhausting activity in the Larger Spotted Dogfish (*Scyliorhinus stellaris*). J Exp Biol 103: 31–46

Holeton GF, Neumann P, Heisler N (1983) Branchial ion exchange and acid–base regulation after strenuous exercise in rainbow trout (*Salmo gairdneri*). Respir Physiol 51: 303–318

Kern DM (1960) The hydration of carbon dioxide. J Chem Educ 37: 14–23

Kerstetter FH, Kirschner LB, Rafuse DD (1970) On the mechanism of sodium ion transport by the irrigated gills of rainbow trout (*Salmo gairdneri*). J Gen Physiol 56: 342–359

Kortüm G (1972) Lehrbuch der Elektrochemie, 5th edn. Verlag Chemie, Weinheim/Bergstraße, p 263

Maetz J (1973) Na^+/NH_4^+, Na^+/H^+ exchanges and NH_3 movement across the gills of *Carassius auratus*. J Exp Biol 58: 255–275

Maetz J (1974) Adaptation to hyper-osmotic environments. Biochem Biophys Perspect Mar Biol 1: 91–149

Payan P, Maetz J (1973) Branchial sodium transport mechanisms in *Scyliorhinus canicula*. Evidence for Na^+/NH_4^+ and Na^+/H^+ exchanges and for a role of carbonic anhydrase. J Exp Biol 58: 487–502

Pequin L (1962) Les teneurs en azote ammoniacal du sang chez la carpe (*Cyprinus carpio L*). C R Hebd Seances Acad Sci 255: 1795–1797

Pequin L, Serfaty A (1963) L'excretion ammoniacale chez un teleosteen dulcicole : *Cyprinus carpio* L. Comp Biochem Physiol 10: 315–324

Radford EP (1964) The physics of gases. In: Fenn WO, Rahn H (eds) Handbook of physiology, respiration, Sect 3, vol I. Am Physiol Soc, Washington, pp 125–152

Rahn H (1966) Aquatic gas exchange: theory. Respir Physiol 1: 1–12

Smith HW (1929b) The excretion of ammonia and urea by the gills of fish. J Biol Chem 81: 727–742

Schwartz JH, Tripolone M (1983) Characteristics of NH_4^+ and NH_3 transport across the isolated turtle urinary bladder. Am J Physiol 245: F210–F216

Tang Y, Lin H, Randall DJ (1992) Compartmental distributions of carbon dioxide and ammonia in rainbow trout at rest and following exercise, and the effect of bicarbonate infusion. J Exp Biol 169: 235–249

Ultsch GR, Gros G (1979) Mucus as a diffusion barrier to oxygen: possbile role in O_2 uptake at low pH in carp (*Cyprinus carpio*) gills. Comp Biochem Physiol 62(a): 685–689

Washburn EW (1929) International critical tables of numerical data. Physics, chemistry and technology, vol 31st ed. Mac Graw-Hill, New York, pp 259

Wood JD (1958) Nitrogen excretion in some marine teleosts. Can J Biochem Physiol 36: 1237–1242

Wood CM (1989) The physiological problems of fish in acid waters. In: Morris R, Taylor EW, Brown DJA, Brown JA (eds) Acid toxicity and aquatic animals. Soc Exp Biol Sem Ser, Uni Press, Cambridge, pp 58–97

Wood CM (1993) Ammonia and urea metabolism and excretion. In: Evans DH (ed) The physiology of fishes. Marine Science Series. CRC Press, Boca Raton, pp 343–378

Wood CM, Munger RS, Toews DP (1989): Ammonia, urea and H^+ distribution and the evolution of ureotelism in amphibians. J Exp Biol 144: 215–233

Wood CM, Bergman HL, Laurent P, Maina JN, Narahara A, Walsh P (1994) Urea production, acid–base regulation and their interactions in the Lake Magadi tilapia, a unique teleost adapted to a highly alkaline environment. J Exp Biol 189: 13–36

Wright PA, Randall DJ (1987) Ammonia distribution and excretion in fish. Fish Physiol Biochem 3: 107–120

Wright PA, Wood CM (1985) An analysis of branchial ammonia excretion in the freshwater rainbow trout: effects of environmental pH change and sodium uptake blockade. J Exp Biol 114: 329–353

Wright PA, Wood CM (1988) Muscle ammonia stores are not determined by pH gradients. Fish Physiol Biochem 5: 159–162

Wright PA, Randall DJ, Wood CM (1988a) An in vitro and in vivo study of the distribution of ammonia between plasma and red cells of rainbow trout (*Salmo gairdneri*). J Exp Biol 134: 423–428

Wright PA, Randall DJ, Wood CM (1988b) The distribution of ammonia and H^+ between tissue compartments in lemon sole (*Parophrys vetulus*) at rest, during hypercapnia and following exercise. J Exp Biol 136: 149–175

Ion Transfer Processes

Chapter 5

Morphological Basis of Acid-Base and Ionic Regulation in Fish

P. Laurent[1] and *S.F. Perry*[2]

Contents

1 Introduction

The fish gill is a complex structure, or, more exactly, an organized system of structures aimed at fulfilling several vital functions.

The primary role is respiratory gas transfer. To perform this function a sophisticated vascular system has progressively evolved from a primitive organization. Complex innervations control this vasculature in order to match perfusion and ventilation.

Transfer of respiratory gas across the gill epithelium requires a thin and large interface which, consequently, leads to significant passive movements of ions and

[1]Laboratoire de Morphologie fonctionnelle et ultrastructurale des Adaptations C.E.P.E – C.N.R.S. Strasbourg, France
[2]Département de Biologie, Université d'Ottawa, 30 Marie Curie Ottawa, Ontario, Canada K1N 6N5

Advances in Comparative and Environmental Physiology, Vol. 22

water according to existing gradients. The involvement of the gills in osmotic and ionic regulation has been the subject of intensive research for several decades. More recently, the focus has been directed additionally towards acid–base regulation owing to its relationships with ionic regulation.

As in other epithelia, the combination of morphological and physiological techniques has provided valuable information in the areas of acid–base and ionic regulation. The goal of this chapter, which concerns freshwater teleosts, is to review some of the classical aspects of this discipline as well as reporting some of the more recent advances pertaining to fish gill structure and function.

2 Ionic Regulation

Teleost fish are able to regulate the ionic composition of their internal body compartments within narrow limits in diverse external environments ranging from dilute freshwater to concentrated seawater. Indeed, certain species (termed euryhaline) can adapt to both freshwater and seawater habitats while maintaining relatively constant ionic composition. At least two significant physiological adaptations permit ionic regulation in variable external environments. These adaptations include modifications in both renal and branchial structure and function. Numerous comprehensive reviews have been written pertaining to fish gill morphology (Laurent and Dunel 1980; Laurent 1984, 1985, 1989; Laurent and Hebibi 1989; Laurent and Perry 1991) or ionic regulation (Girard and Payan 1980; Evans et al. 1982; Evans 1984; DeRenzis and Bornancin 1984; Cameron and Iwama 1989; Shuttleworth 1989; Avella and Bornancin 1990; Perry and Laurent 1990; Wood and Goss 1990a,b; Wood 1991; Goss et al. 1992a). Thus, it is not our intent to provide a detailed description of ionic regulation in fishes or to review general fish gill morphology. Instead, this discussion focuses on the interrelationships between ionic regulation and fish gill morphology while attempting to establish the morphological basis for ionic regulation in variable external environments.

2.1 Gill Epithelial Morphology in the Freshwater Environment

Freshwater-adapted teleosts are hyperosmotic/hyperionic regulators and thus maintain the ionic concentration and the osmolarity of their body fluids greater

Fig. 1. a TEM photo of a group of mucus cells on the trailing edge of gill filament. Note different stages of maturation; *right* a cell which has almost completely emptied its mucus granules; *left*, an immature mucus cell; *MC* mucus cell. **b** SEM photo of filament epithelium of rainbow trout. Note apical surface of progressive stages of chloride cell emergence (*1,2,3,4*). Two mucus cells are visible (*) owing to the granules.

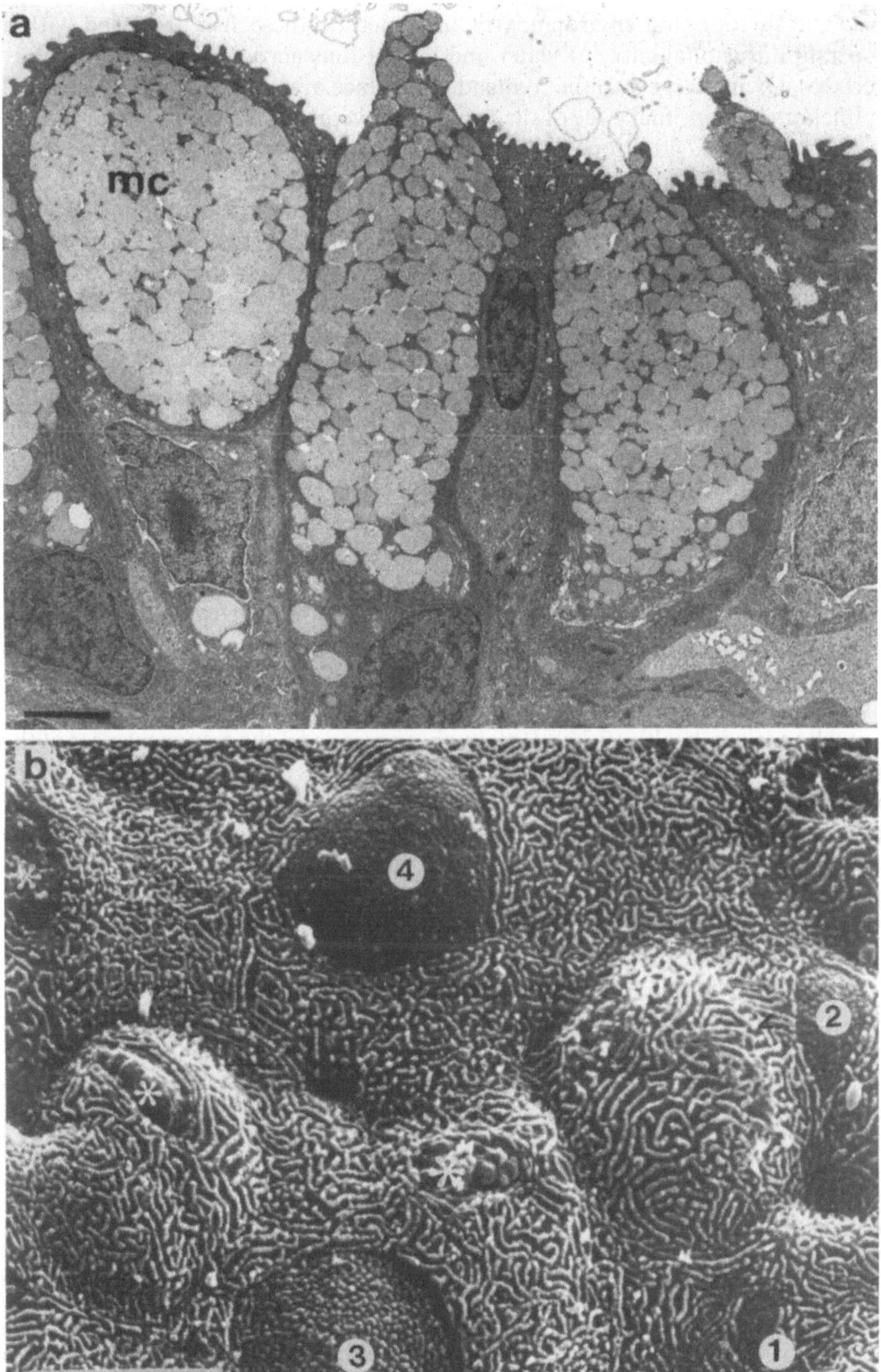
a
mc
b
4
2
3
1

than the surrounding environment. Consequently, these fish are faced with a constant diffusional entry of water and loss of ions across the permeable body surfaces. Owing largely to the tremendous surface area of the gill, the branchial epithelium is the primary site of such passive movements of salts and water. These ionic losses, in addition to those occurring from renal excretion, are counterbalanced by ionic uptake mechanisms located in gill epithelial cells.

The gill epithelial surface contains at least three distinctive epithelial cell types. The pavement cells are by far the predominant cell type on both the filament and lamellar epithelia and generally comprise greater than 90% of the total gill epithelial surface area. In most teleost species, the apical membrane of the pavement cell is ornamented with projecting microridges (Olson and Fromm 1973; Dunel-Erb and Laurent 1980) which may serve to increase the apical surface of these cells and thus enhance the diffusive movements of respiratory gases (Fig. 1b). The apical membrane microridges may also play a role in anchoring mucus to the epithelial surfaces. The microridges are sometimes less developed or entirely absent on the lamellar pavement cells (Hossler et al. 1979), which may be advantageous for decreasing the extent of mucus accumulation on the gas transfer interface. In brown bullhead catfish (*Ictalurus nebulosus*), the pavement apical membrane is characterized by arrays of microvilli rather than microridges (Goss et al. 1992b; Perry et al. 1992). The significance of this observation including further morphological descriptions will be given in the second part of this review (Sect. 3). Classically, the pavement cell has been viewed as a metabolically rather inactive cell based on the relative scarcity of mitochondria in healthy fishes. Recent observations on the responsiveness of the apical membrane microvilli to acid–base disturbances and on the large variations in the density of mitochondria have lead to reopen this question (see below). A particularly intriguing aspect of pavement cell ultrastructure in teleosts is the existence of ordered arrays of rod-shaped particles on the apical membranes of freeze-fractured gill epithelia (Sardet 1977). Although the function of these structures is unknown, a comparison with other epithelia suggests a role both in acid–base and ionic regulation.

Mucus cells are located predominantly on the filament epithelium both on the leading and the trailing edges (Laurent and Hebibi 1989; Fig. 1a). The mucus cells secrete a polyionic glycoprotein known to bind both Na^+ and Cl^- (Handy 1989). Mucus cells are often considered to be more abundant in freshwater-adapted fish (an effect generally attributed to increased prolactin secretion). It is

Fig. 2. a TEM photo of rainbow trout filament chloride cell (*cc*). On the filament, chloride cells are columnar. On the lamellar epithelium, they are flat. These characteristics are due solely to a difference in epithelia thickness. Note the organization of the apical region of the cell: numbers of vesicles (tubulovesicular system), a rather discrete tubular system in loose association with mitochondria. Note that the apical membrane protrudes slightly outward. *pvc* Pavement cell. **b** TEM photo of the apical region of eel chloride cell. Note that microvilli form a tuft covered with glycocalix (*gl*). The apical contact of the cell with the environment is narrow. Two types of vesicles are visible. Most of these characteristics are still unexplained. *pvc* Pavement cell

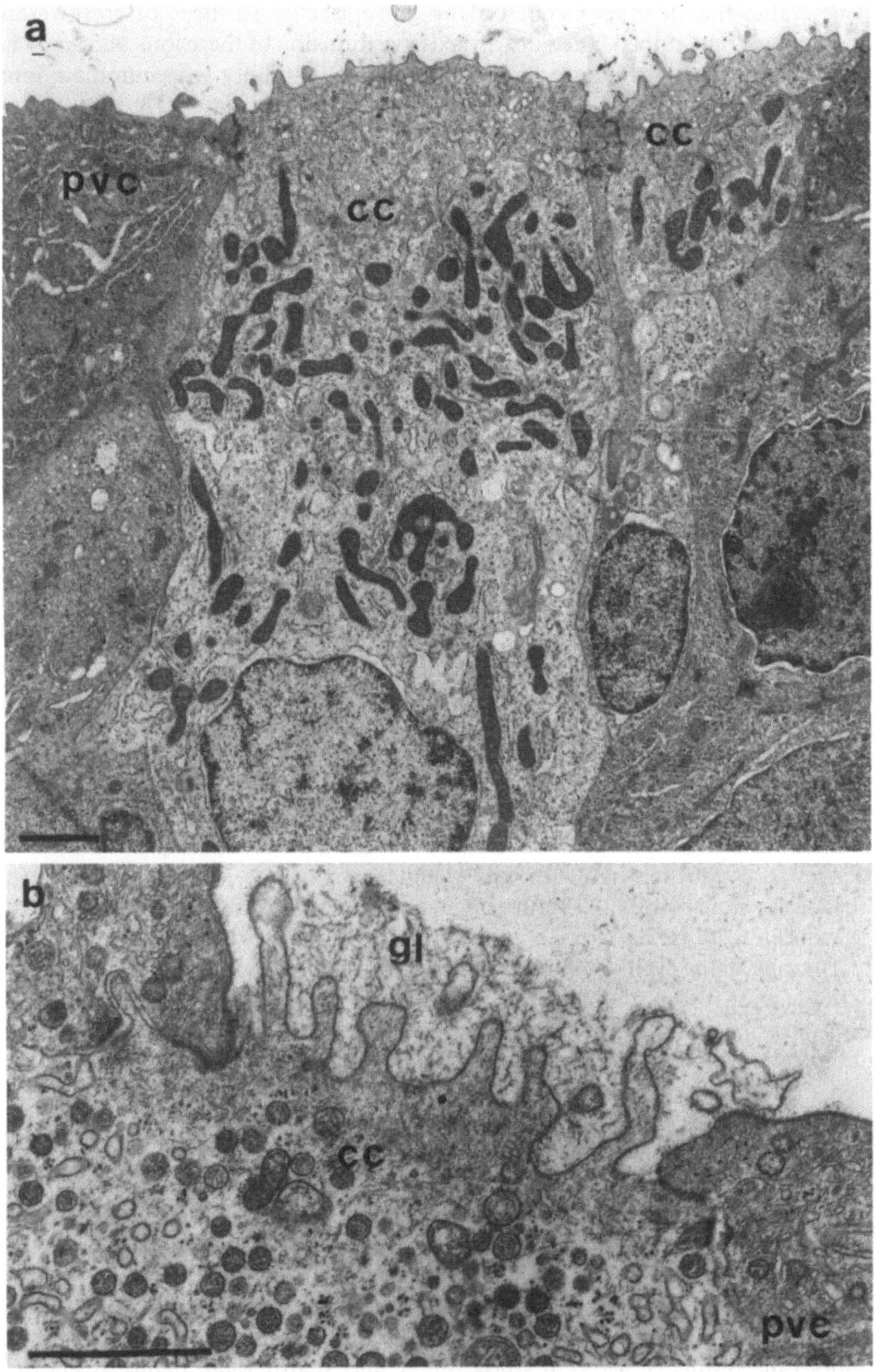
a
pvc
cc
cc
b
gl
cc
pvc

conceivable that a mucus coat on the gill epithelial surfaces of freshwater-adapted fish may impede the outward passive diffusion of these ions and assist in their uptake from the environment by virtue of its ability to accumulate ions thereby effectively decreasing the blood-to-water ionic gradient. However, this point has been recently reconsidered by comparing the density of mucus cells in the same euryhaline species acclimated to different environments. No significant difference was observed between freshwater- and seawater-adapted rainbow trout (Laurent and Hebibi 1989).

Chloride cells exist on both the lamellar and filament epithelium (Laurent and Dunel-Erb 1980; Laurent 1984; Perry and Wood 1985; Avella et al. 1987) although they are normally most concentrated on the afferent side (trailing edge) of the filament or in the interlamellar regions. It is apparent from the various studies that have examined the surface morphology of chloride cells, that there is tremendous diversity in the surface structure of the apical membrane. In most instances, the apical membrane is ornamented with microvilli (Fig. 3c) although the spatial arrangement may differ so greatly that chloride cells among different species may bear little resemblance to each other (e.g. Perry et al. 1992). Indeed, even within any particular species, the surface appearance of the chloride cell may vary markedly (e.g. Franklin and Davieson 1989; Franklin 1990; Laurent and Perry 1990). For example, in rainbow trout (Laurent and Perry 1990) and other salmonids (Franklin and Davieson 1989) the degree of microvillous ornamentation is highly variable. In some cases, the microvilli are densely arranged so as to confer a rough appearance to the apical membrane of the cell, as viewed with the scanning electron microscope, while in others the apical membrane may be totally devoid of microvilli so as to confer a smooth appearance to the surface of the cells (Fig. 1b). It is likely that these different surface appearances reflect different stages in the life cycle of the chloride cell (see Wendelaar Bonga and van der Meij 1989) rather than different types of chloride cell per se. Nevertheless, the primary significance of changes in size and shape of the microvillous structure is probably mostly functional and concerns the size of the apical membrane surface area available for carriers and channels.

Owing to the highly variable surface structure of gill chloride cells, definitive identification often requires the examination of its ultrastructure with the transmission electron microscope. Such studies reveal important internal differences which are far from being functionally elucidated. For instance, two types of chloride cells and their transformation subsequent to seawater adaptation have been described in a freshwater euryhaline fish (*Lebistes reticulatus*; Pisam et al. 1987). Generally, the chloride cell is characterized by numerous mitochondria and an amplification of the basolateral membranes to form an extensive tubular system within the cell (Fig. 2a). Another important compartment is represented by the tubulovesicular system, a structure present in most epithelial cells such as, for example, turtle urinary bladder cells (Stetson 1989). This structure delivers vesicles which finally fuse with the apical membrane (Fig. 2b). The large numbers of mitochondria suggest high metabolic activity and indeed the results of a recent study (Perry and Walsh 1989) demonstrate that chloride cells have the highest rate of oxidative metabolism of the various gill epithelial types.

2.2 Interrelationships Between Gill Morphology and Ionic Uptake in the Freshwater Environment

2.2.1 Sodium and Chloride Uptake

Although it is clear that the gill is the principal location of the trans-epithelial uptake of both Na^+ and Cl^-, there is still considerable uncertainty as to the exact epithelial cell involved. Girard and Payan (1980) suggested that lamellar pavement cells are the sites of Na^+ uptake in freshwater-adapted trout. This idea originated from experiments using a saline-perfused head preparation, a technique which was believed to be capable of distinguishing between ionic fluxes across the lamellar and filament epithelia. However, owing to some technical problems (Perry and Farrel 1989) in addition to the presence of chloride cells on the lamellae (see below), the question was not resolved definitively.

In more recent years, the results of both morphological and physiological studies have tended to support the notion that the chloride cell is the site of Na^+ and Cl^- uptake in freshwater-adapted teleosts. First, various studies have reported proliferation of gill epithelial chloride cells in ion-poor water (Laurent and Dunel 1980; Laurent et al. 1985; Avella et al. 1987; Leino et al. 1987; Perry and Laurent 1989). Indeed, Perry and Laurent (1989) conclude that proliferation of chloride cells in rainbow trout exposed to ion-poor water was an important adaptive response to increase the ionic transporting capacity of the gill. Second, qualitative relationships (Avella et al. 1987) and significant correlations have been established between the fractional surface area of chloride cells exposed to the water and rates of Na^+ and Cl^- uptake both within single species (Perry and Laurent 1989; Laurent and Perry 1990; Perry et al. 1992; Goss et al. 1992b) and among different species (Perry et al. 1992). Third, the characteristic cell ultrastructure, especially the abundance of mitochondria, the high quantities of Na^+/K^+ -ATPase (e.g. Naon and Mayer-Gostan 1983), and the high rate of oxidative metabolism (Perry and Walsh 1989), are all indicative of a role in active (energy-dependent) ionic transport.

Although the indirect evidence provided by the results of recent studies (see above) suggests that the chloride cell is a significant site of both Na^+ and Cl^- uptake in freshwater-adapted teleosts, this does not exclude a role for the predominant epithelial cell type, the pavement cell, in ionic uptake, nor does it eliminate the possibility that Na^+ and Cl^- are transported across the epithelium by different cell types. Unfortunately, a basic limitation to correlating the rates of ionic uptake against gill chloride cell fractional area arises from the usual co-variance between Na^+ and Cl^- uptake under steady state conditions. If Na^+ and Cl^- uptake did not co-vary, severe internal acid–base imbalance would develop (in the absence of efflux modulation) if their rates changed independently of each

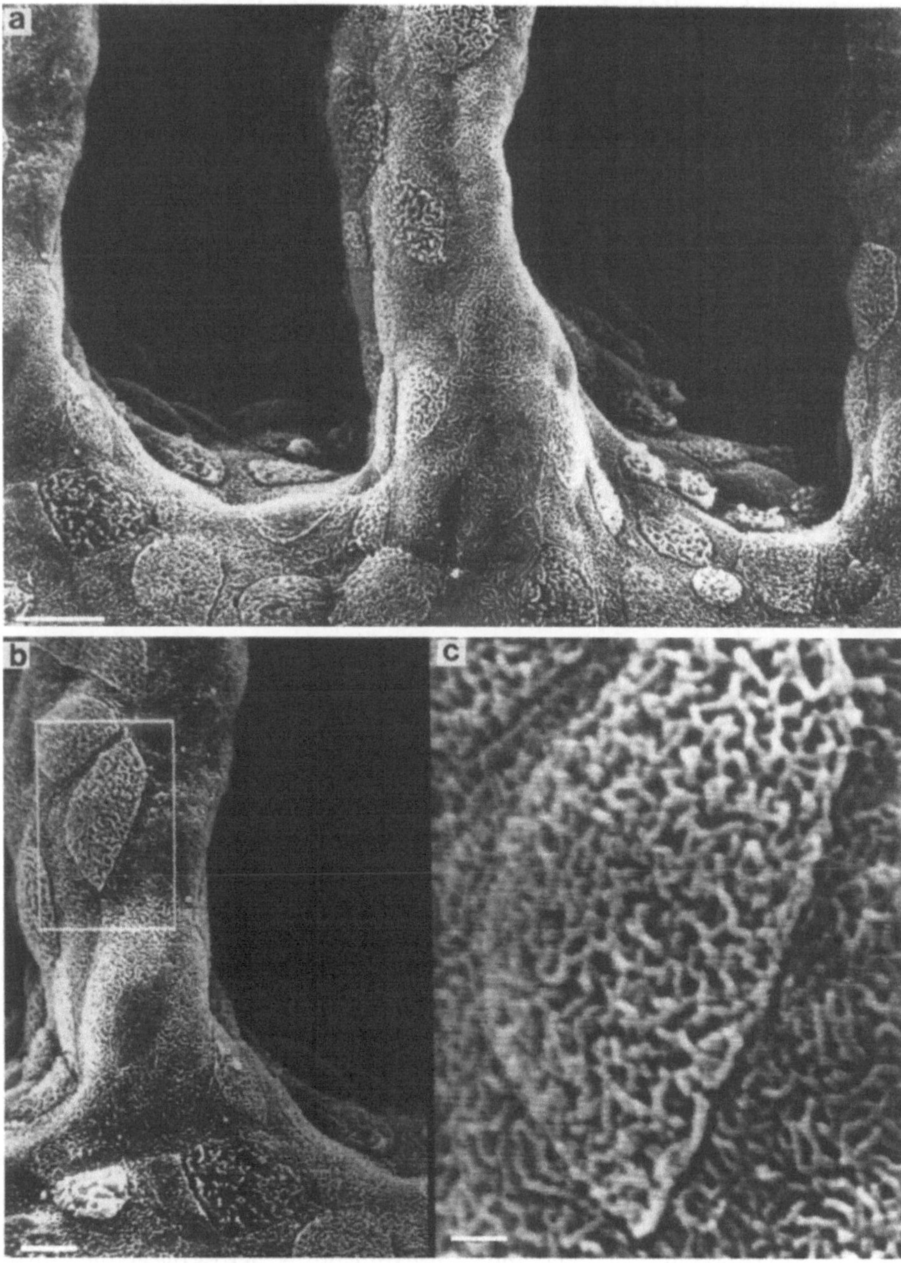

Fig. 3a–c. SEM showing proliferation of chloride cells on gill lamellae of rainbow trout chronically treated with cortisol (see Fig. 8b). *Bars*: **a** = 10; **b** = 5; **c** =1 μm

other. Thus, the rate of one of the two ionic uptake processes might be causally correlated with gill chloride cell surface area while the other might simply be coincidental.

We currently believe that the chloride cell is the sole site of Cl^- uptake, whereas both the chloride cell and the pavement cell are potential sites of Na^+ uptake. The evidence will be given below (see Sect. 3) and concerns the examination of ionic fluxes and gill morphology in brown bullhead catfish under normocapnic and hypercapnic conditions (Goss et al. 1992a,b; see also review by Perry and Laurent 1993). Because the chloride cells are located, to varying degrees, on the filament and lamellar surfaces (Fig. 3a,b), both epithelia are probable locations of Cl^- uptake.

2.2.2 Calcium Uptake

It is generally accepted that the gill is the predominant site of inward Ca^{2+} absorption although the general body surfaces could also be involved (Perry and Wood 1985; see also review by Fenwick 1989). The current model for Ca^{2+} uptake (Perry and Flik 1988) proposes that Ca^{2+} enters gill epithelial cells passively owing to a favourable electrochemical gradient across the apical (water-facing) membranes. The ultimate step in Ca^{2+} uptake involves its active transport across the basolateral membranes via a high-affinity Ca^{2+} ATPase (Flik et al. 1983).

Although direct evidence is lacking, it is generally accepted that the chloride cell is the specific gill epithelial cell type which is responsible for Ca^{2+} uptake. This consensus is based upon several observations including (1) a positive correlation between the apical membrane surface area of exposed (uncovered and contacting the external environment) chloride cells and the rate of Ca^{2+} uptake in rainbow trout, *Oncorhynchus mykiss* (Perry and Wood 1985; Fig. 4), (2) inhibition of Ca^{2+} uptake in rainbow trout by external lanthanum concurrent with its specific

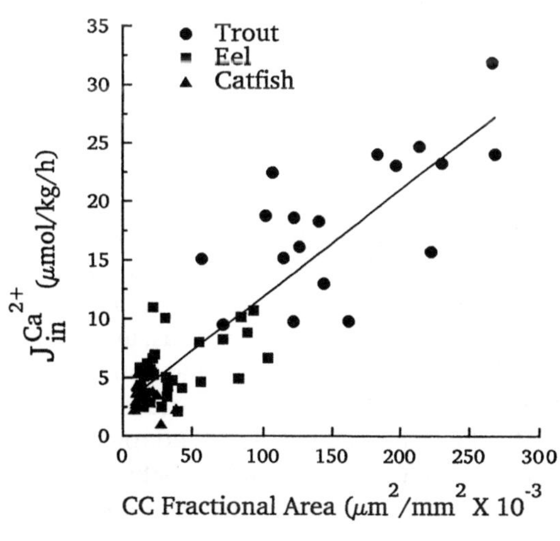

Fig. 4. Interspecific correlation between whole-body calcium uptake ($J_{in}Ca^2$) and gill filament chloride cell (*cc*) fractional area, in three species of freshwater teleost (S.F. Perry, G.G. Goss and J.C. Fenwick, in prep.)

deposition on the apical membrane of chloride cells (Perry and Flik 1988), and (3) oxalate-induced Ca^{2+} precipitates confined to the apical regions of chloride cells in goldfish, *Carassius auratus* (Ishihara and Mugiya 1987). There are data indicating the Ca^{2+} uptake occurs across both the lamellar and filamental epithelial surface (Payan et al. 1981; Perry and Wood 1985).

A survey of the literature reveals marked interspecific differences in the rates of Ca^{2+} uptake among the various teleosts that have been examined (see review by Fenwick 1989). Although some of these differences can be explained by differences in the Ca^{2+} concentration of the acclimation water and/or the growth stage of the fish, it is nevertheless apparent that much of the variability among the species can be explained by similar variability in the surface area of branchial chloride cells exposed to water (S.F. Perry, G.G. Goss and J.C. Fenwick, unpubl. observ.). Indeed, Fig. 4 illustrates the significant correlation between the rate of Ca^{2+} uptake and gill chloride cell fractional surface area among three species of freshwater-adapted teleost, rainbow trout (*Oncorhynchus mykiss*), brown bullhead catfish (*Ictalurus nebulosus*) and American eel (*Anguilla rostrata*). These relationships do not only explain interspecific variability in the rates of Ca^{2+} uptake among different species but further support the general notion that the chloride cell is indeed the site of Ca^{2+} uptake in fish. Nevertheless, in the eel, recent observations have failed to demonstrate a significant correlation between Ca^{2+} uptake and fractional apical surface area of chloride cells (P. Laurent, unpubl.).

2.3 Interrelationships Between Gill Morphology, Ionic Regulation and Gas Transfer

In freshwater-adapted teleosts, a variety of situations are known to induce proliferation and/or enlargement of branchial chloride cells including (1) exposure of fish to water containing low levels of Ca^{2+} (Perry and Wood 1985; Leino et al. 1987; Spry and Wood 1988) or low NaCl levels (Laurent and Dunel-Erb 1980; Laurent et al. 1985; Avella et al. 1987; Thomas et al. 1988; Perry and Laurent 1989; Laurent and Hebibi 1989), and (2) treatment with exogenous hormones including cortisol (Doyle and Epstein 1972; Foskett et al. 1983; Laurent and Perry 1990; McCormick 1990; Perry et al. 1992; Fig. 5a,b) and growth hormone (e.g. Madsen 1990; Laurent et al. 1994). Presumably, chronic stress of fish and the associated elevation of endogenous cortisol levels would also induce proliferation of branchial chloride cells. The proliferation of chloride cells, regardless of the cause, occurs on both the filament and lamellar epithelia (Fig. 3a,b). The proliferation of the chloride cells on the lamellae may markedly reduce the surface area of lamellar pavement cells (see Fig. 3a). Thus, an increase in the branchial surface area of chloride cells might enhance ionic uptake while impairing gas transfer because of the reduction of the lamellar surface occupied by pavement cells which, according to their characteristics, are adapted to gas transfer. Few studies have addressed this potential conflict between ionic regulation and gas transfer. Thomas et al. (1988) compared the responses of two

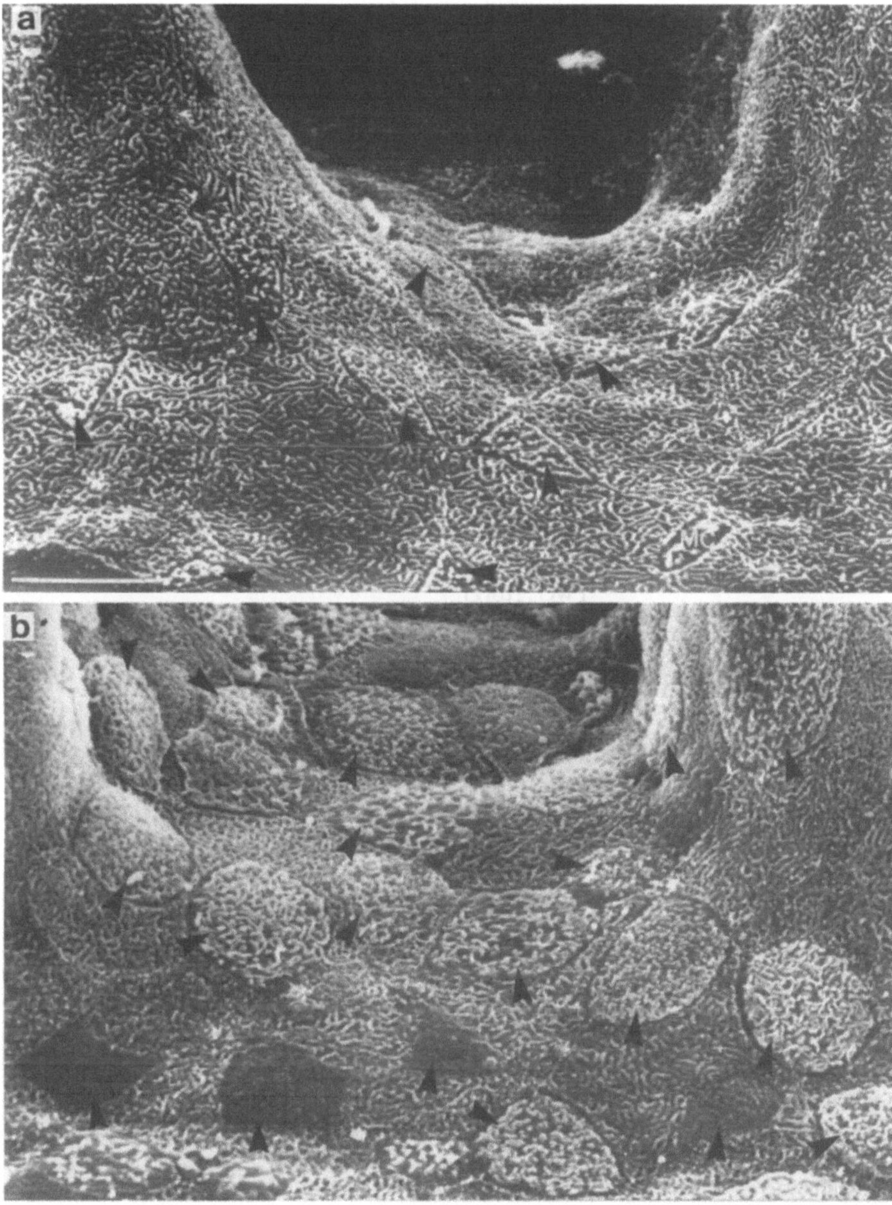

Fig. 5. **a** SEM of filament and interlamellar regions of rainbow trout gill under control conditions. *Arrowheads* point to chloride cells. *Bar* = 10 μm. **b** SEM of filament and interlamellar regions of rainbow trout treated with cortisol (4 mg kg^{-1} intramuscularly for 10 days). Note the pronounced proliferation and expansion of chloride cells on the filament and in the interlamellar region

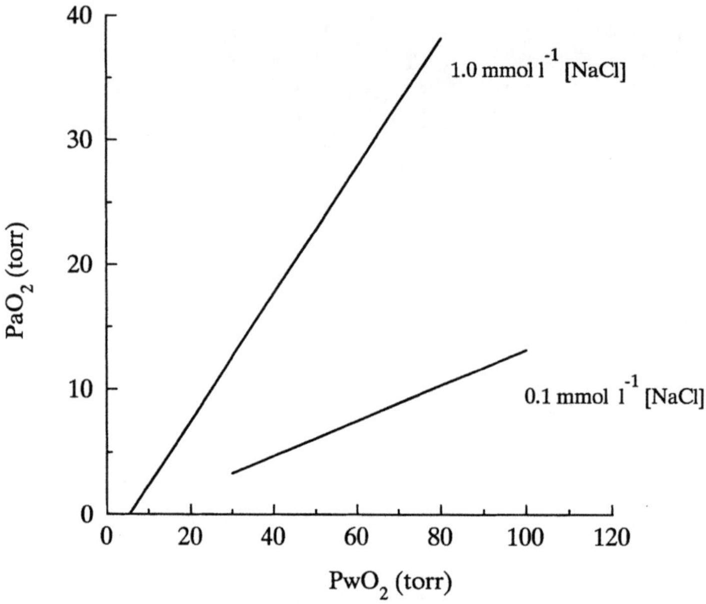

Fig. 6. The relationships between water PO_2 (PwO_2) and arterial PO_2 ($PaPO_2$) during hypoxia in two groups of rainbow trout acclimated to different natural concentrations of NaCl, 1.0 mmol l^{-1} or 0.1 mmol l^{-1}. The fish acclimated to the more dilute medium had lower values of (PaO_2) for any given value of (PwO_2). These low values are presumably related to proliferation of chloride cells (Thomas et al. 1988)

groups of rainbow trout to hypoxia, where one group was acclimated to approximately 1 mmol l^{-1} NaCl and the other to 0.1 mmol l^{-1} NaCl. The results demonstrated that the fish acclimated to the more diluted medium were less tolerant to hypoxia (Fig. 6). This reduced tolerance was attributed to proliferation of chloride cells on the lamellar epithelia, presumably an adaptive response to enhance ionic regulation. However, Laurent and Hebibi (1989) reported a marked increase in the diffusing capacity of the trout gill after acclimation to ion-poor water by virtue of a thinning of the blood-to-water diffusion distance. This may be interpreted as a secondary compensatory response to an initial proliferation of chloride cells, or as suggested by Laurent and Hebibi (1989), as a facilitation of NH_3 excretion.

3 Acid–Base Regulation

Freshwater fish are able to compensate for acid–base disturbances by manipulating branchial ion exchanges in order to modify the relative uptake or excretion of acidic and basic equivalents (Cameron 1976; McDonald and Wood 1981; Cameron and Kormanik 1982; Heisler 1984; Wood et al. 1984; Claiborne

Table 1. Relationships between acid–base satus and ion fluxes MacDonald et al. 1989)

	$J_{net}Cl$	$J_{net}Na$	$J_{in}Cl$	$J^{in}Na$
Blood acidosis	–	+	↘	↗
Blood alkalosis	+	–	↗	↘

and Heisler 1986; Perry et al. 1987; Goss and Wood 1990a,b). Base excretion is clearly linked to chloride uptake via Cl^-/CHO_3^- exchange, whereas acid excretion is thought to be linked to Na^+ uptake via Na^+/H^+ exchange or Na^+ channels and proton-ATPase (see below). Therefore, it is understandable why and how the activity of these exchangers/pumps might play a crucial role in determining the acid–base status of the fish.

The primary site for acid–base regulation is the gill (Cameron 1980; McDonald and Wood 1981; Evans 1982; Wood 1988). The kidney and its associated structures (urinary bladder) play an additional role (e.g. Wood and Caldwell 1978; Wheatly et al. 1984; Wood 1988). The relationships between the acid–base status and ionic fluxes are summarized in Table 1. It has been shown that adjustments in the rates of Cl^- and Na^+ uptake during acid–base disturbances result from changes in the maximal rate of transport. Changes in the affinity of the transporters for these ions: (1/Km), appear to be less important (Goss and Wood 1990a,b). Therefore, from a morphological viewpoint, maximal transport capacity is determined by the numbers of available transport sites (Maetz 1974), which, in turn, are dependent on the surface area of the membrane domain on which these sites are located.

3.1 Modification of Chloride Uptake During Environmental Hypercapnia

Exposure of fish to environmental hypercapnia induces respiratory acidosis which is gradually compensated for by elevation of plasma bicarbonate levels until a new steady-state pH is achieved. When hypercapnia is removed, a phase of metabolic alkalosis develops, because dissolved carbon dioxide dissipates more rapidly than the excess bicarbonate.

In the brown bullhead catfish, exposure to hypercapnia (2% CO_2 in air, Pw CO_2 = 15 torr) resulted in a negative net flux of Cl^- (J_{net} Cl^-) and a positive net flux of Na^+ (J_{net} Na^+; Goss et al. 1992b). The change in J_{net} Cl^- persisted throughtout the hypercapnic period (48 h), whereas $J_{net}Na^+$ returned to the control value within the first 12 h. Exposure to hypercapnia caused a decrease in Cl^- uptake ($J_{in}Cl^-$) and an increase in Na^+ uptake ($J_{in}Na^+$). During recovery from hypercapnia, $J_{net}Cl^-$ and $J_{in}Cl^-$ were greatly stimulated compared to the hypercapnic period.

Our recent experiments using hypercapnia in the catfish *Ictalurus nebulosus* as a model system give strong evidence for a morphological component to acid–base regulation. This species has a significant population of chloride cells and

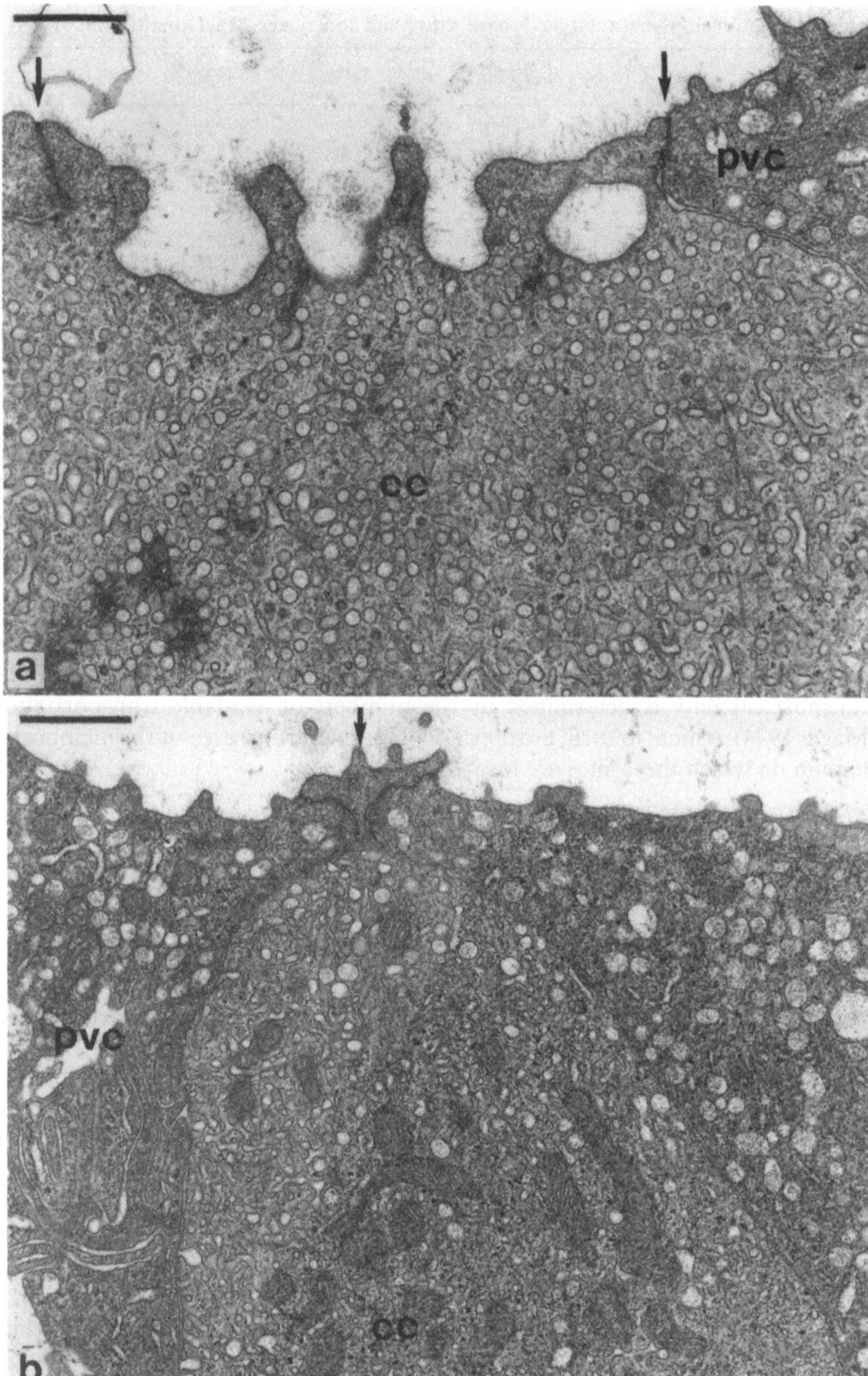
pvc
cc
a
pvc
cc
b

therefore can use them efficiently to regulate the transbranchial fluxes of basic equivalents (Goss et al. 1992b). After 6 h of hypercapnia, the fractional area of the chloride cells (apical membrane surface area per unit area of epithelium in contact with the water) decreased to 50% of the control value. This effect was largely due to a reduction of the individual surface area rather than a change in the numbers of chloride cells per surface area unit (density). By 48 h, chloride cell fractional area was reduced to 5% of its initial value owing to an extreme reduction of both surface area and cell density (Fig. 7a,b). During the post-hypercapnic recovery, there was a rapid increase in individual chloride cell surface and cell numbers.

These results are important, because the cellular process by which a chloride cell is functionally uncoupled from the water is clearly shown. Pavement cells which encircle the apex of a chloride cell progressively stretch themselves until they completely mask the apical membrane. There is no sign of swelling in the pavement cell nor of shrinkage of the chloride cell. Chloride cells possess a rich subapical network of actin filaments (terminal web) associated with the pericellular ring of occluding junctions. This organization has been shown to control membrane channels in some epithelian cells (Hardy and DiBona 1982). In chloride cells, actin filaments might play a role in modifying the apical surface area during acid–base disturbances. However, from our observations, it is more likely that the process of covering of the chloride cell is the result of an activation of the pavement cells rather than an active process of chloride cell apical contraction. From this point of view, possible interactions between pavement cells and chloride cells need further investigation. Similar changes in gill morphology and ionic fluxes have been observed in hypercapnic trout (G.G. Goss, S.F. Perry and P. Laurent, in prep).

3.2 Modification of Chloride Uptake During Environmental Hyperoxia

Morphological adjustments also contribute effectively to acid–base regulation in the trout *Oncorhynchus mykiss*. In the control trout, Na^+ and Cl^- influxes are larger than in the catfish. Consistently, the chloride cells have a larger apical surface area (Perry et al. 1992). Hyperoxia is used widely as a tool to induce respiratory acidosis. In a recent study, trout were exposed to hyperoxic water (water P_{O2} = 500 torr) for 72 h and subsequently returned to normoxia (Goss et al. 1992b). During hyperoxia, Cl^- influx decreased abruptly while Na^+ influx decreased slightly. This response was still appropriate to correct acidosis, because Cl^- loss was greater than that of Na^+ such that complete compensation was

Fig. 7. a TEM photo of the apical part of chloride cell (*cc*) in a normocapnic catfish (*Ictalurus nebulosus*). *Arrows* point to the apical region of the cell in contact with the external milieu. (Note the tight junction of *cc* with pavement cell, *pvc*). Thin filaments anchored in the apical membrane represent the glycocalix. *Bar* = 1 μm. **b** The same as in **a**, here from a catfish exposed to hypercapnia (15 h, 2% CO_2). Note pavement cells covering the apical membrane of a chloride cell. *Bar* = 1 μm

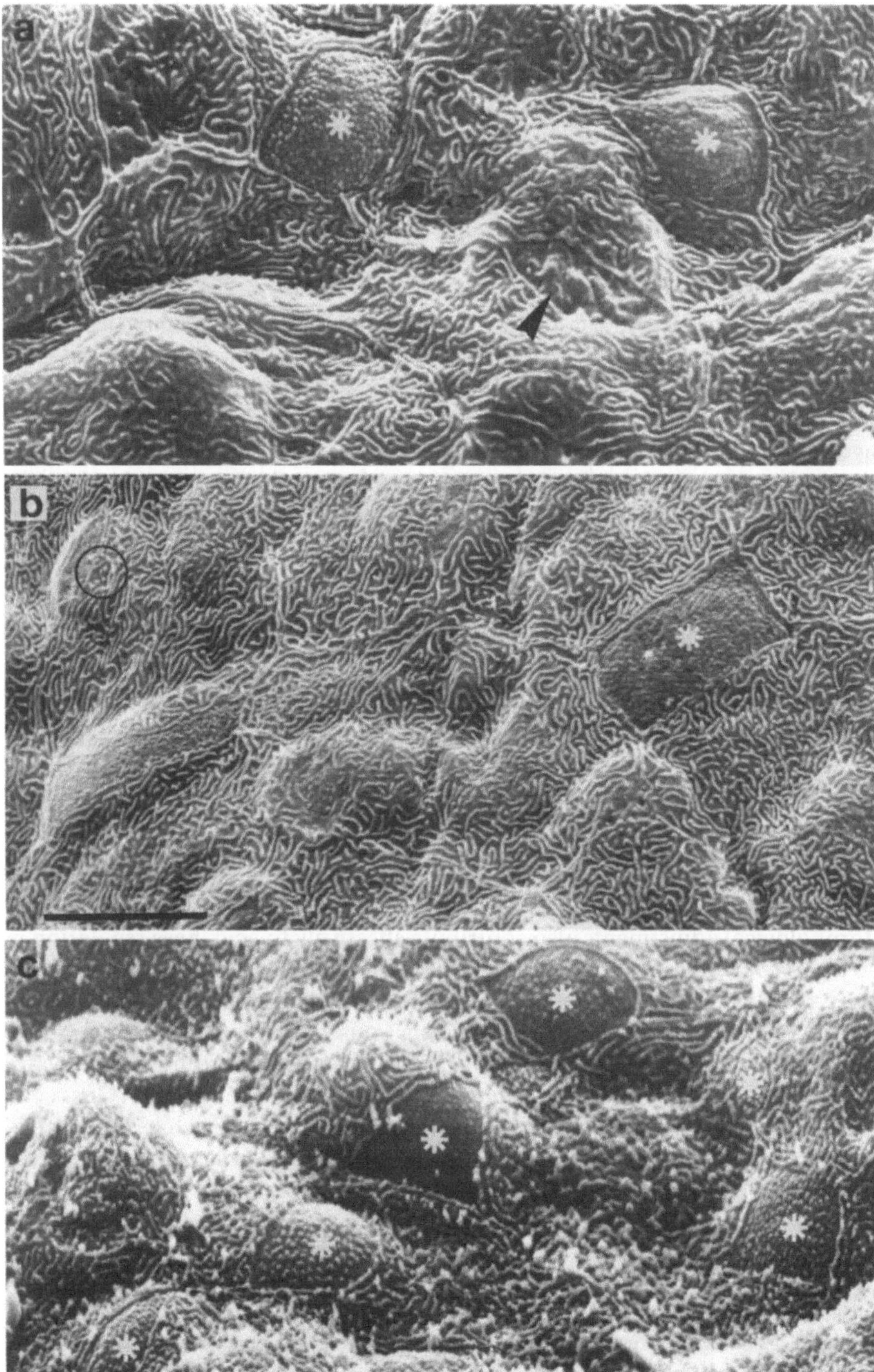
a
b
c

achieved by 72 h. After removal (6 h) of the hyperoxic stimulus (inducing metabolic alkalosis), Cl^- influx was greatly stimulated while Na^+ influx was decreased further. Concomitantly, the covering/uncovering process of chloride cells occurred. Once again, during recovery from hyperoxia, the surface area of the chloride cell apical membrane increased in parallel with the post-hyperoxic increase in Cl^- influx, while Na^+ influx decreased (Fig. 8a–c). The correspondence between Cl^- influx and chloride cell apical membrane surface area reaffirms our contention that the Cl^-/HCO_3^- exchange mechanism is located on the chloride cell. It is interesting that during the phase of recovery from hyperoxia (metabolic alkalosis) the increase in Cl^- influx was greater than that predicted from the increase in chloride cell apical membrane surface area. Presumably, this resulted from an increased substrate (HCO_3^-) availability for the Cl^-/HCO_3^- exchanger owing to the significant rise in plasma HCO_3^- levels (Wood and Goss 1990a,b). Clearly, the physical covering of chloride cells must reduce the stimulatory effect of elevated internal HCO_3^- on Cl^-/HCO_3^- exchange during chronic acidosis.

3.3 Modification of Sodium Uptake and Pavement Cells

Pavement cells are by far the most abundant cell type on the fish gill epithelium. They cover the filament body and the lamellae. It is likely that there is more than a single type of pavement cells, as suggested by the variable apical membrane ornamentations in some species. In addition, as pointed out by Laurent (1984), some pavement cells are squamous as in other protective integuments, while others are more or less columnar. They are provided with a large number of cytoplasmic vesicles and deeply infolded parietal (basolateral) membranes. A common feature is the presence of a conspicuous Golgi apparatus, which presumably is the site of active protein synthesis to supply the external membrane cell coat (glycocalix). Another characteristic property of these cells is the ornamentation of their external surface. This surface, which has been extensively studied with the scanning electron microscope (see review by Laurent 1984), displays a complex arrangement of microridges differing slightly in shape and size between species.

Recently, we have re-examined the external morphology of the gill epithelial pavement cells. In the trout (*Onchorhyncus mykiss*), the external cell surface of the pavement cells located on the filament is ornamented with fingerprint-like microridges. The pavement cells on the lamellar epithelium display microvilli of variable length and density. In the eel, there are microridges on both the filament and lamella, whereas in catfish (*Ictalurus nebulosus*) the pavement cells are

Fig. 8a–c. Representative SEM photos of the gill filament epithelium in rainbow trout, *Oncorhinchus mykiss*. **a** Control fish; **b** acidotic (72 h hyperoxia); **c** 8 h after recovery from hyperoxia. Experimental protocol described in Goss et al. (1992). *Asterisks* point to chloride cells; *arrowhead* points to mucus cell. Note the *encircled* very small apical surface of a chloride cell engaged in a process of covering by pavement cell. *Bar* = 10 μm

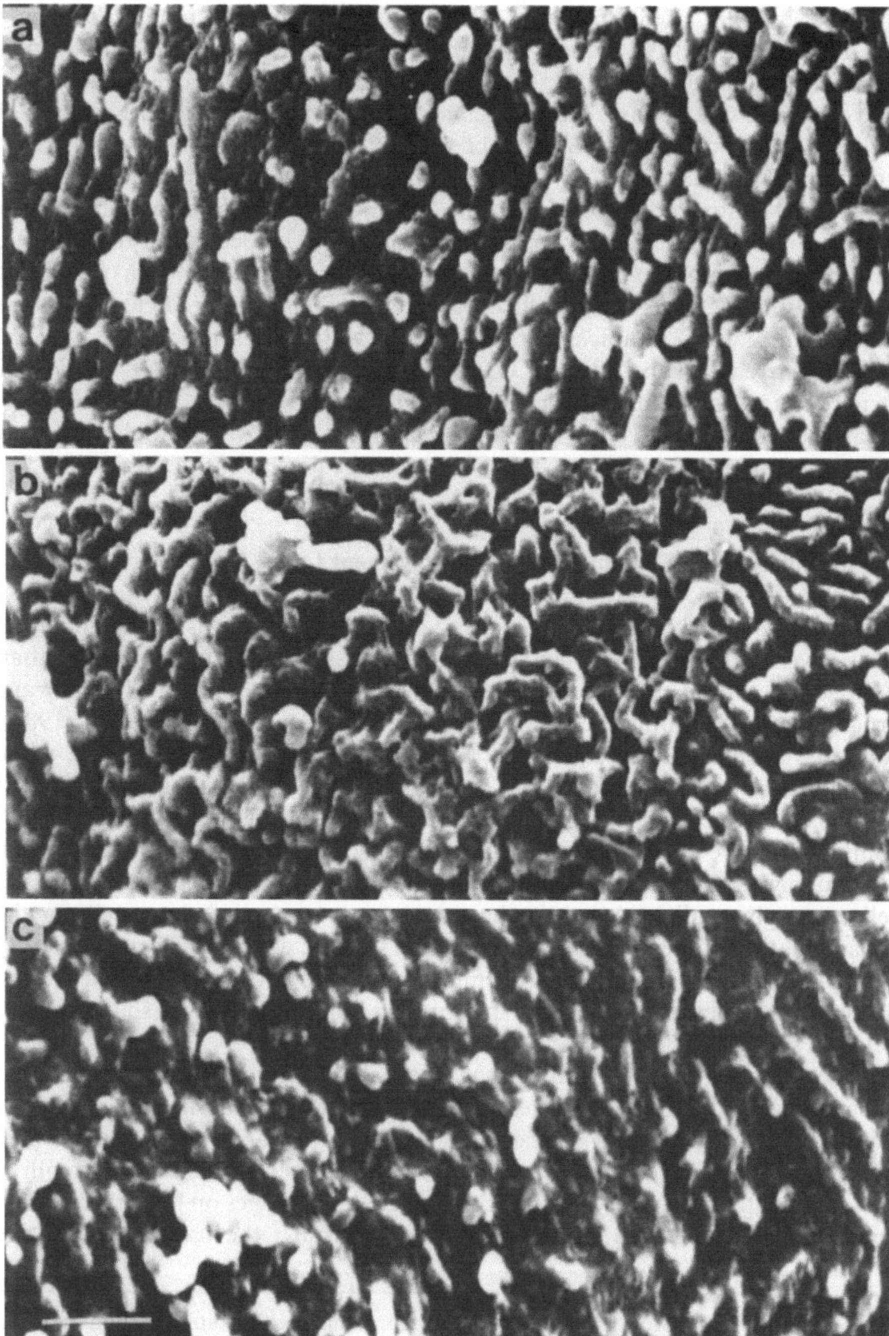
a
b
c

ornamented solely by microvilli. In all species we have studied, it was clear that the density of ornamentation on pavement cells varies markedly. We have recently conducted experiments to quantify, by morphometry, the density of apical membrane microvilli present on pavement cells in the catfish subjected to hypercapnia in order to induce acidosis and stimulate acidic excretion and Na^+ uptake. There was a 30% increase in the density of PVC apical microvilli after 6 h of hypercapnia, reaching a two-fold increase after 24–48 h (Goss et al. 1992a). Return to normocapnia restored the control values. In addition, there were simultaneous ultrastructural changes in pavement cells (Goss et al. 1992a; P. Laurent, in prep.). Similar results have been obtained from rainbow trout exposed to hyperoxia, another acidosis-inducing treatment (Fig. 9; see above for effects on chloride cell apical membrane surface area). In acidotic fish, pavement cells containing numerous mitochondria distributed apically in the cell were frequently observed. This apical distribution of mitochondria is in marked contrast to the situation in control catfish in which the pavement cells have few mitochondria which are located perinuclearly. An interesting finding, as revealed with the transmission electron microscope at high magnification, was the presence of coated vesicles originating as buds from extremities of Golgian cisternae (Laurent et al. 1994; Fig. 10). The vesicle coat is formed from studs inserted into the three-layered membrane of the bud. A stud consists of a spherical head (10 nm) supported by a thin short stalk (about 10 nm long). After their release into the cytoplasm, the vesicles are spherical and appear in cross section to support a crown of 12 to 20 regularly spaced studs. The entire structure is 100 nm in diameter. This description is reminiscent of the H^+ pump vesicles and H^+ ATPase studs described for the H^+ secretory cells of toad and urinary bladder epithelia or kidney epithelial cells (Orci et al. 1986; Brown et al. 1987, 1988). A cycle, including the origin of the vesicle up to its insertion in the apical membrane (fusion), has been observed to occur in pavement cells but, as far as we know, never in chloride cells. The dynamics of this cycle is influenced by the acid–base status of the fish. In control fish, the number of cytoplasmic free vesicles was significantly lower than in acidotic fish, whereas pear-shaped pits (insertion) and flat-coated domains (fusion) were observed more frequently on the apical membrane of acidotic catfish (Laurent et al. 1994).

Thus, the finding of vesicles similar in pavement cells to those described in other H^+ secreting epithelia and confirmed as containing the structural correlate of the H^+ATPase (Brown 1989) suggests that the fish gill pavement cells are equipped with H^+ pumps located on their apical membrane in a way markedly influenced by the acid–base status of the fish. In other words, in acidotic fish, pavement cells might serve to excrete protons. Several additional observations strengthen the hypothesis of H^+ secretion by pavement cells: first, the presence in

Fig. 9a–c. SEM photos of pavement cell surface on the rainbow trout lamellar epithelium. **a** Control condition; **b** after 72 h of hyperoxia; **c** 6 h after recovery from hyperoxia. Note the increased number of ridges and microvilli during hyperoxia and a surface organization in rainbow trout completely different from eel (Fig. 5)

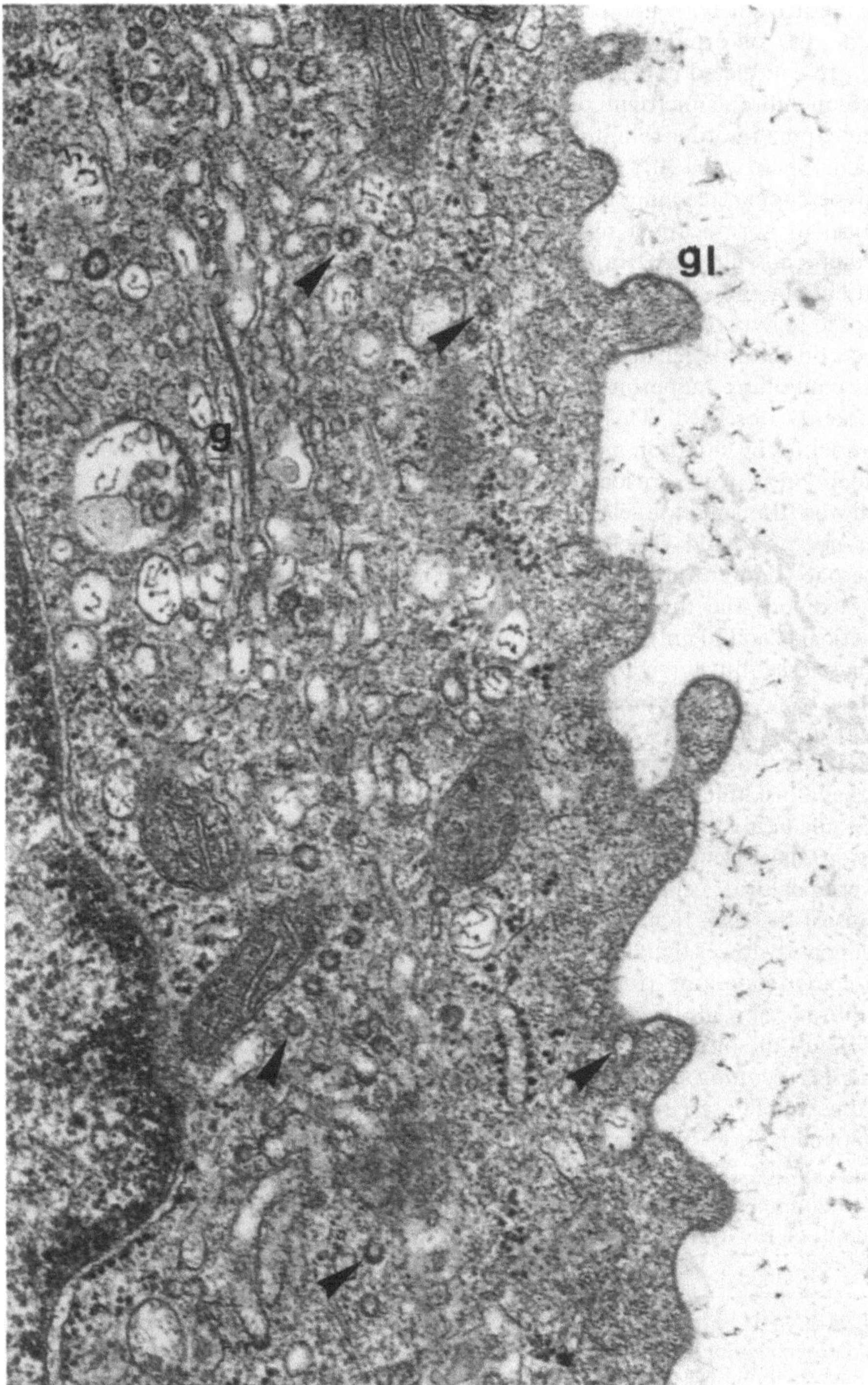
gl
g

pavement cells of carbonic anhydrase (Rahim et al. 1988; see also Perry and Laurent 1990), an enzyme present in acid-secreting cells (see Steinmetz 1986); second, the inability of the chloride cells to carry on H^+ secretion when fish are acidotic and the chloride cells are hidden by pavement cells (Goss et al. 1992b). In addition, recent data have shown that, during acidosis in catfish, considerable amounts of N^+ ions are taken up from the environment (Goss et al. 1992b). This implies the presence in pavement cells of both sodium channels and proton pumps. It has been argued that a Na^+/H^+ antiporter cannot function in the freshwater fish gill based on thermodynamic criteria. The existence of proton pumps in freshwater fish gills instead of Na^+/H^+ antiporter has been suggested (Avella and Bornancin 1989; Lin and Randall 1991; see also review by Avella and Bornancin 1990). According to this concept, Na^+ ions are driven into the gill epithelium through Na^+ channels owing to the electrogenicity of H^+ATPase. Basolateral Na^+/K^+-ATPase would then complete the uptake process by pumping cytosolic Na^+ into the extracellular compartment. As discussed above, based on the absence of functional chloride cells during acidosis in catfish, these channels should be located on the apical membrane of the pavement cell. It is worth mentioning that the eel adapted to low Na^+ water is unable to regulate its extracellular pH during hypercapnia (Hyde and Perry 1989), because of an impossibility of excreting protons (see below).

3.4 Is There a Morphological Basis for Different Strategies of Acid–Base Regulation?

All freshwater species studied to date have chloride cells present on the gills. However, density of cells and the apical surface areas are variable among the species. A recent comparative study has assessed gill chloride cell morphology in eel, catfish, tilapia, and trout (Perry et al. 1992). While there are significant differences between each of them, it is apparent from this study (and others) that the eel displays the lowest apical membrane surface areas by far. Two different interpretations can be offered to explain these differences. First, all fish require chloride cells to maintain its ionic balance which is continuously compromised by ionic losses. In other words, the chloride cell apical surface area is related to the rates of ionic efflux; the "leakier" the gill epithelium, the larger must be the apical membrane areas of the chloride cell. This is probably true, because the diffusive efflux in the eel is the lowest among the species studied (P. Laurent, S.F. Perry and G.G. Goss, in prep.). The second interpretation which, actually, might be complementary to the first one, is that tight epithelia cannot regulate the acid–

Fig. 10. Representative TEM photo of gill pavement cell in acidotic catfish, *Ictalurus nebulosus*; 48 h hypercapnia. Experimental protocol described in Goss et al. (1992). *Arrowheads* point to coated vesicles presumably carrying proton pumps to the apical membrane. Note Golgi cisternae (*g*) and apically located mitochondria; *gl* glycocalix. *Bar* = 0.5 μm base status of the animal as effectively. It opposes the effluxes of acid–base equivalents via the paracellular pathways as suggested by McDonald et al. (1989). Unfortunately, data concerning the morphological correlates of gill epithelial permeability are still very scarce.

What can be functionally expected from a fish possessing a small number of chloride cells or an extremely low fractional apical membrane surface area of the chloride cells? In the eel, it should be extremely difficult for the fish to modify its acid–base status by reduction or activation of Cl^-/HCO_3^- exchangers (located on chloride cells). In fact, contrary to the trout, the eel relies solely on stimulation of Na^+ uptake and its coupling to acid excretion (Hyde and Perry 1989). This probably explains why eels when placed in a Na^+ free medium are unable to compensate for hypercapnic acidosis (Hyde and Perry 1989). On the other hand, acclimation of the eel to increased external [NaCl] caused a significant blood alkalosis: branchial Na^+ uptake was stimulated while Cl^- uptake was maintained at an exceptionally low level. In such an animal, the ability to regulate hypercapnic acidosis was still intact (Hyde and Perry 1989). These data indicate that, unlike the trout, the eel relies only on adjustments of branchial Na^+ uptake for acid–base regulation during hypercapnic acidosis owing to low, inflexible rates of gill unidirectional Cl-fluxes which, in turn, are expected from the quantification of the gill chloride cell surface area.

Recent experiments conducted in eels subjected to hypercapnia have clearly shown that poor, if any, correlation exists between the fractional apical surface area of the chloride cell and experimental sequences of hypercapnia and recovery (P. Laurent, S.F. Perry and G.G. Goss, in prep.). The chloride cell density is quite variable from one specimen to another but never reaches the values measured in other freshwater teleosts. It is worth mentioning that although the apical surface area of the chloride cell is small there are still long microvilli emerging like tufts from a narrow window between pavement cells (Fig. 2b). This clearly indicates that if the chloride cells in the freshwater eel are engaged in an active functional process, this is apparently not related to Cl^-/HCO_3^- exchange.

According to the concept that the eel relies solely on its population of pavement cells to regulate its acid–base balance, one might expect some correlation between the pavement cell morphology and the acid–base status. In addition to a mechanism of apical membrane insertion of proteins specifically involved in proton translocation (see above), cellular responses to acute respiratory acidosis have been demonstrated in the intercalated cell of rat medullary collecting duct (Madsen and Tisher 1983). In this case, morphometric analysis has revealed a significant increase in the surface density of the apical plasma membrane and a decrease in the number of tubulovesicular profiles in the

→

Fig. 11a–c. SEM photos of pavement cell surface on the eel lamellar epithelium. **a** Control condition; **b** after 6 h of hypercapnia; **c** upon 12 h recovery to normocapnic condition. Note that the long microridges seen in control fragment or become very tortuous during hypercapnia. Morphometric analysis would indicate that total surface area is multiplied by a factor of 10

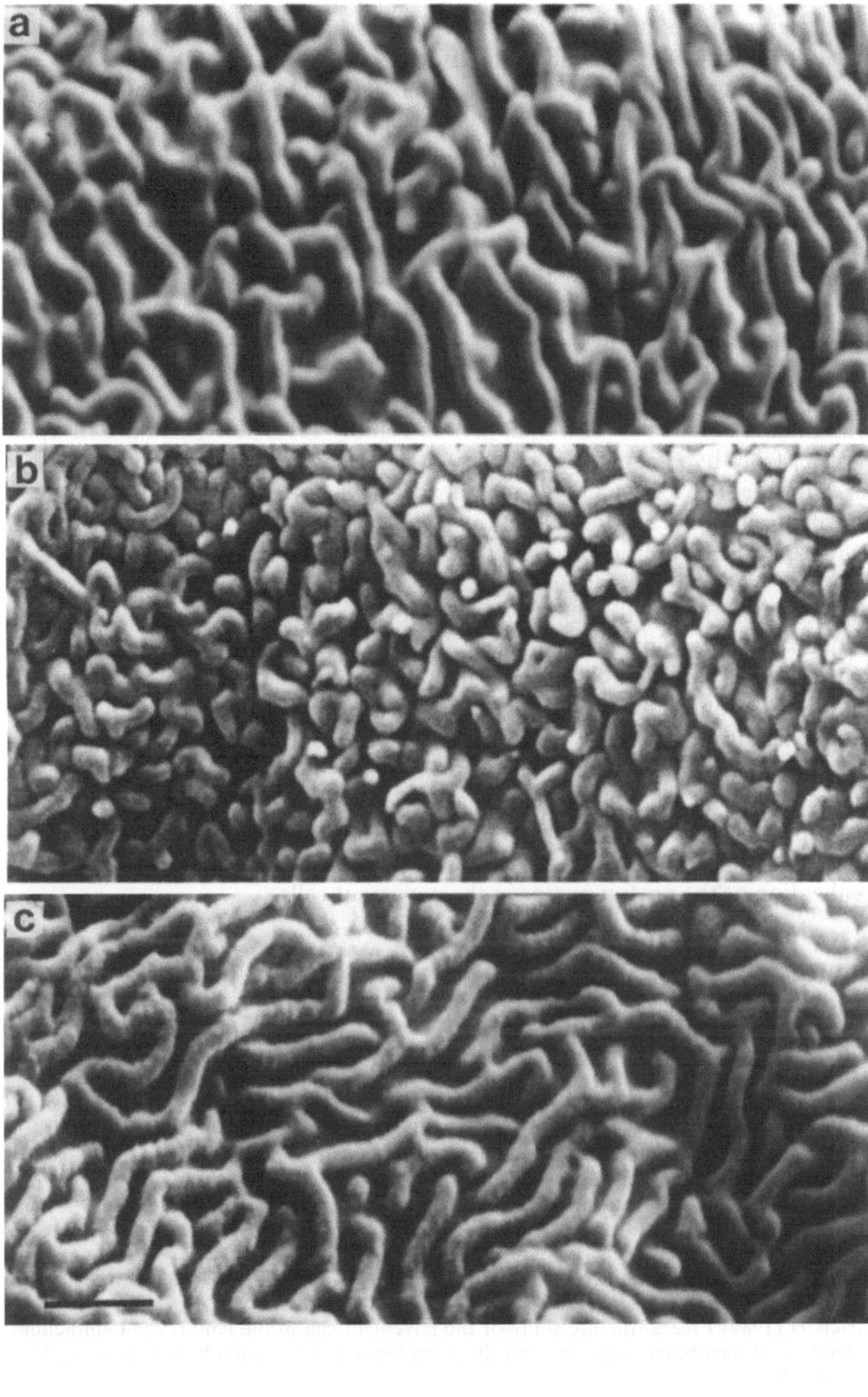
a
b
c

apical region of the intercalated cells with respiratory acidosis. The result is an increased density of apical villi or microridges.

This event was, indeed, observed in the acidotic eel, the pavement cells of which were particularly rich in rough endoplasmic reticulum and Golgian stacks (Laurent et al. 1994). The reversible occurrence of an increased number of villi (Fig. 11) and changes in the internal morphology of pavement cells have been observed in the eel in a way quite similar to that reported in the catfish (Goss et al. 1992b). Recently, Lin and Randall (1991) have suggested that the H^+ATPase (proton pumps) is situated in the chloride cell owing to the presence of a large population of mitochondria necessary to energize the pumps. However, among other morphological changes, the density of mitochondria increased markedly in the apical region of the pavement cells during acidosis in catfish, trout and eel. Under these conditions pavement cells resembled typical mitochondria-rich cells. This change was rapidly reversible (within 6 h) upon recovery from hypercapnia.

Thus there is, today, a body of evidence suggesting a complementary role of pavement cells and chloride cells in freshwater fish gills. Owing to a larger apical membrane surface area in ion-poor water than under control conditions, chloride cells must be the site of Cl^-/HCO_3^- exchange. Conversely, because of the almost complete disappearance of chloride cells from the gill surface during hypercapnia (acidosis), Na^+ ions must be taken up by another way. Morphological studies have produced considerable evidence in favour of the pavement cells as the site of entry of Na^+ ions in acidotic fish (conditions which increase sodium influx and acidic excretion markedly). Concerning the mechanism of H^+ excretion, as in other types of epithelia, the presence in the fish gill of the Na^+/H^+ antiporter is unlikely. According to Ehrenfeld et al. (1985), proton-ATPase pumps would create the driving force to transfer Na^+ ions into the gill epithelia, whereas H^+ ions are excreted. We have now some morphological reasons to postulate the localization of the proton pumps in the pavement cells.

References

Avella M, Bornancin M (1989) A new analysis of ammonia and sodium transport through the gills of the freshwater rainbow trout (*Salmo gairdneri*). J Exp Biol 142: 155–175

Avella M, Bornancin M (1990) Ion fluxes in the gills of freshwater and seawater salmonid fish. In: Truchot JP, Lahlou B (eds) Animal nutrition and transport processes. 2. Transport, respiration and excretion. Comp Physiol, vol. 6 Karger, Basel, pp 1–13

Avella M, Masoni A, Bornancin M, Mayer-Gostan N (1987) Gill morphology and sodium influx in the rainbow trout (*Salmo gairdneri*) acclimated to artificial freshwater environments. J Exp Zool 241: 159–169

Bartels H (1984) Orthogonal arrays of particles in the gill epithelium of the Atlantic hagfish, *Myxine glutinosa*. Cell Tissue Res 238: 657–659

Bartels H (1989) Freeze-fracture study of the pavement cell in the lamprey gill epithelium. Analogy of membrane structure with the granulae cell in the amphibian urinary bladder. Biol Cell 66: 165–171

Brown D (1989) Membrane cell recycling and epithelial cell function. Am J Physiol 256(20): F1–F12
Brown D, Gluck S, Hartwig JJ (1987) Structure of the novel membrane-coating material in proton-secreting epithelial cells and identification as an H^+ATPase. J Cell Biol 105: 1637–1648
Brown D, Hirsch S, Gluck S (1988) Localization of a proton-pumping ATPase in rat kidney. J Clin Invest 82: 2114–2126
Brown D, Sabolic I, Gluck S (1991) Colchicine-induced redistribution of proton pumps in kidney epithelial cells. Kidney Int 40: S79–S83
Cameron JN (1976) Branchial ion uptake in Artic grayling: resting values and the effects of acid–base disturbance. J Exp Biol 123: 145–157
Cameron JN (1980) Body fluids pools, kidney function, and acid–base regulation in the freshwater catfish *Ictalurus punctatus*. J Exp Biol 86: 171–185
Cameron JN, Iwama GK (1989) Compromises between ionic regulation and acid–base regulation in aquatic animals. Can J Zool 67(12): 3078–3084
Cameron JN, Kormanik GA (1982) The acid–base response of gills and kidneys to infused acid and base loads in the channel catfish, *Ictalurus punctatus*. J Exp Biol 99: 143–160
Claiborne JB, Heisler N (1986) Acid–base regulation and ion transfers in the crap (*Cyprinus carpio*): pH compensation during graded long- and short-term environmental hypercapnia, and the effect of bicarbonate infusion. J Exp Biol 113: 465–470
DeRenzis G, Bornancin M (1984) Ion transport and gill ATPase. In: Hoar WS, Randall DJ (eds) Fish physiology, vol XB, Academic Press, New York, pp 65–104
Dunel-Erb S, Laurent P (1980) Ultrastructure of marine teleost gill epithelia: SEM and TEM study of the chloride cell apical membrane. J Morphol 165: 175–186
Ehrenfeld J, Garcia-Romeu F, Harvey BJ (1985) Electrogenic active proton pump in *Rana esculenta* skins and its role in sodium ion transport. J Physiol Lond 359: 331–355
Evans DH (1982) Salt and Water exchange across vertebrate gills. In: Houlihan DF, Rankin JC, Shuttleworth TJ (eds) Gills. Society for Experimental Biology Seminar Series, vol 16. Cambridge Univ Press, Cambridge, pp 149–172
Evans DH (1984) The roles of gill permeability and transport mechanisms in euryhalinity. In: Hoar WS, Randall DJ (eds) Fish physiology, vol XB. Academic Press, New York, pp 239–283
Evans DH, Claiborne JB, Farmer LL, Mallery C, Krasney EJ Jr (1982) Fish gill ionic transport: methods and models. Biol Bull 163: 108–130
Fenwick JC (1989) Calcium exchange across fish gills. In: Pang PKT, Schreibman MP (eds) Vertebrate endocrinology: fundamentals and biomedical implications, vol 3. Academic Press, New York, pp 319–342
Flik G, Wendelaar Bonga SE, Fenwick JC (1983) Ca^{2+}-dependent phosphatase and ATPase activities in eel gill plasma membranes. I. Identification of Ca^{2+}-activated ATPase activities with non-specific phosphatase activities. Comp Biochem Physiol B76: 745–754
Foskett JK, Bern HA, Machen TE, Conner M (1983) Chloride cells and the harmonal control of teleost fish osmoregulation. J Exp Biol 106: 255–281
Franklin GE (1990) Surface ultrastructure changes in the gills of sockeye salmon (Teleostei: *Oncorhynchus nerka*) during seawater transfer: comparison of successful and unsuccessful adaptation. J Morphol 206: 13–23
Franklin GE, Davison W (1989) SEM observations of morphologically different chloride cells in freshwater-adapted sockeye salmon *Oncorhynchus nerka*. J Fish Biol 34: 803–804
Fritsche C, Kleiman JG, Bain JL, Heinen RR, Riley DA (1991) Carbonic anhydrase and proton secretion in turtle bladder mitochondria-rich cells. Am J Physiol 260: F443–F458
Girard JP, Payan P (1980) Ion exchanges through respiratory and chloride cells in freshwater- and seawater-adapted teleosts. Am J Physiol 238: R260–R268
Goss GG, Wood CM (1990a) Na^+ and Cl^- uptake kinetics, diffusive effluxes and acidic equivalent fluxes across the gills of rainbow trout. I. Responses to environmental hyperoxia. J Exp Biol 152: 521–547
Goss GG, Wood CM (1990b) Na^+ and Cl^- uptake kinetics, diffusive effluxes and acidic equivalent fluxes across the gills of rainbow trout. II. Responses to bicarbonate infusion. J Exp Biol 152: 549–571

Goss GG, Laurent P, Perry SF (1992a) Evidence for a morphological component in acid–base regulation during environmental hypercapnia in the brown bullhead *Ictalurus nebulosus.* Cell Tissue Res 268: 539–552
Goss GG, Perry SF, Wood CM, Laurent P (1992b) Relationships between ion and acid–base regulation in freshwater fish. J Exp Zool 263: 143–149
Handy RD (1989) The ionic composition of rainbow trout body mucus. Comp Biochem Physiol A93: 571–575
Hardy MA, DiBona DR (1982) Microfilaments and the hydrosmotic action of vasopressin in toad urinary bladder. Am J Physiol 243: C200–C204
Heisler N (1986) Acid–base regulation in fishes. In: Heisler N (ed) Acid–base regulation in animals. Elsevier, Amsterdam, pp 309–356
Hossler FE, Ruby JR, McIlwain TD (1979) The gill arch of the mullet, *Mugil cephalus.* I. Surface ultrastructure. J Exp Zool 208: 379–398
Hughes JM (1984) General anatomy of the gills. In: Hoar WS, Randall DJ (eds) Fish physiology, vol XA. Academic Press, New York, pp 1–72
Hyde DA, Perry SF (1989) Differential approaches to blood acid–base regulation during exposure to prolonged hypercapnia in two freshwater teleosts: the rainbow trout (*Salmo gairdneri*) and the American eel (*Anguilla rostrata*). Physiol Zool 62: 1164–1186
Ishihara A, Mugiya Y (1987) Ultrastructural evidence of calcium uptake by chloride cells in the gill of the goldfish, *Carassius auratus.* J Exp Biol 242: 121–129
Kleyman TR, Cragoe EJ Jr (1988) Amiloride and its analogs as tools in the study of ion transport. J Membr Biol 105: 1–21
Laurent P (1984) Gill internal morphology. In: Hoar WS, Randall DJ (eds) Fish physiology, vol XA. Academic Press, New York, pp 73–183
Laurent P (1985) A survey of the morpho-functional relationships between epithelia, vasculature and innervation within the fish gill. In: Duncker HR, Fleischer G (eds) Vertebrate morphology. Fischer Stuttgart, pp 340–351
Laurent P (1989) Gill structure and function: Fish. In: Wood SC (ed) Comparative pulmonary physiology: current concepts. Dekker, New York, pp 69–120
Laurent P, Dunel S (1980) Morphology of gill epithelia in fish. Am J Physiol 238: R147–R159
Laurent P, Hebibi N (1989) Gill morphometry and fish osmoregulation. Can J Zool 67: 3055–3063
Laurent P, Perry SF (1990) Effects of cortisol on gill morphology and ionic uptake in the freshwater trout, *Salmo gairdneri.* Cell Tissue Res 259: 429–442
Laurent P, Perry SF (1991) Environmental effects on fish gill morphology. Physiol Zool 64 (1): 4–25
Laurent P, Hobe H, Dunel-Erb S (1985) The role of environmental sodium chloride relative to calcium in gill morphology of freshwater salmonid fish. Cell Tissue Res 240: 675–692
Laurent P, Goss GG, Perry S (1994) Proton pumps in fish gill pavement cells? Arch Int Physiol Biochem 102: 77–79
Laurent P, Dunel-ERB S, Chevalier C, Lignon J (1994) Gill epithelial cells kinetics in a freshwater teleost, *oncorhynchus mykiss* during adaptation to ion-poor water and hormonal treatments. Fish Physiol Biochem 13(5): 353–370
Leino RL, McCormick JH, Jensen KM (1987) Changes in gill histology of fathead minnows and yellow perch transferred to soft or acidified soft water with particular reference to chloride cells. Cell Tissue Res 250: 389–399
Lin H, Randall DJ (1991) Evidence for the presence of an electogenic proton pump on the trout gill epithelium. J Exp Biol 161: 119–134
McCormick SD (1990) Cortisol directly stimulates differentiation of chloride cells in tilapia opercular membrane. Am J Physiol 259: R857–R863
MacDonald DG, Wood CM (1981) Branchial and renal acid and ion fluxes in the rainbow trout, *Salmo gairdneri.* J Exp Biol 98: 273–287
Madsen SS (1990) Enhanced hyposmoregulatory response to growth hormone after cortisol treatment in immature rainbow trout, *Salmo gairdneri.* Fish Physiol Biochem 8: 271–279

Madsen KM, Tisher CC (1983) Cellular response to acute respiratory acidosis in rat medullary collecting duct. Am J Physiol 245: F670–F679
Maetz J (1974) Aspects of adaptation to hypo-osmotic and hyper-osmotic environments. In: Malins DC, Sargent JR (eds) Biochemical and biophysical perspectives in marine biology. Academic Press, New York, vol 1, pp 1–166
Naon R, Mayer-Gostan N (1983) Separation by velocity sedimentation of the gill epithelial cells and their ATPase activities in the seawater adapted eel *Anguilla anguilla*. Comp Biochem Physiol 75A: 541–547
Olson KR, Fromm PO (1973) A scanning electron microscope study of secondary lamellae and chloride cells of rainbow trout (*Salmo gairdneri*). Z Zellforsch 143: 439–449
Orci L, Glick BS, Rotheman JE (1986) A new type of coated vesicular carrier that appears not to contain clathrin: its possible role in protein transport within the Golgi stack Cell 46: 171–184
Payan P, Mayer-Gostan N, Pang PKT (1981) Site of calcium uptake in the freshwater trout gill. J Exp Biol 216: 345–347
Perry SF, Flik G (1988) Characterization of branchial transepithelial calcium fluxes in freshwater trout, *Salmo gairdneri*. Am J Physiol 254: R491–R498
Perry SF, Laurent P (1989) Adaptational responses of rainbow trout to lowered external NaCl concentration: contribution of the branchial chloride cell. J Exp Biol 147: 147–168
Perry SF, Laurent P (1990) The role of carbonic anhydrase in carbon dioxide excretion, acid–base balance and ionic regulation in aquatic gill breathers. In: Truchot JP, Lahlou B (eds) Animal nutrition and transport processes. 2. Transport, respiration and excretion. Comp Physiol, vol. 6. Karger, Basel, pp 39–57
Perry SF, Laurent P (1993) Environmental effects on fish gill structure and function. In: Jensen F, Rankin C (eds) Fish ecophysiology. Chapman & Hall, London, 221–264
Perry SF, Walsh PJ (1989) Metabolism of isolated fish gill cells: contribution of epithelial chloride cells. J Exp Biol 144: 507–520
Perry SF, Wood CM (1985) Kinetics of branchial calcium uptake in the rainbow trout: effects of acclimation to various external calcium levels. J Exp Biol 116: 411–433
Perry SF, Malone S, Ewing D (1987) Hypercapnic acidosis in the rainbow trout, *Salmo gairdneri*. I. Branchial ionic fluxes and blood acid–base status. Can J Zool 65: 888–895
Perry SF, Goss GG, Laurent P (1992) The interrelationships between gill chloride cell morphology and ionic uptake in four freshwater teleosts: the effects of cortisol. Can J Zool 9: 1775–1786
Pisam M, Caroff A, Rambourg A (1987) Two types of chloride cells in the gill epithelium of a freshwater-adapted euryhaline fish: *Lebistes reticulatus*. Their modifications during adaptation to salt water. Am J Anat 179: 40–50
Pisam M, Boeuf G, Prunet P, Rambourg (1990) Ultrastructural features of mitochondria-rich cells in stenohaline freshwater and seawater fishes. Am J Anat 187: 21–31
Rahim SM, Delaunoy JP, Laurent P (1988) Identification and immunocytochemical localization of two different carbonic anhydrase isoenzymes in the teleostean fish erythrocytes and gill epithelia. Histochemistry 89: 451–459
Sardet C (1977) Ordered arrays of intramembrane particles on the surface of fish gills. Cell Biol Int Rep 1: 409–417
Sardet C (1980) Freeze fracture of the gill epithelium of euryhaline teleost fish. Am J Physiol 238: R207–R212
Shuttleworth TJ (1989) Overview of epithelial ion-transport mechanisms. Can J Zool 67 (12): 3032–3038
Spry DJ, Wood CM (1988) Zinc influx across the isolated, perfused head preparation of the rainbow trout (*Salmo gairdneri*) in hard and soft water. Can J Fish Aquat Sci 45: 2206–2215
Steinmetz P (1986) Cellular organization of urinary acidification. Am J Physiol 251: F173–F197
Stetson DL (1989) Turtle urinary bladder: regulation of ion transport by dynamic changes in plasma membrane area. Am J Physiol 257: R973–R981
Thomas S, Fievet B, Motais R (1988) Adaptative respiratory responses of trout to acute hypoxia. Effects of water ionic composition on blood acid–base response and gill morphology. Respir Physiol 74: 77–90

Wendelaar Bonga SE, van der Meij CJM (1989) Degeneration and death, by apoptosis and necrosis, of the pavement and chloride cells in the gills of the teleost *Oreochromis mossambicus*. Cell Tissue Res 255: 235–243

Wheatly MG, Hōbe H, Wood CM (1984) The mechanisms of acid–base and ionoregulation in the freshwater rainbow trout during environmental hyperoxia and subsequent normoxia. II. The role of the kidney. Respir Physiol 55: 155–173

Wood CM (1988) Acid–base and ionic exchanges at gills and kidney after exhaustive exercise in the rainbow trout. J Exp Biol 136: 461–481

Wood CM (1991) Branchial ion and acid–base transfer in freshwater teleost fish. Environmental hyperoxia as a probe. Physiol Zool 64: 68–102

Wood CM, Caldwell FH (1978) Renal regulation of acid-base balance in a freshwater fish. J Exp Zool 205: 301–307

Wood CM, Goss GG (1990a) Kinetic analysis of the relationships between ion exchange and acid–base regulation at the gills of freshwater fish. In: Truchot JP, Lahlou B (eds) Animal nutrition and transport processes. 2. Transport, respiration and excretion. Comp Physiol, vol 6. Karger, Basel, pp 119–136

Wood CM, Goss GG (1990b) Kinetic analysis of the relationships between ion exchange and acid–base regulation at the gills of freshwater fish. Com Physiol 6: 119–136

Wood CM, Wheatly M, Hōbe H (1984) The mechanisms of acid–base and ionoregulation in the freshwater rainbow trout during environmental hyperoxia and subsequent normoxia. II. Branchial exchanges. Respir Physiol 55: 175–192

Chapter 6

The Roles of Natriuretic Peptide Hormones in Fish Osmoregulation and Hemodynamics

D.H. Evans

Contents

Department of Zoology, University of Florida, Gainesville, FL 32611 USA; Mt. Desert Island Biological Laboratory, Salsbury Cove, ME 04672, USA

Advances in Comparative and Environmental Physiology, Vol. 22

1 Introduction

Since the original description of the distinct natriuretic activity of an extract of rat atria (DeBold et al. 1981), it has become apparent that various members of a natriuretic peptide (NP) hormone family[1] are synthesized in a variety of mammalian tissues (Vollmar 1990), but most specifically the heart and brain (Sudoh et al. 1990), and can produce multiple renal, adrenal, cardiovascular, and brain effects (e.g. Needleman et al. 1989; Brenner et al. 1990; Samson and Quirion 1990)[2]. In addition, preliminary results suggest that NPs may be important in the function of such diverse tissues as the thyroid (Ahren 1990), ovary (Tornell et al. 1990), thymus (Vollmar et al. 1990), and gut (Traynor and O'Grady 1991) of mammals. This variety of effects of NPs is mediated both by direct actions on specific tissues and via intermediary hormones (e.g. Samson and Quirion 1990). The intent of this review is to describe the emerging literature in comparative physiology suggesting that NPs may play an important role in fish osmoregulation and hemodynamics. Surprisingly, the bulk of evidence is consistent with a role in adaptation to seawater (and concomitant hypovolemia), rather than adaptation to freshwater (and hypervolemia) as might be expected from the NP-mediated response to hypervolemia as described in mammals (e.g. Brenner et al. 1990). Before reviewing the data supporting this conclusion, a short discussion of the function of NPs in mammals is appropriate. Numerous excellent reviews have been published, most recently: Brenner et al. (1990); Samson and Quirion (1990);

[1] Members of the family appear to be grouped into four types: A-type natriuretic peptide (also termed ANP, ANF, and atriopeptin), B-type natriuretic peptide (also termed BNP), C-type natriuretic peptide (also termed CNP), and V-type natriuretic peptide (VNP; Takei et al. 1991). For a recent review, see Rosenzweig and Seidman (1991).

[2] Most physiological studies have used ANP or truncated versions. However, it is now clear that BNP can produce the same array of cardiovascular and renal responses (e.g. Gunning et al. 1990). Physiological responses to CNP remain relatively unstudied, but preliminary data in mammals suggested that its hypotensive and natriuretic potency is about 1/100th that of ANP or BNP (Sudoh et al. 1990). More recent work indicates that CNP might even be antinatriuretic (Stingo et al. 1992), but more hypotensive than ANP (Clavel et al. 1993).

Rosenzweig and Seidman (1991); Koller and Goeddel (1992); and Ruskoaho (1992)[3].

2 Biochemistry of Natriuretic Peptide Hormones in Mammals

2.1 Structure

For reference, the amino acid sequence of selected members of the NP family, including recently described fish peptides, is presented in Fig. 1. Various nomenclatures and fragments have been described, but it is now generally accepted that the family consists of three, closely related peptides in addition to ANP (Rosenzweig and Seidman 1991). Firstly, BNP was initially described in the brain, but is predominantly found in the heart, including the ventricle (Hosoda et al. 1991). Secondly, CNP was also first described in the brains of mammals (Sudoh et al. 1990) and fishes (Price et al. 1990; Takei et al. 1990a), but has now been found in the hearts of sharks (Schofield et al. 1991; Suzuki et al. 1991, 1992), although not in mammals (Minamino et al. 1991). Finally, VNP has recently been isolated from the ventricle of the Japanese eel (Takei et al. 1991) and appears to

HUMAN ANP	S-L-R-R-S-S C-F-G G-R-M D-R-I-G A-Q S-G-L-G-C N-S-F-R---Y-COOH
RAT ANP	S-L-R-R-S-S C-F-G G-R-I D-R-I-G A-Q S-G-L-G-C N-S-F-R---Y-COOH
FROG ANP	S-S-D C-F-G S-R-I D-R-I-G A-Q S-G M G-C --G---R-R-F-COOH
EEL ANP	S-K-S-S-S-P C-F-G G K-L-D-R-I-G-S Y S-G-L-G-C N-S---R-K-COOH
PIG BNP	S-P-K-T-M-R-D-S G-C-F-G R-R L D-R-I-G-S L S-G-L-G-C N-V-L-R-R-Y-COOH
HUM. BNP	S-P-K-M-V-Q-G-S G-C-F C-R-K-M D-R-I S S S S-G-L-G-C K-V-L-R-R-H-COOH
PIG CNP	G-L-S-K G-C-F-G-L-K-L-D-R-I-G-S-M-S-G-L-G-C COOH
CHICKEN CNP	G-L-S-R-S C-F-G V K-L-D-R-I-G-S-M-S-G-L-G-C COOH
KILLIFISH CNP	G-W-N-R-G-C-F-G-L-K-L-D-R-I-G-S-M-S-G-L-G-C COOH
EEL CNP	G-W-N-R-G-C-F-G-L-K-L-D-R-I-G-S L S-G-L-G-C COOH
DOGFISH CNP-SC	G-P-S R-G-C-F-G V K-L-D-R-I-G A M-S-G-L-G-C COOH
DOGFISH CNP-SA	G-P-S R S C-F-G-L-K-L-D-R-I-G A M-S-G-L-G-C COOH
EEL VNP	K-S-F-N-S C-F-G T-R-M D-R-I-G-S W S-G-L-G-C N-S-L--K-N-G-T-K-K-K-I-F-G-N-COOH

Fig. 1. Amino acid sequences of various members of the natriuretic peptide hormone (NP)-family of hormones. Eel is *Anguilla japonica*, killifish is *Fundulus heteroclitus. Scyliorhinus canicula* and *Squalus acanthias* are dogfish shark species

[3] In most cases, because of space considerations, only those papers not documented in these reviews will be cited.

be the major circulating NP in this species (Takei et al. 1994). Each hormone is apparently cleaved from the carboxyl-terminal of a prohormone during secretion (e.g. Ruskoaho 1992).

2.2 Sites of Synthesis

Early physiological experiments suggested that the heart itself might play a role in the response to volume loading in mammals. For instance, Smith (1957) proposed that a receptor in the atrium might be responsible for the increased natriuresis produced by saline infusions in patients with essential hypertension. Granular vesicles in atrial myocytes that were first described by Kisch and DeBold demonstrated that the granularity was altered in response to changes in water and electrolyte balance in rats (Brenner et al. 1990). DeBold's classic experiment 2 years later demonstrated that injection of granule-enriched atrial extracts elicited a transient, but significant, natriuresis and hypotension in rats (DeBold et al. 1981). Subsequent studies documented immunohistochemical reaction of atrial vesicles with antibodies directed against ANP, confirming the site of synthesis of the hormone (Brenner et al. 1990). The ANP_{ir} in the variety of other tissues (including the ventricles) where it has been demonstrated (e.g. Vollmar 1990) is not associated with vesicles, and is at much lower levels than in the atria (Needleman et al. 1989). Specific sites of ANP_{ir} in the brain include the hypothalamus, paraventricular nuclei, and the anteroventral region of the third ventricle, areas important in osmoregulation in mammals (Brenner et al. 1990). Surprisingly, the kidney itself also has recently been shown to synthesize NPs (e.g. Suzuki et al. 1993).

2.3 Control of Release

It is generally accepted that atrial stretch associated with hypervolemia is the major stimulus for secretion of ANP from the mammalian heart (Brenner et al. 1990), although at least some studies have shown that an increase in Na^+ concentration or osmolarity may stimulate release from isolated atria or dispersed atrial myocytes (Brenner et al. 1990). In addition to these volume or osmotic stimuli, various hormones and neurotransmitters have been implicated in the control of ANP secretion from the heart, including glucocorticoids (Brenner et al. 1990), thyroxine (Mori et al. 1990), calcitonin gene-related peptide (Yamamoto et al. 1988), angiotensin II (ANG II; Volpe et al. 1991), acetylcholine, epinephrine and vasopressin (Brenner et al. 1990), endothelin (Mantymaa et al. 1990), and prolactin (Zarmir et al. 1987). It is important to note that the production of at least four of these putative secretagogues, ANG II, vasopressin, prolactin, and endothelin are, in turn, inhibited by ANP (see below), suggesting negative feedback axes from (*not to*) the heart. Stimuli of secretion of brain NPs remain somewhat less studied. It should be obvious that such neurotransmitters or brain hormones such as catecholamines, acetylcholine, ANG II, and vasopressin probably play a role, but critical experiments are few (e.g. Huang et al. 1991). There is also a single study suggesting that medium Na^+ concentration plays a role in secretion of ANP from isolated hypothalamic fragments (Shibasaki et al. 1988).

2.4 Receptors

Three closely related receptors for NPs have now been identified, cloned, and studied pharmacologically (Maack 1992). The receptor ANPR-A [also termed B (biological)-ANF receptor, ANP-RI, and GC-A] has an apparent molecular weight (M_r) of ~ 130 kDa and (along with ANPR-B; see below) is a transmembrane guanylyl cyclase producing cyclic GMP in most of the target tissues studied (Leitman and Murad 1987). The EC_{50} for cGMP production in COS-7 cells transfected with ANP-A receptor plasmid DNA is 3 nM for ANP and 25 nM for BNP (Schulz et al. 1989).

The receptor ANPR-B (also termed ANP-R3, because it was the third receptor described, and GC-B) has approximately the same M_r as ANPR-A, and shares 88% homology with that receptor in the 252 amino acid residues at the C terminal (Chang et al. 1989). The EC_{50} of cGMP production by this receptor is 25 μM for ANP and 6 μM for BNP (Schulz et al. 1989), suggesting differential, but rather low, sensitivity when compared with ANPR-A. The suggestion that a more potent, natural ligand might be present (Schulz et al. 1989) was corroborated by a recent study demonstrating that ANPR-B is quite sensitive to CNP (Koller et al. 1991).

The receptor ANPR-C (also termed the C or clearance receptor, and ANP-R2, because it was the second receptor described) is a 60–70 kDa glycoprotein that forms a disulfide-bonded homodimer (Schenck et al. 1987) and binds ANP with a very high affinity, but does not display guanylyl-cyclase activity. Contrary to either the ANPR-A or B, this receptor is also sensitive to a variety of truncated peptides (Scarborough et al. 1989; Brenner et al. 1990). This finding, along with the apparent abundance of ANPR-C on vascular endothelium, suggests that this receptor may function to clear ANP-like peptides from the circulation. This proposition is supported by the finding that there are often major discrepancies between the kinetics of ANP-specific binding ($K_D << 1$ nM) and the kinetics of either the physiological response (e.g. vascular relaxation) or ANP-stimulated cGMP production ($EC_{50} > 1$ nM; e.g. Leitman and Murad 1987). For reference, mammalian plasma levels of ANP-like peptides are in the range of 2 to 100 pM (Genest and Cantin 1988), approaching the binding K_D, but far below the usual EC_{50} described. A physiological role for the ANPR-C was established by Maack et al. (1987), who demonstrated that a ring-deleted ANP analog (C-ANF_{4-23}) binds with a high affinity to NP receptors in the isolated perfused kidney, but does not produce any renal effects and is not antagonistic to ANP-stimulated natriuresis. However, injection of C-ANF_{4-23} into intact rats did produce natriuresis and hypotension, associated with an increase in plasma ANP_{ir}, consistent with the proposition that blockade of ANPR-C by C-ANF_{4-23} antagonizes the usual clearance function of ANPR-C, allowing plasma levels of ANP to remain high. There is some evidence that ANPR-C also may mediate some physiological responses via either inhibition of cyclic AMP (Anand-Srivastava et al. 1990; Rashed et al. 1993) or stimulation of phosphatidylinositol metabolism (Hirata et al. 1989).

3 Physiology of Natriuretic Peptide Hormones in Mammals

Although there is emerging evidence that NPs may function in a variety of physiological responses, the bulk of evidence supports initial conclusions that the primary focus of action is a reduction in blood pressure mediated by both direct actions on the cardiovascular system and stimulation of salt and water excretion by the kidney.

3.1 Effects on Heart and Peripheral Resistance

The NP receptors have been identified autoradiographically on cardiac tissue, although it is unclear if the sites of action are myocytes or glial cells (e.g. James et al. 1990), or if the response in vivo is due to systemic ANP or release from adjacent myocytes (e.g. Gu 1991). Infusion of ANP decreases cardiac output, associated with a fall in stroke volume and, sometimes, rate. Current evidence suggests that this response is mediated by direct actions on the heart tissue or coronary perfusion, and, indirectly, by alteration of autonomic input and/or venous return (e.g. Brenner et al. 1990), but the relative importance of each control mechanism is not known.

The ANP has been shown to dilate or relax a variety of mammalian blood vessels, especially when they have been preconstricted with effectors such as angiotensin II and norepinephrine. The vascular endothelium is not necessary for the response, and ANP is not antagonistic to contraction produced by membrane depolarization when extracellular K^+ concentrations are increased (e.g. Winquist and Hintze 1990). Thus, ANP is generally considered to be an antagonistic hormone, rather than a controller of vascular tone, and there is some debate on its relative role in the control of peripheral resistance or tissue perfusion, especially since the hypotension produced by an ANP-induced fall in cardiac output and intravascular volume (see above) would normally result in sympathetic stimulation of peripheral vasconstriction (Brenner et al. 1990). However, it has recently been shown that injection of antiserum against rANP into rats produced a rise in mean blood pressure and cardiac output (as well as antinatriuresis), suggesting "an important physiological role in the regulation of hemodynamics and water-electrolyte balance" (Sasaki et al. 1989).

3.2 Effects on Kidney

DeBold's pioneering experiment (DeBold et al. 1981) and the accepted terminology of this family of peptides highlight the ability of ANP to stimulate both an increase in urine output and an increase in urinary Na^+ concentration. Given the complex architecture of the mammalian (indeed, vertebrate) kidney, there are a multitude of putative sites of action for these responses.

It is still debated if ANP infusion produces an alteration in total or regional renal blood flow. However, it is clear that ANP stimulates glomerular filtration by an increase in hydraulic pressure (produced by afferent arteriolar dilation and efferent arteriolar constriction) and an increase in the glomerular permeability coefficient secondary to mesangial cell relaxation (e.g. Cogan 1990). Interestingly, CNP has recently been found to be antinatriuretic in the dog (Stingo et al. 1992). In addition, there is still some debate on the relative roles of glomerular (or other hemodynamic) vs tubular transport effects in the natriuretic response to ANP (Cogan 1990; Zeidel 1990; Knepper et al. 1991).

Because the proximal tubule is the site of reabsorption of the majority of filtered urine, it seems intuitive that this might be the site of ANP action. However, receptors have not been identified kinetically or autoradiographically in segments of the proximal tubule, and some studies have been unable to demonstrate ANP inhibition of solute or water transport by the proximal tubule (Brenner et al. 1990). Nevertheless, other studies have demonstrated that, at least under conditions of angiotensin II or norepinephrine stimulation, fluid absorption by perfused proximal tubules may be significantly inhibited by ANP (e.g. Garvin 1992). There seems to be general agreement that inhibition of sodium and water uptake in the inner medullary collecting duct (IMCD) is a major site of action of ANP (e.g. Brenner et al. 1990). Contrary to more proximal segments, ANP receptors have been localized via autoradiography in the IMCD, and various studies have established that ANP can inhibit Na^+ (and presumably water) reabsorption by the IMCD epithelium by inhibiting Na^+ uptake via apical channels (Zeidel 1993). A recent study demonstrated that ANP could also stimulate basolateral NaCl uptake (via a Na/K/2Cl cotransporter) and secretion by perfused IMCD (Rocha and Kudo 1990), indicating that stimulation of secretion as well as inhibition of reabsorption in the IMCD plays a role in ANP-induced natriuresis.

3.3 Interactions with Other Endocrine Systems

In addition to the direct cardiac, vasomotor, and renal epithelial effects (see above), the hypotension and natriuresis produced by NPs may also be produced via interactions with other hormone systems. A variety of studies have now suggested that ANP can produce its cardiovascular and renal effects via inhibition of the renin-angiotensin system (e.g. Brenner et al. 1990; Johnson et al. 1991), endothelin (Kohno et al. 1991; Ota et al. 1992), vasopressin (Brenner et al.

1990), prolactin (via stimulation of dopamine; Samson 1990), and adrenocorticotropic hormone (ACTH) (Dayanithi and Antoni 1990; Ibanez-Santos et al. 1990). Interestingly, it appears that ANP may stimulate growth hormone release by inhibition of somatostatin (Samson 1990), and CNP may stimulate prolactin secretion directly in the hypothalamus, contrary to the effect of ANP (Huang et al. 1992).

4 Physiology of Fish Osmoregulation

The fact that the entire cardiac output from the fish heart passes through the branchial vasculature, which is separated from a noniso-osmotic environment by a thin epithelium, dictates that nearly all species of fishes face substantial osmoregulatory problems (see Evans 1979, 1984, and 1993 for more complete review and literature citations for much of the following).

4.1 Teleosts

Teleosts maintain a plasma osmotic concentration in the range of 300–450 mOsm, approximately 300 times as concentrated as freshwater and one-third as concentrated as seawater (Fig. 2). Because the branchial epithelium is relatively permeable to both ions and water, teleosts in either environment must osmoregulate. In freshwater, these fishes must balance the constant influx of water (hypervolemia) and loss of ions (predominantly NaCl). In seawater, the gradients are reversed; hence, they must compensate for the efflux of water (hypovolemia) and influx of NaCl.

4.1.1 Osmoregulation in Freshwater

Teleosts maintain body fluid volume and concentration in freshwater by a combination of branchial and renal mechanisms. Glomerular filtration rates are high (ca. 400 $\mu l\ 100\ g^{-1}\ h^{-1}$) and tubular reabsorption of NaCl is substantial. The resulting urine is significantly more dilute than the plasma, approaching 20 mOsm l^{-1}; urinary output is approximately 300 $\mu l\ 100\ g^{-1}\ h^{-1}$. Renal tubules in freshwater fishes include the proximal and distal tubules, and the collecting duct (Hentschel and Elger 1989; Evans 1993). Interestingly, the distal tubule is missing in some euryhaline, marine species that enter freshwater (e.g. the killifish *Fundulus heteroclitus* and the toadfish *Opsanus tau*), as well as the freshwater stickleback (*Gasterosteus aculeatus*), suggesting that this tubule is not a vital site of salt reabsorption. However, the collecting duct is present in all species that enter freshwater, indicating an important role for this segment in the production of a dilute urine (Evans 1979).

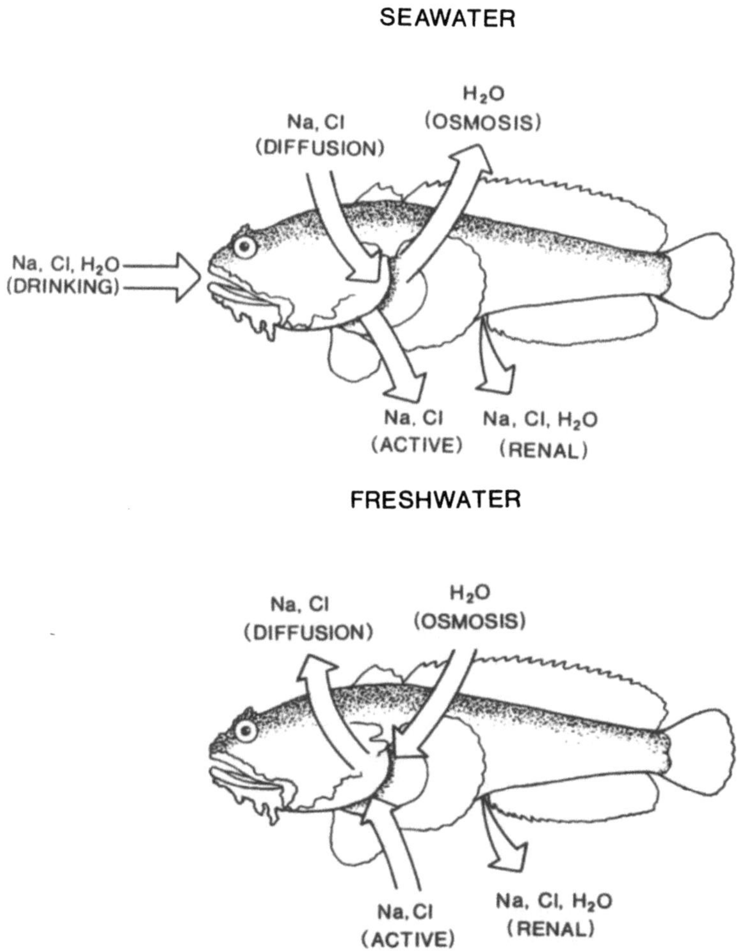

Fig. 2. Osmoregulation in a typical marine or freshwater teleost fish. In seawater, branchial loss of water and gain of NaCl is balanced by oral ingestion of the medium, excretion of nearly iso-osmotic urine, and active branchial extrusion of NaCl. In freshwater, branchial gain of water and loss of NaCl is balanced by excretion of relatively large volumes of hypo-osmotic urine and active branchial extraction of NaCl from the medium (see text for details)

Branchial and renal loss of NaCl is balanced by active uptake of these ions from the external medium, generally considered to be via parallel antiports of Na^+/H^+ (NH_4^+) and Cl^-/HCO_3^- (e.g. Marshall 1988; Heisler 1989; McDonald et al. 1989; Shuttleworth 1989). However, recent evidence suggests that the coupling between Na^+ and H^+ may be electrical, rather than chemical (Avella and Bornancin 1989; Lin and Randall 1991), and it appears that most ammonia excretion is via both nonionic and ionic diffusion (Evans and Cameron 1986; Evans et al. 1989a). Recent evidence suggests that the parallel extrusion of H^+ and HCO_3^- are the major pathways for fish acid/base regulation (e.g. Heisler

1993). Indeed, the antiports are present in marine species, including the elasmobranch and hagfish, suggesting that the acid/base function of these antiports evolved before their ion-regulatory function (Evans 1982, 1984, 1986). Recent evidence also suggests that the chloride cell of the branchial epithelium is the site of these antiports (Perry and Laurent 1989; Laurent and Perry 1990), as well as the NaCl extrusion sites in seawater teleosts (see below).

4.1.2 Osmoregulation in Seawater

The gut, as well as the gills and kidney, control osmoregulation in marine teleost fishes. To balance the osmotic loss of water across the branchial epithelium, marine teleost fishes ingest the medium. Some esophageal uptake of salt and water takes place (e.g. Kirsch and Meister 1982), but the bulk of the ingested salt and water is absorbed in the small intestine via coupled Na/K/2CI transport, with water following passively (e.g. Evans 1979; Groot and Bakker 1988). Renal glomerular filtration rates (GFRs) and flows are extremely low (ca. 10% those in freshwater teleosts) in marine teleosts, and the urine NaCl concentration is usually approximately isotonic to the plasma. Various species of marine teleosts are actually aglomerular, and it was initially considered that urine formation in these species is secondary to divalent ion secretion in the proximal tubule (e.g. Evans 1979). However, recent studies (reviewed by Beyenbach and Baustian 1989) have demonstrated that glomerular marine and freshwater teleosts, as well as elasmobranchs, can secrete NaCl into the lumen of the proximal tubule via Na/K/2CI transport, although urine hypertonic to the plasma has only been described once (Stanley and Fleming 1964). Whether NaCl secretion plays a role in urine formation by aglomerular kidneys remains to be determined, but it has been calculated that approximately 41% of ureteral Na^+ in the trout (*Oncorhynchus mykiss*, formerly *Salmo gairdneri*) may have been secreted in the proximal tubule (Beyenbach and Baustian 1989).

The plasma NaCl load in marine teleosts that is produced by diffusional influx and intestinal uptake is balanced by branchial extrusion via a Na/K/2Cl transport system in the chloride cell of the branchial epithelium (see reviews by Zadunaisky 1984; Karnaky 1986; Foskett 1987). Analogous to the model for secretion of NaCl by the fish renal proximal tubule (Beyenbach and Baustian 1989), the transporter in the teleost gill chloride cell is on the basolateral membrane, along with Na/K-ATPase, and final secretion of Cl^- takes place across apical Cl^- channels; Na^+ follows passively across leaky paracellular junctions (Karnaky 1986).

4.2 Elasmobranchs

The sharks, skates, and rays are primarily marine, although the fact that their plasma NaCl concentrations are significantly below those of seawater suggests that they arose in a freshwater environment (e.g. Shuttleworth 1988). Elasmobranch plasma is actually slightly hypertonic to seawater, secondary to the retention of two organic solutes: urea (ca. 400 mM l^{-1}) and trimethylamine oxide (TMAO; ca. 80mM l^{-1}). In contrast to the relatively low branchial water permeability of teleosts (Evans 1979), elasmobranchs maintain an extremely high branchial water permeability (e.g. Shuttleworth 1988). The resulting osmotic influx of water is balanced by a relatively high GFR (ca. 350 μl 100 g^{-1} h^{-1} ≈ freshwater teleosts; Henderson et al. 1988) and urine flow (ca. 100 μl 100 g^{-1} h^{-1}; Evans 1979). Urinary NaCl concentrations are significantly above those of marine teleost fishes, but still below plasma levels, despite the presence of NaCl secretory mechanisms in the proximal tubule (Beyenbach and Baustian 1989). Therefore, extrarenal NaCl secretion must take place to balance the diffusion influx of salt, which is relatively small, because of a very low branchial permeability to salt (Evans 1979; Shuttleworth 1988). It is now clear that the rectal gland is the dominant, if not sole, site for extrarenal NaCl secretion (Fig. 3). A variety of studies (reviewed by Shuttleworth 1988) have demonstrated that the secretory cells of the rectal gland function in much the same way as the chloride cell of teleosts, utilizing basolateral cotransport of Na/K/2Cl, coupled with basolateral Na/K-ATPase. The resulting solution has a significantly higher NaCl concentration than the plasma (Evans 1979; Shuttleworth 1988). However, the rectal gland is apparently not a vital organ (e.g. Evans et al. 1982), suggesting that branchial extrusion may also play a role. No direct studies of NaCl extrusion by the elasmobranch branchial epithelium have been published, but a recent study of ammonia transport across the shark gill suggested that the Na/K/2Cl transporter may be present, and therefore presumably involved in net NaCl secretion (Evans and More 1988). Chloride cells are found in the shark gill epithelium, but their number and Na/K-ATPase content are less than 20% of that found in teleost gills (Shuttleworth 1988).

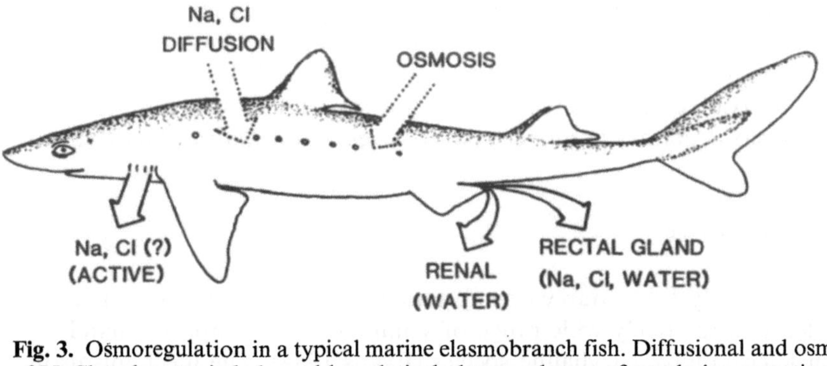

Fig. 3. Osmoregulation in a typical marine elasmobranch fish. Diffusional and osmotic gain of NaCl and water is balanced by relatively large volumes of nearly iso-osmotic urine and extrusion of NaCl via the rectal gland (and possibly the branchial epithelium; see text for details)

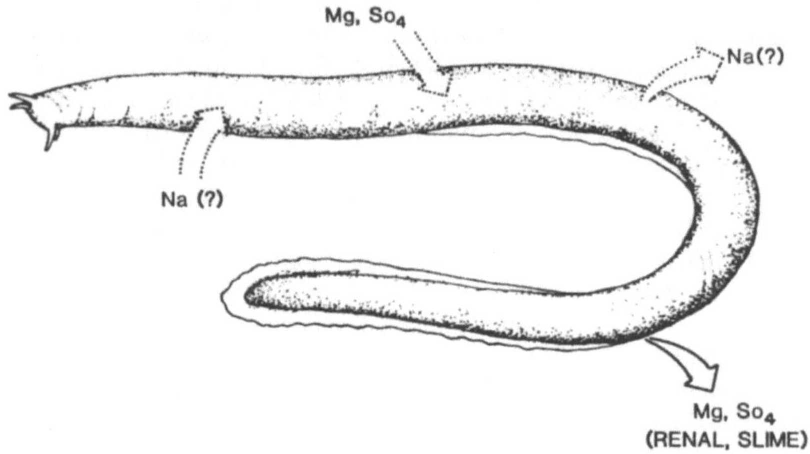

Fig. 4. Osmoregulation by the marine hagfish. Slight diffusional gain of divalent salts is balanced by renal and slime-gland extrusion, and slight diffusional loss of Na^+ may be balanced by active extraction of Na^+ from the medium (see text for details)

4.3 Agnatha

The two orders of jawless fishes, the lampreys and hagfish, have been distinct for approximately 500 million years (Hardisty 1979). Lampreys osmoregulate in a basically teleostean pattern, but hagfish are unique in the subphylum Vertebrata in having plasma NaCl concentrations and total osmolarity nearly equivalent to those of their marine environment (e.g. Evans 1979; Hardisty 1979). Some ionic gradients exist, however, with plasma Na^+ concentrations generally slightly above those of the medium (Evans 1979). Because the urinary Na^+ urine/plasma ratio is nearly at unity, it is apparent that this hypernatremia may be maintained by branchial Na^+ uptake mechanisms. In fact, there is now some evidence that Na^+/H^+ exchange is present in the hagfish gill, suggesting that this transport system evolved in the marine environment for acid/base regulation, rather than ionic regulation in freshwater (Fig. 4; Evans 1984). Nevertheless, it is clear that, contrary to teleosts, elasmobranchs, and lampreys, hagfish do not suffer from substantial osmoregulatory or ionoregulatory stress in the marine environment.

4.4 Euryhaline Species

Teleosts are the only fish group with substantial numbers of members that are able to tolerate a relatively wide range of salinities. In fact, approximately 84 families of teleosts have euryhaline species (Evans 1984). The mechanisms involved have been discussed elsewhere (Evans 1984); suffice it to say that the limiting factor in teleost euryhalinity appears to be branchial ionic permeability relative to the affinity of the NaCl uptake systems. Euryhaline marine species move from a condition of volume depletion and salt loading to a condition of volume loading and salt depletion. They, therefore, must turn off the extrusion of NaCl and oral ingestion, and increase glomerular filtration and tubular ion reabsorption when entering brackish or freshwater environments. The reverse is true of euryhaline, freshwater species, which are apparently much rarer (with the notable exceptions of killifish and salmonids; Evans 1984). It is apparent that euryhaline lampreys (e.g. *Petromyzon marinus*) osmoregulate in much the same way as euryhaline teleosts (Evans 1979).

Euryhaline elasmobranchs are relatively rare, and there is only a single family of rays (Potamotrygonidae) that are actually freshwater residents. Because it appears that even marine elasmobranchs have the transport steps (i.e. Na^+/H^+ and Cl^-/HCO_3^- exchangers) necessary for ionic uptake in dilute salinities, as well as renal mechanisms for the excretion of water and reabsorption of salts, the actual factor(s) limiting elasmobranch euryhalinity are unknown (Evans 1984). Nevertheless, it should be noted that at least in the spiny dogfish (*Squalus acanthias*) volume loading, rather than salt loading, appears to be the major stimulus for rectal gland secretion (Solomon et al. 1985b), despite its dominant role in the excretion of salt (see above).

Hagfish are very stenohaline, presumably, because of their extremely high permeability to water (Evans 1979). However, at least *Eptatretus stoutii* can volume regulate in 80% seawater, but it fails in 122% seawater (McFarland and Munz 1965).

5 Endocrine Control of Fish Osmoregulation

As might be expected, the relatively complex physiology that characterizes fish osmoregulation is under the control of a variety of hormones and local effectors. Recent, more comprehensive reviews include: Lahlou (1980); Foskett et al. (1983); Evans (1984); Rankin and Bolis (1984); Foskett (1987); Hirano et al. (1987); Mayer-Gostan et al. (1987); Evans (1990); Wendelaar Bonga (1993).

5.1 Freshwater Teleosts

There is still general agreement that prolactin is the major effector controlling a variety of the physiological steps that characterize osmoregulation in freshwater teleost fishes (e.g. Loretz and Bern 1982; Bern 1983; Foskett et al. 1983; Mayer-Gostan et al. 1987; Herndon et al. 1991). Its major effects appear to be: reduction in branchial water and salt permeability; inhibition of salt extrusion, probably by dedifferentiation of chloride cells; reduction in esophageal salt permeability; reduction in intestinal salt uptake and water permeability, and reduction in urinary bladder water permeability, but enhancement of urinary bladder salt uptake. All of these responses would enhance the retention of plasma NaCl and reduce the osmotic influx of water or enhance renal water excretion, processes which are critical for acclimation to freshwater. These responses are generally thought to be long-term after prolonged exposure to reduced salinities (e.g. Foskett et al. 1983). It appears that cortisol, which is usually considered to be a major hormone in seawater acclimation of teleosts (see below), may also play a role in acclimation to freshwater. Laurent and Perry (1990) have recently shown that daily intramuscular injection of cortisol stimulated chloride cell proliferation and NaCl influxes in freshwater trout.

More acute responses to lowered salinities may be partially mediated via cathecholamines, because it has been shown that NaCl extrusion by the chloride cells in seawater-acclimated *F. heteroclitus* opercular epithelium (which models the gill chloride cell) is inhibited by ∂-adrenergic stimulation (Zadunaisky 1984). Interestingly, it has also been shown recently that arginine vasotocin (AVT) reduces gill adenylate cyclase activity in gill tissue from seawater-acclimated trout (Guibbolini and Lahlou 1987). Because cyclic AMP is known to be involved in secretagogue stimulation of gill chloride cell NaCl extrusion (e.g. Zadunaisky 1984), these data suggest that neurohypophyseal peptides may be involved in the inhibition of this salt extrusion. Such inhibition is critical for acclimation to freshwater. They may also be involved in the stimulation of urine flow, because AVT is generally diuretic in fishes (e.g. Pang et al. 1983), rather than antidiuretic, as it is in mammals. Recent work has demonstrated that plasma levels of AVT (estimated by bioassay) are increased in freshwater-acclimated flounder and trout (Perrott et al. 1991).

5.2 Seawater Teleosts

Cortisol appears to be the major hormone controlling acclimation to seawater. Its plasma concentrations increase in seawater (e.g. Forrest et al. 1973), and cortisol stimulates the differentiation of chloride cells (McCormick 1990). In addition, cortisol increases the NaCl permeability of the esophagus, the intestinal salt uptake and water permeability, and the urinary bladder water permeability (e.g. Foskett et al. 1983; Mayer-Gostan et al. 1987), which are all responses critical for acclimation to seawater. As in tetrapods, pituitary ACTH stimulates adrenal cortical release of cortisol (e.g. Lahlou 1980).

The renin/angiotensin system is apparently also critical to seawater acclimation. Both renin and ANG II levels in plasma are elevated in seawater-acclimated fishes (e.g. Wilson 1984). Injection of ANG II into various species of teleosts stimulates drinking of the medium, but other factors may be involved, at least in the eel (*Anguilla japonica*), because infusion of hypertonic saline into either freshwater- or seawater-acclimated eels increased plasma ANG II concentration, but did not stimulate drinking (Nishimura 1987). Analogous to the axis in mammals, ANG II can stimulate cortisol secretion in flounder (*Platichthys flesus*; Perrott and Balment 1990). A recent study suggests that ACTH and ANG II may act synergistically to stimulate cortisol secretion in the trout (Decourt and

Lahlou 1987). Infusion of ANG II is associated with a significant increase in dorsal aortic blood pressure and heart rate (Gray and Brown 1985). Such pressor doses of ANG II produce glomerular diuresis in teleosts, but nonpressor doses reduce urine flow secondary to morphological changes in the glomerulus (Brown et al. 1990).

Recent work, primarily on salmonid fishes, has suggested that growth hormone (GH) may play a major role in acclimation to increased salinities, independent of its promotion of growth (e.g. Sakamoto et al. 1990). For instance, both plasma GH and gill Na/K-ATPase levels increase during acclimation to seawater (e.g. Prunet et al. 1989). A functional connection between cortisol and GH is indicated by the finding that cortisol stimulates the secretion of GH from the pituitary in vitro (Nishioka et al. 1985) and GH enhances the release of cortisol from interrenal tissue both in vivo and in vitro (Young 1988). However, other hormones in addition to cortisol and GH must be involved in acclimation to seawater, because injection of either or both hormones into hypophysectomized coho salmon (*Oncorhynchus kisutch*) did not restore sea-water gill Na/K-ATPase levels, despite enhanced tolerance of seawater (Richman et al. 1987).

5.3 Elasmobranchs

Investigation of the hormonal control of elasmobranch osmoregulation has focused on the rectal gland. Some evidence suggests that vasoactive intestinal peptide (VIP) and somatostatin may be released from neurons in the gland and stimulate and inhibit, respectively, secretions (e.g. Stoff et al. 1988), but the role played by at least VIP has been questioned, and it appears that at least in *Scyliorhinus canicula,* a non-VIP-like peptide may be the dominant secretagogue (Shuttleworth 1988). The role of corticosteroids is unclear. One study found that interrenalectomy decreased rectal gland flow; another found that injection of cortisol or deoxycorticosterone inhibited flow (Shuttleworth 1988). Surprisingly, there is some evidence that volume expansion, rather than salt load, is the main stimulus for rectal gland secretion (Solomon et al. 1985b).

5.4 Agnatha

Despite the presence of most of the hormonal systems found in other fishes, hormonal control of agnathan osmoregulation remains nearly unstudied (e.g. Hardisty 1979). No clear picture of the hormones controlling osmoregulation in lampreys or Na^+ balance in hagfish has emerged.

6 Evidence for Role of Natriuretic Peptide Hormones in Fish Osmoregulation

In the past 5 years a relatively small body of research has suggested that NPs may play a role in osmoregulation and/or hemodynamics in teleosts, elasmobranchs, and hagfish. For brief, previous reviews, see Evans (1990) and Evans and Takei (1992).

6.1 Immunological Localization

Several studies have documented the presence of NP-like immunoreactivity in the atrium, ventricle, and brain of teleosts, elasmobranchs, and agnatha (Table 1). All of these studies utilized antibodies raised against mammalian NPs, or some truncated peptide. No immunohistochemical data have been published using antibodies raised against fish NPs. Despite some negative data, the large number of species in Table 1 suggests that NP-like peptides are found throughout the three extant groups of fishes. The presence of NP immunoreactivity (NP_{ir}) in the ventricle of many species indicates that, contrary to mammals, the ventricles of fishes may be a significant site of synthesis. Thus, the fact that the neonatal mammalian ventricle is a significant source of ANP (e.g. Brenner et al. 1990) may be an interesting case of "ontogeny recapitulating phylogeny". The relative lack of data on NP_{ir} in the brain of fishes is striking, but recent studies of brain distribution in *Opsanus beta* and *S. acanthias* indicate that NP-containing neurons are present in the diencephalon, suggesting a role in osmoregulation as in mammals (see above). However, NP_{ir} is also present in a variety of other areas of the brain [i.e. telencephalon, mesencephalon (including the tectum), and the medulla], supporting the conclusion that, as in mammals, NPs may serve as brain neuropeptides (Donald and Evans 1992; Donald et al. 1992).

Various studies using heterologous radioimmunoassay (RIA), have also determined that ANP_{ir} is present in the plasma of several species of teleosts, elasmobranchs, and agnatha, with concentrations equivalent to those described in mammalian plasma (ca. 100 pg $ml^{-1} \approx 30$ pM; see Evans 1990). A recent study using a homologous eel RIA demonstrated that plasma ANP concentration in this species is actually above that described for mammals (247 pM; Takei et al. 1992). Despite the relative uncertainty of heterologous RIAs, studies have demonstrated that acclimation of two euryhaline marine teleosts (the flounder *Pseudopleuronectes americanus* and the sculpin *Myoxocephalus octodecimspinosus*) to dilute salinity is correlated with a significant reduction in plasma ANP_{ir} (Evans et al. 1989a), whereas acclimation of a euryhaline, freshwater teleost (the chub, *Gila atraria*) to an increased salinity (Westenfelder et al. 1988), or the salmon *Salmo salar* or trout *O. mykiss* (Smith et al. 1991) to seawater, is associated with an increase in plasma ANP_{ir}. These data suggest that fish NPs are involved in osmoregulation, and that, contrary to what might be expected from

Table 1. Immunohistochemical localization of natriuretic peptide hormone (NP)-like peptides in fishes

Species	Atrium	Ventricle	Brain	Reference
Freshwater teleosts				
Cyprinus carpio	+	+	ND[a]	Chapeau et al. (1985)
	+	+	ND	Uemura et al. (1990)
	+	+	ND	Hirohama et al. (1988)
Cichlosoma biocellatum	+	+	ND	Chapeau et al. (1985)
Carassius auratus	+	+	ND	Chapeau et al. (1985)
Puntius schwanefeldi	0	0	ND	Chapeau et al. (1985)
Anguilla japonica	+	+	ND	Uemura et al. (1990)
	+	+	ND	Takei et al. (1990)
Channa maculata	+	+	ND	Uemura et al. (1990)
Lepomis macrochirus	+	+	ND	Uemura et al. (1990)
Salmo gairdneri	+	0	ND	Uemura et al. (1990)
	+	+	ND	Reinecke et al. (1985)
Oryzias latipes	0	0	ND	Hirohama et al. (1988)
Tribolodon hakonensis	+	+	ND	Hirohama et al. (1988)
Seawater teleosts				
Conger myriaster	+	+	ND	Uemura et al. (1990)
Oplegnathus fasciatus	+	0	ND	Uemura et al. (1990)
Pagrus major	+	+	ND	Uemura et al. (1990)
Tachurus japonicus	+	0	ND	Uemura et al. (1990)
Hexagrammos otakii	0	0	ND	Uemura et al. (1990)
	0	0	ND	Hirohama et al. (1988)
Chrysiptera cyanea	0	0	ND	Hirohama et al. (1988)
Sebastes inermis	0	0	ND	Hirohama et al. (1988)
Opsanus beta	+	+	+	Donald and Evans (1992)
Pseudopleuronectes americanus	+	+	ND	Donald (unpubl.)
Myoxocephalus octadecimspnisus	+	+	ND	Donald (unpubl.)
Chondrichthyes				
Triakis scyllia	+	+	ND	Donald (unpubl.)
	+	+	ND	Uemura et al. (1990)
Narke japonica	0	0	ND	Uemura et al. (1990)
Squalus acanthias	+	+	ND	Reinecke et al. (1987b)
	+	+	+	Donald et al. (1992)
Raja erinacea	+	+	ND	Donald et al. (1992)
Raja clavata	+	+	ND	Reinecke et al. (1987b)
Chimera monstrosa	+	+	ND	Reinecke et al. (1987b)
Scyliorhinus canicula	ND	ND	+	Vallarino et al. (1990)
Agnatha				
Eptatretus burgeri	+	0	ND	Uemura et al. (1990)
Lampetra japonica	0	0	ND	Uemura et al. (1990)
Myxine glutinosa	+	+	+	Reinecke et al. (1987a)
	0	0	+	Donald et al. (1992)

[a] ND = not determined.

analogy with mammals, salt- rather than volume-loading appears to be the stimulus for ANP secretion in at least teleosts (Evans 1990). The eel (*A. japonica*) may be an exception to this proposition, because its plasma ANP levels (in this case measured with a homologous RIA) decline in higher salinities (Takei and Balment 1993). Whether the eel is indeed an exception awaits further studies on other species using homologous RIAs.

A recent preliminary study has found that acclimation of the sea lamprey *Petromyzon marinus* to seawater was correlated with a threefold increase in plasma ANP_{ir} (Freeman and Bernard 1990), suggesting that, as in teleosts, a salt rather than volume load may be the stimulus for secretion of ANP.

6.2 Localization of Natriuretic Peptide Hormone Receptor Sites

Surprisingly, the first published study visualizing putative NP receptor sites in fishes utilized hagfish. Kloas et al. (1988) found ANP binding to the ventral aorta and the kidney (glomerulus and archinephric duct) of *Myxine glutinosa*. Our demonstration that rat ANP vasodilates the isolated ventral aorta of the same species indicates that the aortic NP receptors must play a role in gill hemodynamics (Evans 1991). More recently, Broadhead et al. (1992) have analyzed ANP receptors in gill cells from the eel, *A. anguilla* also demonstrating that activation of the receptors stimulated the production of cyclic GMP. Similar studies have analyzed NP receptors in the trout kidney, conger eel heart, and shark rectal gland (Cerra et al. 1992; Gunning et al. 1993; Perrott et al. 1993). We (Donald et al. 1995; Toop et al. 1995; J.A. Donald, T. Toop, and D.H. Evans, in prep.) have recently localized NP receptors in the gills of *O. beta, M. glutinosa,* and *S. acanthias* autoradiographically and have determined, by competitive binding assays, affinity cross-linking, and cyclic GMP production that both biological and clearance receptors are present. Recently, Duff and Olson (1992) have shown that infusion of an ANPR-C antagonist into *O. mykiss* produces physiological responses consistent with decreased clearance of plasma NPs, suggesting a physiological role for clearance receptors in vivo. Interestingly,we have been unable to localize any NP receptors in the kidney of *O. beta*, a congeneric to *O. tau* that has been shown to become natriuretic after infusion of high doses of rANP (Lee and Malvin 1987), suggesting that the natriuresis may have been the result of secondary effects.

6.3 Extraction and Sequencing of Fish Natriuretic Peptide Hormones

Fish NPs have only been isolated and sequenced since 1989. Their amino acid sequences are included in Fig. 1 for comparison with relevant mammalian NPs. Currently, NPs have been sequenced from only two species of teleosts (*A. japonica* and *F. heteroclitus*) and three species of sharks (*S. acanthias, S. canicula*, and *Triakis scyllia*). In addition to eel ANP (eANP; Takei et al. 1989), we now have the sequences for eCNP (Takei et al. 1990a), killifish CNP (kCNP; Price et al. 1990), and three shark CNPs (sCNP; Schofield et al. 1991; Suzuki et al. 1991, 1992). Although both eCNP and kCNP were initially isolated from teleost brain tissue, the sCNP from *S. canicula* was isolated from cardiac extracts and the amino acid sequence of sCNP from *S. acanthias* was deduced from the gene cloned from a cardiac cDNA library, after initial partial cloning using the polymerase chain reaction. Thus, at least in sharks CNP appears to be a major cardiac peptide, in direct contrast to its apparent rarity in cardiac tissue in mammals (e.g. Minamino et al. 1991; Vollmar et al. 1993). Finally, Takei and colleagues (Takei et al. 1991) have recently isolated a novel member of the NP family from the eel ventricle and termed it VNP. Presently, this is the only VNP that has been isolated, but immunoreactivity in the ventricle of various vertebrates, and physiological responses to ventricular extracts (see above and below) in fishes, may be due to a VNP, rather than an ANP.

6.4 Physiological Responses to Natriuretic Peptide Hormones in Teleosts

Because nearly all of the currently published data on physiological responses to ANP predate the extraction of fish NPs, mammalian ANP (usually rat, human, or some truncated peptides) has been the usual stimulant in these studies. It is clear that future studies should utilize fish NPs when possible, with the proviso that species-specific, rather than fish-specific peptides may be necessary (see below).

6.4.1 Kidney

Initial studies attempting to demonstrate a role for NPs in fish osmoregulation centered on renal responses, because of ANP's classic ability to produce natriuresis in mammals. Infusion of rather substantial doses of rANP (10–60 μg kg^{-1}; theoretically producing a plasma concentration ≈ 0.1–1 μM) into either freshwater trout (Duff and Olson 1986) or seawater toadfish (*O. tau*: Lee and Malvin 1987) produced a slight diuresis (200 and 120%, respectively, of control), but more substantial natriuresis (500% and 133%, respectively). Because *O. tau* is aglomerular, it is apparent that the ANP-induced diuresis/natiuresis was via tubular, rather than glomerular, mechanisms, although our inability to localize renal receptors in the congener *O. beta* suggests that the diuresis may have been produced indirectly (see above). Nevertheless, this extremely small data base suggests that teleosts, like mammals, respond to an increase in plasma ANP with renal diuresis and natriuresis in either freshwater or seawater. Eel ANP and VNP have recently been shown to be antidiuretic in the freshwater-acclimated eel (*A. japonica;* Takei and Balment 1993), indicating that much more work is necessary with homologous peptides before a clear picture of the renal effects of ANP in teleosts is possible.

6.4.2 Intestine

Because the intestine is an important osmoregulatory organ, at least in marine teleosts, it might be expected that NPs would play some role in controlling salt and water absorption across this epithelium. This appears to be the case, because application of ANP to the serosal side of isolated, intestinal epithelium from the flounder (*P. americanus*) inhibited the short-circuit current, as well as the influxes of both Na^+ and Cl^-, in a dose-dependent manner (EC_{50} = 7 nM; O'Grady et al. 1985). Similar data have recently been published for the eel intestine (Ando et al. 1992), where it was shown that eel ANP was significantly more efficacious than mammalian ANP. Because it has been shown that NaCl uptake across this epithelium is via coupled Na/K/2Cl transport (see Sect. 4.1.2), the authors propose that ANP inhibits some step in this transepithelial transport. In addition to effects on intestinal transport, it appears that NPs may also alter intake of seawater by teleosts. Takei and Balment (1993) have recently demonstrated that eel ANP and VNP inhibit the ingestion of seawater by acclimated eels. It is important to note that inhibition of ingestion and salt uptake in the seawater teleost intestine would not only reduce a salt load, but also inhibit the only means of regulating water balance. Thus, this response to ANP would be ionically, but not osmotically, appropriate.

6.4.3 Gill

Gill epithelial transport of NaCl is also affected by ANP, at least in vitro. In direct contrast to what has been demonstrated in the intestine, application of mammalian ANP to the isolated opercular epithelium from the killifish *F. heteroclitus* stimulated NaCl transport with an EC_{50} of < 10 nM (Scheide and Zadunaisky 1988). Because coupled Na/K/2Cl transport is also the carrier in this system, these data suggest that ANP can stimulate some component of this system in the gill, but inhibit some component of the same system in the intestine. In either case, an apparent effect on coupled Na/K/2Cl transport is of some interest, because this transporter is also present in the thick ascending limb of the mammalian loop of Henle, where neither ANP receptors nor ANP responses have been demonstrated (e.g. Brenner et al. 1990). However, the importance of this apparent stimulation is unclear, because maximal stimulation (at 10 μM ANP) was only 19%, far below the 50–100% stimulation produced by β-adrenergic agonists (e.g. Zadunaisky 1984). Moreover, this stimulation of the seawater gill salt extrusion mechanism by ANP may be species-specific, because O'Grady et al. (1985) did not find any effect of even 1 μM ANP on transport (measured as a transepithelial electrical potential difference) across the isolated opercular epithelium of *P. americanus*, although it responded to both epinephrine and bumetanide (a specific inhibitor of coupled Na/K/2Cl), indicating that the tissue was transporting NaCl.

6.4.4 Cardiovascular System

Much more extensive literature exists on the effects of mammalian, and even fish ANP on the fish cardiovascular system. As the branchial resistance vessels fishes are in series with the heart and the aortic and systemic vasculature, alteration of gill blood flow can be controlled by: (1) cardiac output; (2) ventral aortic resistance; (3) branchial vessel resistance; (4) dorsal aortic resistance; and (5) systemic vessel resistance, either individually or in concert. For instance, an increase in dorsal aortic blood pressure could be secondary to any or all of the following: an increase in cardiac output, a fall in branchial vessel resistance, or an increase in systemic vessel resistance. Alterations in branchial bloodflow could be secondary to changes in cardiac output, ventral aortic resistance, branchial resistance, and dorsal aortic or systemic vessel resistance. Because alterations in branchial blood flow could impact on the surface area and diffusional thickness of the gill epithelium (e.g. Laurent and Hebibi 1989), ANP's putative cardiovascular effects could have secondary consequences on fish osmoregulation.

Initial studies found that infusion of mammalian ANP (1–10 $\mu g\ kg^{-1}$) into the trout (*O. mykiss*) produced a significant increase in dorsal aortic pressure (Duff and Olson 1986), but a subsequent study of the trout found no change in dorsal aortic pressure after human ANP infusion (10 $\mu g\ kg^{-1}$; Eddy et al. 1990). In addition, rat ANP infusion ($\approx$60 $\mu g\ kg^{-1}$) produced no change in celiac artery pressure in *O. tau* (Lee and Malvin 1987) . A more recent study showed that rat ANP (3 $\mu g\ kg^{-1}$) produced hypotension in both the ventral and dorsal aorta of the Atlantic cod (*Gadus morhua*; Acierno et al. 1991) and Arnold-Reed et al. (1991) found a fall in dorsal aortic pressure in the flounder (*P. flesus*) after injection of human ANP (10 $\mu g\ kg^{-1}$). Fish cardiac responses to mammalian ANP vary. There was no effect on heart rate in *O. tau* (Lee and Malvin 1987) or *O. mykiss* (Eddy et al. 1990), but ANP produced tachycardia in *G. morhua* (Acierno et al. 1991). Olson and Meisheri (1989) had previously suggested that the dorsal aortic hypertension produced by ANP in the trout (Duff and Olson 1986) is mediated by central or secondary effects, because ANP produced dilation in isolated celiacomesenteric arterial rings after preconstriction with a variety of agonists including norepinephrine, ADH, acetylcholine, U-44069 (a thromboxane A_2 agonist), and serotonin. Despite obvious dilatory effects on an isolated vessel, they found only a slight fall in resistance of the perfused trunk. However, epinephrine-constricted perfused gills responded with a significant fall in resistance when ANP was applied (Olson and Meisheri 1989). These gill data are consistent with our finding that mammalian ANP produced significant, dose-dependent reduction in the total resistance of perfused heads of *O. beta* (Evans et al. 1989a), with an EC_{50} of 3 nM. The sensitivity of the perfused head vasculature to ANP did not change when this species was acclimated to ca. 5% seawater (Evans et al. 1989a).

In an effort to avoid the obvious pitfalls of utilizing the complex vasculature of a perfused head with an intact central and peripheral nervous system, we have recently used vascular rings from the ventral aorta from *O. beta*. The vascular smooth muscle (VSM) responds to mammalian ANP with relaxation (EC_{50} = 4 nM; Evans et al. 1989a). Interestingly, VSM rings from low-salinity acclimated *O. beta* show a much higher sensitivity (EC_{50} = 0.3 nM) to ANP, suggesting some up-regulation of receptors, consistent with our finding that plasma ANP levels are reduced in lower salinities in other euryhaline teleosts (Evans et al. 1989a).

It is clear that despite some uncertainty about the effect of, and pathways involved in, ANP in intact teleosts, the peptide does produce vasodilation of isolated vessels associated with the branchial vasculature and the perfused gill itself. Increased gill perfusion could exacerbate the osmotic and diffusional problems of teleosts in any salinity. However, Olson and Meisheri (1989) found no change in the epinephrine-stimulated water influx into the perfused (freshwater) trout gill when ANP was added to the perfusate, suggesting that osmoregulation may not be compromised by ANP-induced changes in branchial perfusion.

Data on the effect of homologous ANPs on fish physiology are only now emerging, subsequent to the first isolations of teleost fish peptides. Takei showed that eel ANP and eel CNP are much more hypotensive than rat ANP when infused into the ventral aorta of the eel ($EC_{50} \approx 0.3\ \mu g\ kg^{-1}$; Takei et al. 1989, 1990b). However, the fact that both eel ANP and killifish CNP are even slightly less potent than rat ANP in producing relaxation of aortic rings from *O. beta* (Price et al. 1990) suggests that there may be as much difference in ANPs between teleost groups as there is between mammals and fishes. Moreover, some species of teleosts may be exceptionally refractive to mammalian ANP. For instance, eel ANP inhibits the transport of NaCl across the eel intestine (measured as short-circuit current) with an EC_{50} of approximately 10 nM (Ando et al. 1992), virtually identical to that described for the effect of rat ANP on flounder intestine (O'Grady et al. 1985). However, inhibition of transport across the eel intestine by human ANP has an EC_{50} of nearly 1 μM (Ando et al. 1992).

6.4.5 Interactions with Other Teleost Endocrine or Neural Systems

Since prolactin and cortisol (and possibly AVT) are so critical to teleost osmoregulation (see Sect. 5) it is surprising that presently only two studies of ANP-hormonal interaction have been published. Arnold-Reed and Balment (1991) demonstrated that injection of human ANP into seawater-acclimated flounder (*P. flesus*) is followed by a doubling of plasma cortisol levels. Moreover, ANP stimulated secretion of cortisol by seawater-acclimated trout interrenal tissue in vitro; tissue from freshwater-acclimated trout did not respond (Arnold-Reed and Balment 1991). Takei and Balment (1993) have recently corroborated these data by demonstrating that eel ANP is steroidogenic in seawater, but not freshwater eel (*A. japonica*). This stimulation of cortisol release in a seawater-acclimated teleost makes osmoregulatory sense (see above), but is in direct contrast to the usual finding in mammals that cortisol secretion by the adrenal gland is inhibited by ANP (e.g. Brenner et al. 1990).

As has been found in mammals, cortisol appears to be a secretagogue for cardiac ANP release. A recent study has demonstrated that incubation of single ventricular trout cardiocytes with dexamethazone stimulates ANP release (Powell and Miller 1992). Because ANP has been shown to stimulate cortisol release in flounder (see above), it appears that teleosts may have a positive-feedback axis, rather than a negative-feedback axis between the heart and interrenal gland.

6.5 Physiological Responses to Natriuretic Peptide Hormones in Elasmobranchs

6.5.1 Kidney

Infusion of 10 μg kg^{-1} ANP into *S. acanthias* did not affect renal flow in an initial study (Solomon et al. 1985a), but a subsequent investigation of the same species found that an even lower dose (2 μg kg^{-1}) reduced both glomerular filtration and urine flow rate significantly (Benyajati Yokota 1990), with no change in fractional Na^+ excretion, even after the concomitant hypotension had been reversed. This interesting finding suggests direct glomerular effects, exclusive of both vasodepression and tubular Na^+ handling. However, the response was glomerular antidiuresis, in direct contrast to the diuresis described in all other vertebrates (see Sect. 3.2.1), with the exception of the antidiuresis produced in the eel, *A. japonica* (see Sect. 6.5.1). As the authors suggest, "the observed antidiuretic and antinatriuretic effect of synthetic ANP in the dogfish contrasts with its putative role as a hormone mediating hypervolemic regulation" (Benyajati and Yokota 1990). Clearly, further studies using more physiologically relevant plasma concentrations, as well as homologous peptides, are necessary before the role of NPs in elasmobranch renal physiology is known with more certainty.

6.5.2 Rectal Gland

Infusion of mammalian ANP (10 μg kg^{-1}) into the intact dogfish shark (*S. acanthias*) stimulated rectal gland secretion of Cl^- by approximately 300% (Solomon et al. 1985a). The response to ANP was also demonstrated in the isolated perfused gland, but the response was only 50% of that produced by VIP, which has been shown to be a major stimulant of rectal gland salt secretion in this species (e.g. Stoff et al. 1988), and the concentrations of ANP used to elicit a, response were substantial (100 nM). Subsequent studies determined that the ANP-induced secretion is actually mediated by release of VIP from nerve endings in the gland itself (Silva et al. 1987); however, once again, the ANP concentrations utilized were extraordinarily high (minimally, 0.5 μg ANP ml^{-1}

or ~ 150 nM). A recent study has shown that ANP can also stimulate (EC_{50} ~ 100 nM) salt transport across cultured rectal gland epithelium (Karnaky et al. 1991), suggesting that direct stimulation of the gland by NPs probably plays a role in the response as well. Recent studies have shown that shark CNP (Schofield et al. 1991) is about four times as stimulatory of shark rectal gland Cl^- secretion as rANP (both at 10 nM concentration; Forrest et al. 1992), with a threshold of < 0.1 nM (Solomon et al. 1992), in the range of plasma ANP_{ir} levels measured in this species (Evans et al. 1989). In addition, Karnaky et al. (1992) found that 0.1 nM sCNP stimulated salt transport across the cultured rectal gland epithelium fourfold, whereas 1 nM rANP was only slightly stimulatory. Thus, shark CNP appears to be a cardiac, circulating hormone that is very effective in stimulating salt extrusion. This finding is in sharp contrast to the fact that CNP appears to be mainly a neurotransmitter in the mammalian brain with limited natriuretic ability (Sudoh et al. 1990; Minamino et al. 1991; Stingo et al. 1992).

6.5.3 Cardiovascular System

Infusion of mammalian ANP (10 $\mu g\ kg^{-1}$) into intact *S. acanthias* was correlated with a significant fall in dorsal aortic blood pressure (Solomon et al. 1985a). Blood flow to the rectal gland increased by 360%, consistent with dilation of gland blood vessels. However, ANP could stimulate secretion by the isolated perfused rectal gland without any changes in perfusate flow, suggesting that hemodynamic effects were not the cause of the stimulated rectal gland salt secretion in vivo (Solomon et al. 1985a). The fact that ANP has now been shown to stimulate salt transport across cultured rectal gland cells (Karnaky et al. 1991) supports this conclusion. An ANP-induced hypotension in *S. acanthias* was corroborated by a subsequent study that also found that the renal effects could be dissociated from the hemodynamic effects of the peptide (Benyajati and Yokota 1990), and Bjenning et al. (1992) have recently shown that shark CNP is potently hypotensive when injected into *S. canicula*. We found that rat ANP vasodilates *S. acanthias* ventral aortic rings with an EC_{50} of 7 nM (Evans 1991), but recently demonstrated that homologous shark CNP (as well as other killifish and porcine CNP) is 15-fold more effective (EC_{50} = 0.5 nM; Evans et al. 1993). Our work corroborated an earlier study by Bjenning et al. (1992) on *S. canicula*, which showed that shark CNP was much more effective in dilating aortic rings than human ANP, but suggests that it may be the CNP molecule per se, rather than species specificity that is important in this response. Nevertheless, the efficacy of CNP on the shark VSM supports the conclusion, as indicated previously (see Sect. 6.5.2), that CNP is a circulating hormone in sharks, whereas it is generally thought to be solely a neurotransmitter in the mammalian brain.

6.5.4 Interactions with Other Elasmobranch Endocrine or Neural Systems

Except for the apparent interaction with VIP-ergic neurons in the rectal gland (Silva et al. 1987), there are no other published studies indicating an interaction of NPs with any other hormonal systems in the elasmobranchs. On the other hand, the literature on the hormonal control of shark osmoregulation is especially sparse (see Sect. 5.3).

6.6 Physiological Responses to Natriuretic Peptide Hormones in Agnatha

The data are especially sparse in this interesting group, despite the recent visualization of ANP receptor sites in the ventral aorta and kidney of *M. glutinosa* (Kloas et al. 1988). We have recently shown that the rings of ventral aortic VSM of this species relax when mammalian ANP is applied, with an EC_{50} of 4 nM (Evans 1991), identical to that described for the marine teleost (*O. beta*) aortic VSM (Evans et al. 1989). Because *M. glutinosa* is iso-osmotic to seawater, and maintains slight ionic gradients of only Na^+, it is unclear what role NPs may play in this group.

7 Conclusions

Apparently, various members of the NP family of hormones are present in the heart, brain, and plasma of various fish groups, and current data are that NPs may play a role in seawater vs freshwater, in opposition to its role in response to hypervolemia in mammals. Arguments for and against a role for NPs in acclimation to seawater are presented in Table 2. They are separated into data-based and a priori, suggested from known interactions between NPs and hormones in mammals. In addition, it is important to note that the ability to dilate aortic and branchial blood vessels seems to be a common finding in all three fish groups. If this decrease in branchial resistance is associated with an increase in perfusion of the water- and salt-permeable branchial epithelium, one might propose that one of the NPs' consistent effects in fishes is inappropriate in terms of osmoregulation. This is potentially true, because teleosts, elasmobranchs, and agnatha all face osmotic or ionic gradients across these epithelium. Thus, the selective advantage of this consistent, hemodynamic response to NPs in fishes is unclear. It is clear that significantly more data are needed before a definitive role for NPs in fish osmoregulation and branchial hemodynamics can be determined.

Table 2. Matched arguments for and against a role for NPs in seawater fish osmoregulation (see text for details)

For:

1. Acclimation of two species of teleosts to seawater, and a euryhaline, freshwater teleost to increased salinities is associated with a significant increase in plasma ANP_{ir}. Acclimation of the marine lamprey to seawater is associated with an increase in plasma ANP_{ir}.
2. ANP infusion produces natriuresis in a marine teleost, and antidiuresis in the freshwater acclimated eel. The stimulation of natriuresis in the freshwater trout is greater than the stimulation of diuresis in the same species.
3. ANP inhibits NaCl uptake by the seawater flounder and eel intestine. It also inhibits drinking in the eel.
4. ANP stimulates NaCl extrusion by the killifish gill epithelium, as modeled by the opercular epithelium.
5. ANP stimulates the release of cortisol by the interrenal tissue of the seawater, but not freshwater-acclimated flounder and eel.
6. Cortisol stimulates ANP release from cardiac tissue.
7. ANP is apparently antidiuretic in sharks.
8. ANP stimulates salt secretion by the shark rectal gland.
9. ANP inhibits prolactin release in mammals, and prolactin is a major freshwater hormone in teleosts.
10. ANP stimulates growth hormone (GH) in mammals, and GH is a major seawater hormone in teleosts (especially salmonids).
11. ANP inhibits vasopressin release in mammals and arginine vasotocin appears to inhibit salt-extrusion mechanisms in seawater teleosts, and may be diuretic in teleosts. AVT concentrations normally increase in freshwater-acclimated teleosts.

Against:

1. Acclimation of the Japanese eel to seawater is associated with a fall in plasma ANP_{ir}.
2. ANP produces diuresis in the freshwater trout.
3. ANP inhibition of intestinal NaCl influx would also inhibit needed water influx in the seawater flounder and eel. Inhibition of drinking in the eel would have the same effect.
4. ANP apparently does not stimulate NaCl extrusion by the flounder opercular epithelium, although this tissue responds to other secretagogues such as epinephrine.

5, 6, and 7. N/A.

8. The major stimulus for rectal gland secretion appears to be volume loading.

9, 10, and 11. N/A.

12. Angiontensin II (ANG II) production and stimulation of drinking is inhibited by ANP in mammals. ANG II stimulates oral ingestion in fish.
13. ANG II stimulates cortisol release by teleost interrenal tissue.
14. ANG II may be antidiuretic in teleosts.
15. NPs produce increased perfusion of the gill vasculature, which would presumably exacerbate osmoregulatory problems in either sea- or freshwater.

Acknowledgments. The writing of this review and the author's current research was funded by NSF grants DCB 8916413 and IBN 9219122, as well as NIH EHS-P30-ESO3828 to the Center for Membrane Toxicity Studies at the MDIBL. The recent assistance of John Payne, Evan Chipouras, Tes Toop, Archie Vomachka, and John Donald is gratefully acknowledged.

References

Acierno R, Axelsson M, Tota B, Nilsson S (1991) Hypotensive effect of atrial natriuretic factor (ANF) in the Atlantic cod, *Gadus morhua*. Comp Biochem Physiol 99c: 11–14

Ahren B (1990) Atrial natriuretic factor (ANF) inhibits thyroid hormone secretion in the mouse. Life Sci 47: 1973–1977

Anand-Srivastava MB, Sairam MR, Cantin M (1990) Ring-deleted analogs of atrial natriuretic factor inhibit adenylate cyclase/cAMP system. J Biol Chem 265: 8566–8572

Ando M, Kondo K, Takei Y (1992) Effects of eel atrial natriuretic peptide on NaCl and water transport across the intestine of seawater eel. J Comp Physiol 162b: 436–439

Arnold-Reed DE, Balment RJ (1991) Atrial natriuretic factor stimulates in-vivo and in-vitro secretion of cortisol in teleosts. J Endocrinol 128: R17–R20

Arnold-Reed D, Hazon N, Balment RJ (1991) Biological actions of atrial natriuretic factor in flatfish. fish Physiol Biochem 9: 271–277

Avella M, Bornancin M (1989) A new analysis of ammonia and sodium transport through the gills of the freshwater rainbow trout (*Salmo gairdneri*). J Exp Biol 142: 155–176

Benyajati S, Yokota SD (1990) Renal effects of atrial natriuretic peptide in a marine elasmobranch. Am J Physiol 258: R1201–R1206

Bern HA (1983) Fuctional evolution of prolactin and growth hormone in lower vertebrates. Am Zool 23: 663–671

Beyenbach KW, Baustian MD (1989) Comparative physiology of the proximal tubule. In: Kinne RKH (ed) Structure and function of the kidney, vol 1. Karger, Basel, pp 103–142

Brenner BM, Ballermann BJ, Gunning ME, Zeidel ML (1990) Diverse biological actions of atrial natriuretic peptide. Physiol Rev 70: 665–699

Broadhead CL, O'Sullivan UT, Deacon CF, Henderson IW (1992) Atrial natriuretic peptide in the eel, *Anguilla anguilla* L.: its cardiac distribution, receptors and actions on isolated branchial cells. J Mol Endocrinol 9: 103–114

Brown JA, Gray CJ, Taylor SM (1990) Direct effects of angiotensin II on glomerular ultrastructure in the rainbow trout, *Salmo gairdneri*. Cell Tissue Res 260: 315–319

Cerra MC, Canonaco M, Tota B (1992) A quantitative autoradiographic study of ^{125}I atrial natriuretic factor in the heart of a teleost fish (*Conger conger*). J Exp Zool 263: 215–219

Chang M-S, Lowe DG, Lewis M, Hellmiss R, Chen E, Goeddel DV (1989) Differential activation by atrial and brain natriuretic peptides of two different receptor guanylate cyclases. Nature 341: 68–72

Chapeau C, Gutkowska J, Schiller PW, Milne RW, Thibault G, Garcia R, Genest J, Cantin M (1985) Localization of immunoreactive synthetic atrial natriuretic factor (ANF) in the heart of various animal species. J Histochem Cytochem 33: 541–550

Clavell AL, Stingo AJ, Wei CM, Heublein DM, Burnett J Jr (1993) C-type natriuretic peptide: a selective cardiovascular peptide. Am J Physiol 264: R290–R295

Cogan MG (1990) Renal effects of atrial natriuretic factor. Annu Rev Physiol 52: 699–708

Dayanithi G, Antoni FA (1990) Atriopeptins are potent inhibitors of ACTH secretion by rat anterior pituitary cells in vitro: involvement of the atrial natriuretic factor receptor domain of membrane-bound gyanylyl cyclase. J Endocrinol 125: 39–44

DeBold AJ, Borenstein HB, Veress AT, Sonnenberg H (1981) A rapid and potent natriuretic response to intravenous injection of atrial myocardial extract in rats. Life Sci 39: 89–94

Decourt C, Lahlou B (1987) Evidence for the direct intervention of angiotensin II in the release of cortisol in teleost fishes. Life Sci 41: 1517–1524

Donald JA, Evans DH (1992) Immunohistochemical localisation of natriuretic peptides in the heart and brain of the gulf toadfish, *Opsanus beta*. Cell Tissue Res 269: 151–158

Donald JA, Vomachka AJ, Evans DH (1992) Immunohistochemical localisation of natriuretic peptides in the brains and hearts of the spiny dogfish *Squalus acanthias* and the Atlantic hagfish *Myxine glutinosa*. Cell Tissue Res 270: 535–545

Donald JA, Toop T, Evans DH (1995) Localization and analysis of natriuretic peptide receptors in the gills of the gulf toadfish, *Opsanus beta* (Teleostei). Am J Physiol 267:R1437–R1444

Duff DW, Olson KR (1986) Trout vascular and renal responses to atrial natriuretic factor and heart extracts. Am J Physiol 251: R639–R642

Duff DW, Olson KR (1992) Atrial natriuretic peptide clearance receptors in trout: effects of receptor inhibition in vivo. J Exp Zool 262: 343–346

Eddy FB, Smith NF, Hazon N, Grierson C (1990) Circulatory and ionoregulatory effects of atrial natriuretic peptide on rainbow trout (*Salmo gairdneri* Richardson) fed normal or high levels of dietary salt. Fish Physiol Biochem 8: 321–327

Evans DH (1979) Fish. In: Maloiy GMO (ed) Comparative physiology of osmoregulation in animals, vol 1, Academic Press, Orlando, pp 305–390

Evans DH (1982) Mechanisms of acid extrusion by two marine fishes: the teleost, *Opsanus beta,* and the elasmobranch, *Squalus acanthias.* J Exp Biol 97: 289–300

Evans DH (1984) The roles of gill permeability and transport mechanisms in euryhalinity. In: Hoar WS, Randall DJ (eds) Fish physiology, vol XB, Academic Press, Orlando, pp 239–283

Evans DH (1986) The role of branchial and dermal epithelia in acid–base regulation in aquatic vertebrates. In: Heisler N (ed) Acid–base regulation in animals, Elsevier, Amsterdam, pp 139–172

Evans DH (1990) An emerging role for a cardiac peptide hormone in fish osmoregulation. Annu Rev Physiol 52: 43–60

Evans DH (1991) Rat atriopeptin dilates vascular smooth muscle of the ventral aorta from the shark (*Squalus acanthias*) and the hagfish (*Myxine glutinosa*). J Exp Biol 157: 551–555

Evans DH (1993) Osmotic and ionic regulation. In: Evans DH (ed) The physiology of fishes, CRC Press, Boca Raton, pp 315–341

Evans DH, Cameron JN (1986) Gill ammonia transport. J Exp Zool 239: 17–23

Evans DH, More K (1988) Modes of ammonia transport across the gill epithelium of the dogfish pup (*Squalus acanthias*). J Exp Biol 138: 375–397

Evans DH, Takei Y (1992) A putative role for natriuretic peptides in fish osmoregulation. NIPS 7: 15–19

Evans DH, Oikari A, Kormanik GA, Mansberger L (1982) Osmoregulation by the prenatal spiny dogfish, *Squalus acanthias.* J Exp Biol 101: 295–305

Evans DH, Chipouras E, Payne JA (1989a) Immunoreactive atriopeptin in the plasma of fishes: a potential role in gill hemodynamics. Am J Physiol 257: R939–R945

Evans DH, More K, Robbins SL (1989b) Modes of ammonia transport across the gill epithelium of the marine teleost fish *Opsanus beta.* J Exp Biol 144: 339–356

Evans DH, Toop T, Donald J, Forrest JN Jr (1993) C-type natriuretic peptides are potent dilators of shark vascular smooth muscle. J Exp Zool 265: 84–87

Forrest JN Jr, Mackay WC, Gallagher B, Epstein FH (1973) Plasma cortisol response to saltwater adaptation in the American eel, *Anguilla anguilla.* Am J Physiol 224: 714–717

Forrest JN Jr, Kelley GG, Opdyke D, Schofield JP, Aller C (1992) Synthetic shark CNP based on the amino acid sequence of cloned pre-pro shark heart CNP potentially stimulates chloride secretion in the perfused shark rectal gland. Bull Mt Desert Isl Biol Lab 31: 71–72

Foskett JK (1987) The chloride cell. Multiple hormonal control of gill salt secretion. In : Kirsch R, Lahlou B (eds) Comparative physiology of environmental adaptations, vol 1. Karger, Basel, pp 83–91

Foskett JK, Bern HA, Machen TE, Conner M (1983) Chloride cells and the hormonal control of fish osmoregulation. J Exp Biol 106: 255–281

Freeman JD, Bernard RA (1990) Atrial natriuretic peptide and salt adaptation in the sea lamprey *Petromyzon marinus*. Physiologist 33: A-38 (Abstr)

Garvin JL (1992) ANF inhibits norepinephrine-stimulated fluid absorption in rat proximal straight tubules. Am J Physiol 263: F581–F585

Genest J, Cantin M (1988) The atrial natriuretic factor: its physiology and biochemistry. Rev Physiol Biochem Pharmacol 110: 2–145

Gray CJ, Brown JA (1985) Renal and cardiovascular effects of angiotensin II in the rainbow trout, *Salmo gairdneri.* Gen Comp Endocrinol 59: 375–381

Greenwald JE, Ritter D, Tetens E, Rotwein PS (1992) Renal expression of the gene for atrial natriuretic factor. Am J Physiol 263: F974–F978

Groot JA, Bakker R (1988) NaCl transport in vertebrate intestine. In: Greger R (ed) Advances in comparative and environmental physiology, vol 1. Springer, Berlin Heidelberg New York, pp 103–152
Gu J (1991) Immunocytochemical detection of atrial natriuretic peptide in periarterial myocardiocytes of cardiac ventricles of adult normotensive rat. J Histochem Cytochem 39: 829–834
Guibbolini ME, Lahlou B (1987) Neurohypophyseal peptide inhibition of adenylate cyclase activity in fish gills. The effect of environmental salinity. FEBS Lett 220: 98–102
Gunning M, Ballermann BJ, Silva P, Brenner BM, Zeidel ML (1990) Brain natriuretic peptide: interaction with renal ANP system. Am J Physiol 258: F467–F472
Gunning M, Cuero C, Solomon R, Silva P (1993) C-type natriuretic peptide receptors and signaling in rectal gland of *Squalus acanthias*. Am J Physiol 264: F300–F305
Hardisty MW (1979) Biology of the cyclostomes. Chapman and Hall, London
Heisler N (1989) Interactions between gas exchange, metabolism, and ion transport in animals: an overview. Can J Zool 67: 2923–2935
Heisler N (1993) Acid–base regulation. In: Evans DH (ed) The physiology of fishes. CRC Press, Boca Raton, pp 343–378
Henderson IW, O'Toole LB, Hazon N (1988) Kidney function. In: Shuttleworth TJ (ed) Physiology of elasmobranch fishes. Springer, Berlin Heidelberg New York, pp 201–214
Hentschel H, Elger M (1989) Morphology of glomerular and aglomerular kidneys. In: Kinne RKH (ed) Structure and function of the kidney, vol 1. Karger, Basel, pp 1–72
Herndon TM, McCormick SD, Bern HA (1991) Effects of prolactin on chloride cells in opercular membrane of seawater-adapted tilapia. Gen Comp Endocrinol 83: 283–289
Hirano T, Ogasawara T, Bolton JP, Collie NL, Hasegawa S, Iwata M (1987) Osmoregulatory role of prolactin in lower vertebrates. In: Kirsch R, Lahlou B (eds) Comparative physiology of environmental adaptations. Karger, Basel, pp 112–124
Hirata M, Chang CH, Murad F (1989) Stimulatory effects of atrial natriuretic factor on phosphoinositide hydrolysis in cultured bovine aortic smooth muscle cells. Biochim Biophys Acta 1010: 346–351
Hirohama T, Uemura H, Nakamura S, Aotao T (1988) Atrial natriuretic peptide (ANP)-immunoreactivity and ultrastructures of cardiocytes in fish. Zool Sci 5: 833–845
Hosoda K, Nakao K, Mukoyama M, Saito Y, Jougasaki M, Shirakami G, Suga S, Ogawa Y, Yasue H, Imura H (1991) Expression of brain natriuretic peptide gene in human heart: production in the ventricle. Hypertension 17: 1152–1155
Huang FL, Skala KD, Samson WK (1992) C-type natriuretic peptide stimulates prolactin secretion by a hypothalamic site of action. J Neuroendocrinol 4: 593–597
Huang W, Choi CL, Yang Z, Copolov DL, Lim AT (1991) Forskolin-induced immunoreactive atrial natriuretic peptide (ANP) secretion and pro-ANP messenger ribonucleic acid expression of hypothalamic neurons in culture: modulation by glucocorticoids. Endocrinology 128: 2591–2600
Ibanez-Santos J, Tsagarakis S, Rees LH, Besser GM, Grossman A (1990) Atrial natriuretic peptides inhibit the release of corticotrophin-releasing factor-41 from the rat hypothalamus in vitro. J Endocrinol 126: 223–228
James S, Hassall CJS, Polak JM, Burnstock G (1990) Autoradiographical localization of specific atrial natriuretic peptide binding sites on immunocytochemically identified cells in cultures from rat and guinea-pig hearts. Cell Tissue Res 261: 301–312
Johnson BG, Trachte GJ, Drewett JG (1991) Neuromodulatory effect of the atrial natriuretic factor clearance receptor blinding peptide, cANF(4-23)-NH2 in rabbit isolated vasa deferentia. J Pharmacol Exp Ther 257: 720–726
Karnaky KJ Jr (1986) Structure and function of the chloride cell of *Fundulus heteroclitus* and other teleosts. Am Zool 26: 209–224
Karnaky KJ Jr, Valentich JD, Currie MG, Oehlenschlager WF, Kennedy MP (1991) Atriopeptin stimulates chloride secretion in cultured shark rectal gland cells. Am J Physiol 260: C1125–C1130
Karnaky KJ Jr, Stidham JD, Nelson DS, McCraw AS, Valentich JD, Kennedy MP, Currie MG (1992) C-type natriuretic peptide is a potent secretagogue for the cultured shark (*Squalus acanthias*) rectal gland. Bull Mt Desert Isl Biol Lab 31: 122–123

Kim SH, Cho KW, Koh GY, Seul KH, So JN, Ryu H (1989) Phylogenetic study on the immunoreactive atrial natriuretic peptide in the heart. Gen Comp Endocrinol 74: 127–131

Kirsch R, Meister MF (1982) Progressive processing of ingested water in the gut of sea-water teleosts. J Exp Biol 98: 67–81

Kloas W, flugge G, Fuchs E, Stolte H (1988) Binding sites for atrial natriuretic peptide in the kidney and aorta of the hagfish (*Myxine glutinosa*). Comp Biochem Physiol 91A: 685–688

Knepper MA, Lankford SP, Terada Y (1991) Renal tubular actions of ANF. Can J Physiol Pharmacol 69: 1537–1545

Kohno M, Yasunari K, Yokokawa K, Murakawa K, Horio T, Takeda T, Johchi M (1991) Inhibition by atrial and brain natriuretic peptides of endothelin-1 secretion after stimulation with angiotensin II and thrombin of cultured human enfothelial cells. J Clin Invest 87: 1999–2004

Koller KJ, Goeddel DV (1992) Molecular biology of the natriuretic peptides and their receptors. Circulation 86: 1081–1088

Koller KJ, Lowe DG, Bennett GL, Minamino N, Kangawa K, Matsuo H, Goeddel DV (1991) Selective activation of the B natriuretc peptide receptor by C-type natriuretic peptide (CNP). Science 252: 120–123

Lahlou B (1980) Les hormones dans l'osmoregulation des poissons. In: Ali MA (ed) Environmental physiology of fishes. Plenum, New York, pp 201–240

Laurent P, Hebibi N (1989) Gill morphometry and fish osmoregulation. Can J Zool 67: 3055–3063

Laurent P, Perry SF (1990) Effects of cortisol on gill chloride cell morphology and ionic uptake in the freshwater trout, *Salmo gairdneri*. Cell Tissue Res 259: 429–442

Lee J, Malvin RL (1987) Natriuretic response to homologous heart extract in aglomerular toadfish. Am J Physiol 252: R1055–R1058

Leitman DC, Murad F (1987) Atrial natriuretic factor receptor heterogeneity and stimulation of particulate guanylate cyclase and cyclic GMP accumulation. In: Rosenblatt M, Jacobs JW (eds) Atrial natriuretic factor, vol 16 (1). Saunders, Philadelphia, pp 79–105

Lin H, Randall D (1991) Evidence for the presence of an electrogenic proton pump on the trout gill epithelium. J Exp Biol 161: 119–134

Loretz CA, Bern HA (1982) Prolactin and osmoregulation in vertebrates. An update. Neuroendocrinology 35: 292–304

Maack T (1992) Receptors of atrial natriuretic factor. Annu Rev Physiol 54: 11–27

Maack T, Suzuki M, Almeida FA, Nussenzveig D, Scarborough RM, McEnroe GA, Lewicki JA (1987) Physiological role of silent receptors of atrial natriuretic factor. Science 238: 675–678

Mantymaa P, Leppaluoto J, Ruskoaho H (1990) Endothelin stimulates basal and stretch-induced atrial natriuretic peptide secretion from the perfused rat heart. Endocrinology 126: 587–595

Marshall WS (1988) NaCl transport in gills and related structures-vertebrates. In: Gregor R (ed) Comparative and environmental physiology. NaCl transport in epithelia, vol 1. Springer, Berlin Heidelberg New York, pp 48–73

Martin ER, Ballermann BJ (1989) Atrial natriuretic peptide receptors. In: Brenner BM, Stein JH (eds) Contemporary issues in nephrology. Atrial natriuretic peptide, vol 21. Churchill-Livingstone, New York, pp 105–136

Mayer-Gostan N, Wendelaar Bonga SJ, Balm PHM (1987) Mechanisms of hormone actions on gill transport. In : Pang PKT, Schreibman MP (eds) Vertebrate endocrinology: fundamentals and biomedical implications, vol 2. Regulation of water and electrolytes. Academic Press, San Diego, pp 211–238

McCormick SD (1990) Cortisol directly stimulates differentiation of chloride cells in tilapia opercular membrane. Am J Physiol 259: R857–R863

McDonald DG, Tang Y, Boutilier RG (1989) Acid and ion transfer across the gills of fish: mechanisms and regulation. Can J Zool 67: 3046–3054

McFarland WN, Munz FW (1965) Regulation of body weight and serum composition by hagfish in various media. Comp Biochem Physiol 14: 393–398

Minamino N, Makino Y, Tateyama H, Kangawa K, Matsuo H (1991) Characterization of immunoreactive human C-type natriuretic peptide in brain and heart. Biochem Biophys Res Commun 179: 535–542

Mori Y, Nishikawa M, Matsubara H, Takagi T, Toyoda N, Oikawa S, Inada M (1990) Stimulation of rat atrial natriuretic peptide (rANP) synthesis by triodothyronine and thyroxine (T_4): T_4 as a prohormone in synthesizing rANP. Endocrinology 126: 466–471

Needleman P, Blaine EH, Greenwald JE, Michener ML, Saper CB, Stockman PT, Tolunay HE (1989) The biochemical pharmacology of atrial peptides. Annu Rev Pharm Toxicol 29: 23–54

Nishimura H (1987) Role of the renin-angioitensin system in osmoregulation. In : Pang PKT, Schreibman MP (eds) Vertebrate endocrinology: fundamentals and biomedical implications, vol 2. Regulation of water and electrolytes. Academic Press, San Diego, pp 157–187

Nishioka RS, Brau EG, Bern HA (1985) In vitro release of growth hormone from the pituitary gland of tilapia *Oreochromis mossambicus*. Gen Comp Endocrinol 60: 90–94

O'Grady SM (1989) Cyclic nucleotide-mediated effects of ANF and VIP on flounder intestinal ion transport. Am J Physiol 256: C142–C146

O'Grady SM, field N, Nash NT, Rao MC (1985) Atrial natriuretic factor inhibits Na–K–Cl cotransport in teleost intestine. Am J Physiol 249: C531–C534

Olson KR, Meisheri KD (1989) Effects of atrial natriuretic factor on isolated arteries and perfused organs of trout. Am J Physiol 256: R10–R18

Ota K, Kimura T, Shoji M, Inoue M, Sato K, Ohta M, Yamamoto T, Tsunoda K, Abe K, Yoshinaga K (1992) Interaction of ANP with endothelin on cardiovascular, renal, and endocrine function. Am J Physiol 262: E135–E141

Pang PKT, Furspan PB, Sawyer WH (1983) Evolution of neurohypophyseal hormone actions in vertebrates. Am Zool 23: 655–662

Perrott MN, Balment RJ (1990) The renin-angiotensin system and the regulation of plasma cortisol in the flounder, *Platichthys flesus*. Gen Comp Endocrinol 78: 414–420

Perrott MN, Carrick S, Balment RJ (1991) Pituitary and plasma arginine vasotocin levels in teleost fish. Gen Comp Endocrinol 83: 68–74

Perrott MN, Sainsbury RJ, Balment RJ (1993) Peptide hormone-stimulated second messenger production in the teleostean nephron. Gen Comp Endocrinol 89: 387–395

Perry SF, Laurent P (1989) Adaptational responses of rainbow trout to lowered external NaCl concentration: contribution of the branchial chloride cell. J Exp Biol 147: 147–168

Powell WH, Miller HAI (1992) Dexamethasone stimulates release of an ANP-like substance from rainbow trout cardiocytes. Am J Physiol 263: R447–R451

Price DA, Doble KE, Lee TD, Galli S, Dunn BM, Parten B, Evans DH (1990) The sequencing synthesis and biological activity of an ANP-like peptide isolated from the brain of the killifish, *Fundulus heteroclitus*. Biol Bull 178: 279–285

Prunet P, Boeur G, Bolton JP, Young G (1989) Smoltification and seawater adaptation in Atlantic salmon (*Salmo salar*): plasma prolactin, growth hormone, and thyroid hormones. Gen Comp Endocrinol 174: 355–364

Rankin JC, Bolis L (1984) Hormonal control of water movement across the gills. In: Hoar WS, Randall DJ (eds) fish physiology, vol XB. Academic Press, Orlando, pp 177–201

Rashed HM, Sun H, Patel TB (1993) Atrial natriuretic peptide inhibits growth of hepatoblastoma (HEP G2) cells by means of activation of clearance receptors. Hepatology 17: 677–684

Reinecke M, Nehls M, Forssmann WB (1985) Phylogenetic aspects of cardiac hormones as revealed by immunocytochemistry, electronmicroscopy, and bioassay. Peptides 6 (Suppl 3): 321–331

Reinecke M, Betzler D, Forssmann WG (1987a) Immunocytochemistry of cardiac polypeptide hormones (Cardiodilatin/atrial natriuretic polypeptide) in brain and hearts of *Myxine glutinosa* (Cyclostomata). Histochemistry 86: 233–239

Reinecke M, Betzler D, Forssmann WG, Thorndyke M, Askensten U, Falkmer S (1987a) Electronmicroscopical, immunohistochemical, immunocytochemical and biological evi-

dence for the occurrence of cardiac hormones (ANP/CDD) in chondrichthyes. Histochemistry 87: 531–538
Richman NHI, Nishioka RS, Young G, Bern HA (1987) Effects of cortisol and growth hormone replacement on osmoregulation in hypophysectomized coho salmon (*Oncorhynchus kisutch*). Gen Comp Endocrinol 67: 194–201
Rocha AS, Kudo LH (1990) Effect of atrial natriuretic factor and cyclic guanosine monophosphate on water and urea transport in the inner medullary collecting duct. Pflügers Arch Eur J Physiol 417: 84–90
Rosenzweig A, Seidman CE (1991) Atrial natriuretic factor and related peptide hormones. Annu Rev Biochem 60: 229–255
Ruskoaho H (1992) Atrial natriuretic peptide: synthesis, release, and metabolism. Pharmacol Rev 44: 479–602
Sakamoto T, Ogasawara T, Hirano T (1990) Growth hormone kinetics during adaptation to a hyperosmotic environment in rainbow trout. J Comp Physiol (B) 160: 1–6
Samson WK (1990) Neuroendocrine actions of the atrial natriuretic peptides. In: Samson WK Quirion R (ed) Atrial natriuretic peptides, CRC Press, Boca Raton, pp 231–241
Samson WK, Quirion R (eds) (1990) Atrial natriuretic peptides, CRC Press, Boca Raton
Sasaki A, Kida O, Kita T, Kato J, Nakamura S, Kodama K, Miyata A, Kangawa K, Matsuo H, Tanaka K (1989) Effect of antiserum against ∂-rat atrial natriuretic polypeptide in spontaneously hypertensive rats. Am J Physiol 257: H1104–H1109
Scarborough RM, Hsu MA, Kang L-L, McEnroe GA, Schwartz K, Arfsten A, Lewicki JA (1989) Structural and conformational requirements of ANP clearance receptors. In: Brenner BM, Laragh JH (eds) Progress in atrial peptide research. Raven, New York, pp 23–29
Scheide JI, Zadunaisky JA (1988) Effect of atriopeptin II on isolated opercular epithelium of *Fundulus heteroclitus*. Am J Physiol 254: R27–R32
Schenck DB, Phelps MN, Porter JG, Fuller F, Cordell B, Lewicki JA (1987) Purification and subunit composition of atrial natriuretic peptide receptor. Proc Natl Acad Sci USA 84: 1521–1525
Schofield JP, Jones D, Forrest JJ (1991) Identification of C-type natriuretic peptide in heart of spiny dogfish shark (*Squalus acanthias*). Am J Physiol 261: F734–F739
Schulz S, Singh S, Bellet RA, Singh G, Tubb DJ, Chin H, Garbers KL (1989) The primary structure of a plasma membrane guanylate cyclase demonstrates diversity within this new receptor family. Cell 58: 1155–1162
Shibasaki T, Naruse M, Narus K, Yamauchi N, Sim YS, Masuda A, Imaki T, Demura H, Ling N, Inagami T, Shizume K (1988) Effect of sodium ion on atrial natriuretic factor release from rat hypothalamic fragments. Life Sci 42: 1173–1180
Shuttleworth TJ (1988) Salt and water balance—extrarenal mechanisms. In: Shuttleworth TJ (ed) Physiology of elasmobranch fishes. Springer, Berlin Heidelberg New York, pp 171–199
Shuttleworth TJ (1989) Overview of epithelial ion-transport mechanisms. Can J Zool 67: 3032–3038
Silva P, Stoff JS, Solomon RJ, Lear S, Kniaz D, Greger R, Epstein FH (1987) Atrial natriuretic peptide stimulates salt secretion by shark rectal gland by releasing VIP. Am J Physiol 252: F99–F103
Smith HW (1957) Salt and water volume receptors. Am J Med 23: 623–652
Smith NF, Eddy FW, Struthers AD, Talbot C (1991) Renin, atrial natriuretic peptide and blood plasma ions in parr and smolts of Atlantic salmon *Salmo salar* L. and rainbow trout *Oncorhynchus mykiss* (Walbaum) in fresh water and after short-term exposure to sea water. J Exp Biol 157: 63–74
Solomon R, Taylor M, Dorsey D, Silva P, Epstein FH (1985a) Atriopeptin stimulation of rectal gland in *Squalus acanthias*. Am J Physiol 249: R348–R354
Solomon R, Taylor M, Sheth S, Silva P, Epstein FH (1985b) Primary role of volume expansion in stimulation of rectal gland function. Am J Physiol 248: R638–R640
Solomon R, Protter A, McEnroe G, Porter JG, Silva P (1992) C-type natriuretic peptides stimulate chloride secretion in the rectal gland of *Squalus acanthias*. Am J Physiol 262: R707–R711

Stanley JG, fleming WR (1964) Excretion of a hypertonic urine by a teleost. Science 144: 63–64
Stingo AJ, Clavell AL, Aarhus LL, Burnett JJ (1992) Cardiovascular and renal actions of C-type natriuretic peptide. Am J Physiol 262: H308–H312
Stoff JS, Silva P, Lechan R, Solomon R, Epstein FH (1988) Neural control of shark rectal gland. Am J Physiol 255: R212–R216
Sudoh T, Minamino N, Kangawa N, Matsuo H (1990) C-type natriuretic peptide (CNP): a new member of natriuretic peptide family identified in porcine brain. Biochem Biophys Res Commun 168: 863–870
Suzuki E, Hirata Y, Hayakawa H, Omata M, Kojima M, Kangawa K, Minamino N, Matsuo H (1993) Evidence for C-type natriuretic peptide production in the rat kidney. Biochem Biophys Res Commun 192: 532–538
Suzuki R, Takahashi A, Hazon N, Takei Y (1991) Isolation of high-molecular-weight C-type natriuretic peptide from the heart of a cartilaginous fish (European dogfish, *Scyliorhinus canicula*). FEBS Lett 282: 321–325
Suzuki R, Takahashi A, Takei Y (1992) Different molecular forms of C-type natriuretic peptide isolated from the brain and heart of an elasmobranch, *Triakis scyllia*. J Endocrinol 135: 317–323
Takei Y, Balment RJ (1993) Biochemistry and physiology of a family of eel natriuretic peptides. Fish Biochem Physiol 11: 183–188
Takei Y, Takahashi A, Watanabe TX, Nakajima K, Sakakibara S (1989) Amino acid sequence and relative biological activity of eel atrial natriuretic peptide. Biochem Biophys Res Commun 164: 537–543
Takei Y, Tamaki H, Ando K (1990a) Identification and partial characterization of immunoreactive and bioactive atrial natriuretic peptide from eel heart. J Comp Physiol (B) 160: 119–126
Takei Y, Takahashi A, Watanabe TX, Nakajima K, Sakakibara S, Takao T, Shimonishi Y (1990b) Amino acid sequence and relative biological activity of a natriuretic peptide isolated from eel brain. Biochem Biophys Res Commun 170: 883–891
Takei Y, Takahashi A, Watanabe TX, Nakajima K, Sakakibara S (1991) A novel natriuretic peptide isolated from eel cardiac ventricles. FEBS Lett 282: 317–320
Takei Y, Ando K, Kawakami M (1992) Atrial natriuretic peptide in eel plasma, heart and brain characterized by homologous radioimmunoassay. J Endocrinol 135: 325–331
Takei Y, Takahashi A, Watanabe TX, Nakajima K, Ando K (1994) Eel ventricular natriuretic peptide: isolation of a low molecular size form and characterization of plasma form by homologous radiioimmunoassay. J Endocrinol 141: 81–89
Toop T, Donald JA, Evans DH (1995) Localization and characteristics of natriuretic peptide receptors in the gills of the Atlantic hagfish, *Myxine glutinosa* (Agnatha). J Exp Biol 198: 117–126
Tornell J, Carlsson B, Billig H (1990) Atrial natriuretic peptide inhibits spontaneous rat oocyte maturation. Endocrinology 126: 1504–1508
Traynor TR, O'Grady SM (1991) Brain natriuretic peptide stimulates K and Cl secretion across porcine distal colon epithelium. Am J Physiol 260: C750–C755
Uemura H, Naruse M, Hirohama T, Nakamura S, Kasuya Y, Aoto T (1990) Immunoreactive atreal natriuretic peptide in the fish heart and blood plasma examined by electron microscopy, immunohistochemistry and radioimmunoassay. Cell Tissue Res 260: 235–247
Vallarino M, Feuilloley M, Gutkowska J, Cantin M, Vaudry H (1990) Localization of atrial natriuretic factor (ANF)-related peptides in the central nervous system of the elasmobranch fish *Scyliorhinus canicula*. Peptides 11: 1175–1181
Vollmar AM (1990) Atrial natriuretic peptide in peripheral organs other than the heart. Klin Wochenschr 68: 699–708
Vollmar AM , Lang RE, Hanze J, Schulz R (1990) The rat thymus—a site of atrial natriuretic peptide synthesis. Peptides 11: 33–37
Vollmar AM, Gerbes AL, Nemer M, Schulz R (1993) Detection of C-type natriuretic peptide (CNP) transcript in the rat heart and immune organs. Endocrinology 132: 1872–1874
Volpe M, Pepino P, Lembo G, Pignalosa S, Mele AF, Rubattu S, Condorelli G, Covino E, Trimarco B (1991) Modulatory role of angiotensin-II in the secretion of atrial natriuretic factor in rabbits. Endocrinology 128: 2427–2431

Wendelaar Bonga SE (1993) Endocrinology. In: Evans DH (ed) The physiology offishes. CRC Press, Boca Raton, pp 469–502
Westenfelder C, Birch FM, Baranowski RL, Rosenfeld MJ, Shiozawa DK, Kablitz C (1988) Atrial natriuretic factor and salt adaptation in the teleost fish *Gila atraria*. Am J Physiol 255: F1281–1286
Wilson JX (1984) The renin-angiotensin system in nonmammalian vertebrates. Endocrinol Rev 5: 45–61
Winquist RJ, Hintze TH (1990) Mechanisms of atrial natriuretic factor-induced vasodilation. Pharmacol Ther 48: 417–426
Yamamoto A, Kimura S, Hasui K, Fujisawa Y, Tamaki T, Fukui K, Iwao H, Abe Y (1988) Calcitonin gene-related peptide (CGRP) stimulates the release of atrial natriuretic peptide (ANP) from isolated rat atria. Biochem Biophys Res Commun 155: 1452-1458
Young G (1988) Enhanced response of the interrenal of coho salmon (*Oncorhynchus kisutch*) to ACTH after growth hormone treatment in vivo and in vitro. Gen Comp Endocrinol 71: 85–92
Zadunaisky J (1984) The chloride cell: the active transport of chloride and the paracellular pathways. In: Hoar WS, Randall DJ (eds) Fish physiology, vol XB. Academic Press, Orlando, pp 129–176
Zarmir N, Haass M, Dave JR, Zukowska-Grojec Z (1987) Anterior pituitary gland modulates the release of anterior natriuretic peptides from cardiac atria. Proc Natl Acad Sci USA 84: 541–545
Zeidel ML (1990) Renal actions of atrial natriuretic peptide: regulation of collecting duct sodium and water transport. Annu Rev Physiol 52: 747–759
Zeidel ML (1993) Hormonal regulation of inner medullary collecting duct sodium transport. Am J Physiol 265: F159–F173

Chapter 7

Regulation of Ion and Water Transport by Hydrogen Ions in High Resistance Epithelia

B.J. Harvey and *V. Urbach*

Contents

What a wonderful bird the frog are—
When he sit, he stand almost;
When he hop, he fly almost.
He ain't got no sense hardly;
He ain't got no tail hardly either.
When he sit, he sit on what he ain't got—almost.

Anonymous

The Wellcome Trust Cellular Physiology Research Unit, Department of Physiology, University College, Cork, Ireland

Advances in Comparative and Environmental Physiology, Vol. 22

1 Introduction

The barrier epithelia of amphibians, such as frog skin and toad urinary bladder play an essential role in whole-body electrolyte, acid–base and water balance. These epithelia have tight junctions of high electrical resistance between the outermost living cells that render the epithelium "tight" to paracellular flow of Na^+ or K^+, thus allowing vectorial transepithelial transport of these ions. The Na absorption function is carried out by granular cells via Na^+ channels in the apical membranes and an Na/K adenosine triphosphatase (ATPase) pump in the basolateral membranes; K^+ is recycled across this barrier via K^+-selective channels. The multilayered epithelium of frog skin functions as a syncytial Na transport compartment (Farquhar and Palade 1964; Nagel 1976; Mills et al. 1977; Nagel et al. 1981; Harvey and Kernan 1984a; Rick et al. 1978). The granular cells, along with the deeper spinosum and germinativum cells, form the syncytium and are termed principal cells. The principal cells in deeper layers are coupled to the first reactive cell layer by gap junctions (Farquhar and Palade 1965; Sharin and Blankemeyer 1989). Although cells in the different layers share the Na transport load, only the granular cell layer has a polarized apical membrane containing amiloride-sensitive Na^+ channels.

The amphibian skin and urinary bladder actively secrete hydrogen ions under appropriate electrochemical gradients (Frazier and Vanatta 1971; Ludens and Fanestil 1972; Machen and Erlij 1975; Ehrenfeld and Garcia-Romeu 1977; Ramsey 1982). Active H^+ secretion is restricted to mitochondria-rich cells (Harvey and Ehrenfeld 1988c; Ehrenfeld et al. 1989; Larsen et al. 1992). The amphibian skin and urinary bladder share common Na^+ and H^+ transport properties with distal renal tubule, and have been used for the past 50 years as highly successful models in studies of the mechanisms and regulation of ion transport across tight-junction epithelia (Koefoed-Johnsen and Ussing 1958). A vast amount of information is available on the mechanisms and regulation of Na^+, K^+, Cl^-, H^+ and water transport systems in amphibian skin and urinary bladder (suggested reviews; Macknight et al 1980; Kristensen and Ussing 1985; Handler 1988; Beauwens et al. 1990; Harris et al. 1991; Larsen 1991). Many of the first descriptions of transport mechanisms and the cellular action of natriferic and anti-diuretic hormones and diuretics were first described in these epithelia. The application of a wide range of electrophysiological techniques, such as transepithelial voltage clamp, intracellular ion-sensitive microelectrodes, noise and impedance analyses and patch clamp have greatly advanced our knowledge of cellular and molecular control of ion channels, exchangers, and pumps in epithelia.

1.1 Epithelial "Cross-Talk"

A major topic in transport physiology concerns the description of cellular control of ion and water movements across epithelia. "Cross-talk" phenomena help to maintain equilibrium between ion permeabilities at opposite apical and basolateral cell membranes (Schultz 1992; Diamond 1982; Davis and finn 1982). High resistance epithelia transfer large quantities of Na and K across their cells, and the transported load can fluctuate over a wide range and exceed the total cytosolic pool of these ions within minutes. The cross-talk signals involved in cellular osmotic equilibrium must have the characteristics of producing instantaneous, simultaneous and parallel changes in ion permeabilities at opposite apical and basolateral membranes. Cross-talk phenomena have been described between apical Na channels and the Na/K pump (MacRobbie and Ussing 1961), as well as between Na channels and basolateral K channels (Davis and finn 1982; Dawson and Richards 1990). Cross-talk is also important between ion transporters within the basolateral membrane. For example, equilibrium must be maintained between Na/K pump rate and K recyling (Messner et al. 1985); between pH-activated Na/H exchange and the Na/K pump (Ehrenfeld et al. 1992); and between volume-activated K and Cl channels (Ussing 1987).

Cross-talk in epithelia is most likely the result of a concerted action of multiple cytosolic and membrane-bound ion transport regulators. The primary cause of transport modulation may determine the type of signal brought into action and the interplay among signals. Some of the most likely candidates for cross-talk signals include Ca^{2+} (Taylor and Windhager 1979, 1985), G-protein (Cantiello and Aussielo 1991), actin filaments (Cantiello and Ausiello 1991), protein kinase C (Ling and Eaton 1989; Shearman et al. 1989; Civan et al. 1991), and membrane shuttling (Loo et al. 1983; Els and Helman 1991; Schaerer et al. 1991). In the spectrum of cross-talk signals, however, only intracellular pH has been satisfactorily shown to produce simultaneous, instantaneous, and parallel modulation of Na and K channels at opposite cell membranes in an intact epithelium (Harvey et al. 1988). Here we review evidence for the role of intracellular pH in coordinating the activity of Na^+ and K^+ channels and the Na/K pump and we discuss recent studies concerning the metabolic control of basolateral K channels by the adenosine triphosphate (ATP), adenosine diphosphate (ADP) ratio and by cellular Ca^{2+}. We briefly describe novel applications of the patch-clamp technique to study ion channel and pump modulation in intact epithelia isolated in Ussing chambers. For a general description of the regulation of ion channels in high resistance epithelia, the reader is referred to recent excellent review papers by Garty and Benos (1988), Beauwens et al. (1990), Smith and Benos (1991), Turnheim (1991), Dawson (1991), Larsen (1991), Schafer and Hawk (1992) and Lang and Rehwald (1992).

1.2 pH and Ca Effects on Na Absorption

One of the major difficulties in identifying cellular signals of cross-talk has been the requirement for simultaneous measurements of ion transport rate and signal level. Vesicle preparations of purified apical or basolateral membranes provide one possible solution. This technique has been used by Garty and colleagues to investigate the effects of intravesicular Ca and pH on amiloride-sensitive Na flux in apical membrane vesicles prepared from toad urinary bladder (Garty et al. 1985, 1987). Ca^{2+} was shown to decrease Na influx, whereas lowering intravesicular pH from 8 to 7 had no effect. Curiously, sodium uptake was found to be very sensitive to external pH changes between 8.25 and 6.0. These results could indicate that the vesicles were orientated inside-out, because in intact epithelia lowering external pH from 8.0 to 5.5 stimulates Na uptake (Leaf et al. 1964; Funder et al. 1967; Mandel 1978), and lowering intracellular pH from 8 to 7 drastically reduces apical Na permeability (Harvey et al. 1988). The role of Ca in regulating Na uptake in intact epithelia is supported by data showing inhibition of Na permeability by experimental manoeuvers designed to raise cell Ca^{2+} (Grinstein and Erlij 1978; Taylor and Windhager 1979; Chase and Al-Awqati 1983; Chase 1984; Palmer 1985; Frindt et al. 1988). The inhibitory effects of Ca^{2+} on Na absorption reported in these studies are in contrast with results obtained from patch-clamp analysis of Na^+ channel activity. The open probability of Na^+ channels excised from apical membranes of rat collecting tubule was found to be insensitive to changes in Ca^{2+} between 0.1 μM and 1 μM, whereas a slight decrease in pH_i produced prolonged closure of the channels (Palmer and Frindt 1987). In cell-attached patches, Na^+-channel open probability was decreased after a delay of > 10 min following exposure to the Ca^{2+} ionophore ionomycin. Presently, there is no satisfactory explanation for the discrepancy between Na flux and patch-clamp studies, although the available data point to an indirect inhibitory action of Ca involving a cytosolic or membrane-bound mediator that is lost or inactivated on excision of the patch. Some possibilities include a role for Ca^{2+} in the regulation of amiloride-sensitive Na^+ channel activity by methylation of cytosolic proteins (Minuth et al. 1987), exocytosis (Almers 1990), actin polymerization (Cantiello et al. 1991), guanosine nucleotides or protein kinase C (Cantiello et al. 1989; Garty et al. 1989; Cantiello and Ausiello 1991; Civan et al. 1991; Das et al. 1991).

2 Intracellular pH: A Cross-Talk Signal in Principal Cells

2.1 Electrophysiology

In an effort to understand the role of H^+ as a cross-talk signal, we approached this problem in the intact tissue by simultaneously measuring intracellular pH as well as Na^+ and K^+ conductances in principal cells using double-barrelled H^+-sensitive microelectrodes. The application of this technique in isolated frog-skin epithelium, mounted in an Ussing chamber, is illustrated in Fig. 1. The intracellular reference barrel of the microelectrode is used to measure the voltage across apical and basolateral membranes, and the calibrated differential output of reference and ion-sensitive barrels provides a direct reading of intracellular pH. The

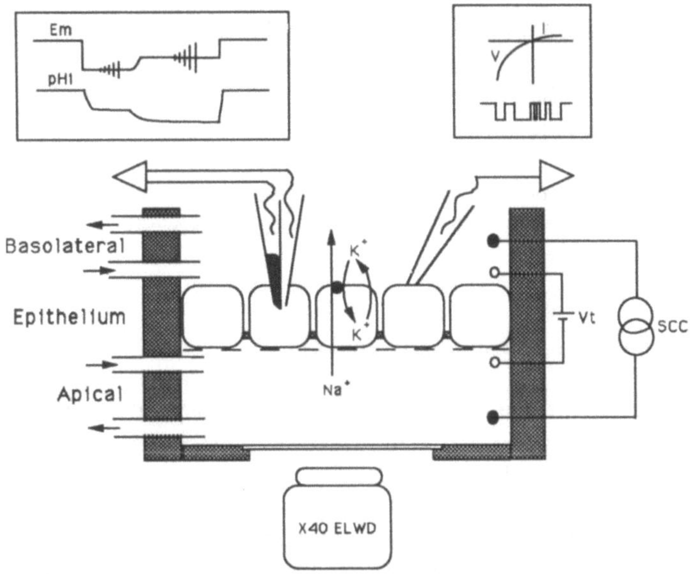

Fig. 1. Experimental setup used to simultaneously measure intracellular pH, membrane potentials and conductances in principal cells of isolated frog-skin epithelium mounted in a miniature Ussing chamber on the stage of an inverted microscope. Transepithelial potential (V_t) and short-circuit current (*SCC*) can be monitored throughout the experiment. Single-channel activity can also be recorded using a patch pipette sealed against the basolateral cell membrane. This arrangement allows comparison of macroscopic transepithelial transport rates and single-channel activity

epithelium can be voltage-clamped to generate current-voltage relations of apical and basolateral membranes. The amount of transepithelial current that actually passes through the cells can be determined using amiloride on the apical side to block transcellular Na transport. The specific sodium conductance of apical membranes (G_a) can be measured from amiloride-sensitive current-voltage (I_a–V_a) relations of this membrane. In high resistance epithelia the Goldman-Hodgkin-Katz (GHK) Na flux equation provides a near-perfect fit to I_a–V_a curves (Eq. 1). In addition to specific Na^+ conductance values, the GHK equation provides estimates of apical Na permeability (P_{Na}) and intracellular Na concentration (Na_i):

$$I_a = \left(\frac{-V_a \cdot P_{Na} \cdot F^2}{R \cdot T}\right) \cdot \left(\frac{Na_o - Na_i \cdot \exp(-V_a \cdot F / R \cdot T)}{1 - \exp(-V_a \cdot F / R \cdot T)}\right) \qquad (1)$$

where Na_o is the external Na concentration, R, F, and T are the gas constant, Faraday constant and absolute temperature, respectively. At zero membrane current, the Na^+ -specific conductance of the apical membrane is given by

$$G_a = \left(\frac{P_{Na} \cdot F^2}{R \cdot T}\right) \cdot \left(\frac{Na_o \cdot Na_i.}{Na_o - Na_i}\right) \ln(Na_o / Na_i). \tag{2}$$

The specific potassium conductance of the basolateral membranes (G_b) can be determined from the slope of current-voltage relations and the specific K transference number for this membrane.

2.2 Changing Intracellular pH at Constant External pH

When examining the effects of intracellular pH (pH_i) on ion transport, it is of the utmost importance to perform these experiments at constant external pH. This is because epithelial Na and K channels are also sensitive to external pH (Funder et al. 1967; Palmer 1985; Onken et al. 1990; Harvey et al. 1988). Some of the experimental strategies used in frog skin to produce acid–base disturbances at constant extracellular pH are illustrated in Fig. 2. These manoeuvers involve asymmetrical application of weak acid–base and take advantage of the permselectivity of apical and basolateral membranes to HCO_3^- and NH_4^+. A rapid and deep intracellular acidification can be produced when the solution bathing the apical side of an isolated epithelium is changed from N-2-hydroxyethyl piperazine –N′-2-ethane sulforic acid (HEPES) pH 7.4-buffered Ringer solution to 5% CO_2/24 mM HCO_3^- Ringer solution buffered at the same pH (Fig. 2A). Because of the relative impermeability of granular cell apical membranes to anions, bicarbonate ions cannot cross this barrier, and the weak acid CO_2 diffuses alone into the cell to produce H^+ and HCO_3^-. The hydration of CO_2 is catalyzed by carbonic anhydrase, but principal cells have very low activity of this enzyme (Rosen and Friedley 1973). HCO_3^- may diffuse rapidly out of the cell across the basolateral membranes through Cl^- channels (down a large electrochemical gradient) or via Cl^- /HCO_3^- exchange. Hydrogen ions are trapped in the cell, because the electrochemical driving force for protons favours H^+ entry until pH_i reaches 6.5. The continued loss of HCO_3^- produces an intracellular "sink" for CO_2 and reinforces the acid load. The CO_2-induced intracellular acidification is rapidly reversible on return to HEPES-buffered Ringer solution, and the rate of recovery of pH_i depends on the activity of Na/H exchange in the basolateral membrane (Harvey and Ehrenfeld 1988b). The pH_i response to CO_2/HCO_3^- is very different when performed from the basolateral side (Fig. 2B). Anion permeability of basolateral membranes is relatively high, and HCO_3^- may penetrate into the cell either through Cl^- channels or through the Band-3 Cl^- /HCO_3^- exchanger (Duranti et al. 1986; Harvey and Ehrenfeld 1988c; Lacoste et al. 1991). The high rate of transfer of HCO_3^-($\approx 10^6$ ions/s) allows the buffer pair to enter the cell simultaneously and to produce a doubling of the intracellular buffering power (Harvey et al. 1988), which dampens the effect of intrinsic acid production on pH_i and results in a moderate intracellular alkalinization. This experimental

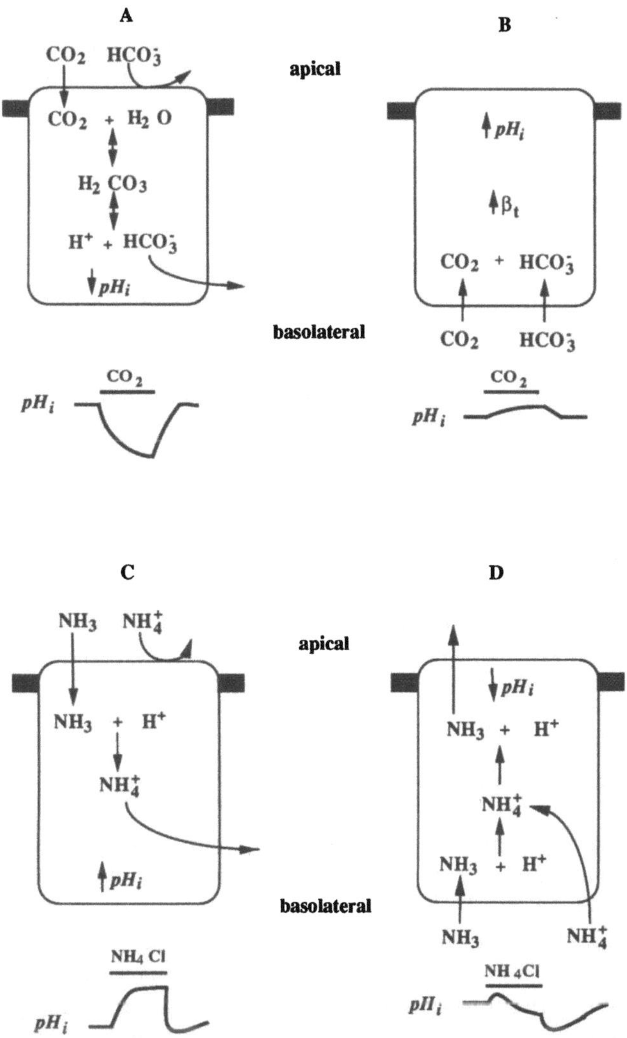

Fig. 2A–D. CO_2/HCO_3 and NH_4Cl loading techniques used to produce intracellular acid–base changes at constant extracellular pH. The mechanisms behind the pH_i changes are explained in the text

condition is similar to the in vivo buffer state and provides a useful model to determine the role of Na/H and Cl^-/HCO_3^- exchangers in regulating pH_i (Harvey et al. 1988).

Loading cells with the weak base NH_4Cl is a well-tried means of producing intracellular acid–base changes (Thomas 1984). Exposure of frog-skin epithelium to 15 mM NH_4Cl in HEPES Ringer buffered at pH 7.4 produces an initial alkalinization followed by acidification, which is deepened immediately following removal of external NH_4Cl. The duration and magnitude of the initial alkalinization phase and the degree of acidification following washout of the base depends on the side of addition of NH_4Cl. At pH 7.4, 15 mM of NH_4^+ solution will contain ≈150 μM uncharged NH_3, which readily penetrates cell membranes and traps intracellular protons. NH_4^+ can enter cells via K^+ channels (Latorre and Miller 1983) or by substituting for K^+ on the Na/K pump. The absence of these transporters from apical membranes means that the proton acceptor NH_3 will preferentially diffuse into the cells and produce a rapid and large alkalinization when the epithelium is exposed to NH_4Cl from the apical side (Fig. 2C). The NH_4^+ thus produced will diffuse out of the cell via K^+ channels in the basolateral membrane. This produces an intracellular "sink" for NH_3 and serves to maintain a constant plateau-phase alkalinization. When NH_4Cl is removed from the apical solution, NH_3 will diffuse faster than NH_4^+ across both apical and basolateral membranes. This drives the reaction $NH_4^+ \rightarrow NH_3 + H^+$, resulting in an intracellular acidification. When the cells are exposed to NH_4Cl from the basolateral side (Fig. 2D), the same profile of pH_i changes is produced, except that NH_4^+ can now diffuse rapidly into the cell through K^+ channels down a large electrochemical gradient until equilibrium is reached at an intracellular pH 6.5. The near simultaneous entry of NH_3 and NH_4^+ produces an initial transient alkalinization followed by a plateau-phase acidification, which is deepened on washout of external NH_4Cl. The advantage of loading principal cells with NH_4Cl, rather than CO_2, is that the effects of both an alkalinization and an acidification can be followed in the same cell; the disadvantages are the possibility of blocking K^+ conductance by NH_4^+, and of cell swelling due to entry of NH_4^+ across the basolateral membranes.

2.3 pH_i Sensitivity of Na Transport Components

Transepithelial Na transport, measured as short-circuit current, is inhibited in frog skin by experimental manoeuvers designed to reduce intracellular pH (Funder et al. 1967; Mandel 1978). The absorption of Na depends on the operation of three essential components; apical Na^+ channels, basolateral Na/K pumps and K^+ channels. In order to fully appreciate the role of pH_i in modulating cross-talk, it is important to determine which transport elements are affected by intracellular H^+. The pH_i sensitivity of total apical membrane Na^+ conductance, basolateral membrane K^+ conductance and Na/K pump current were analyzed using the techniques described in Figs. 1 and 2 and are shown in Fig. 3. The simultaneous recording of apical and basolateral membrane current-voltage relations during acid–base challenges demonstrates that an alkali load increases Na^+ and K^+ conductances, whereas an acid load has the opposite effect (Fig. 3A,B). In contrast, the current-voltage relations of the K^+-dependent "Na/K pump" current is relatively insensitive to an acid load (Fig. 3C) between pH_i 7.27 and 6.95, which is sufficient to reduce the macroscopic Na^+ and K^+ conductances. These findings are comparable to the relative insensitivity, over a physiological pH range, of ouabain-sensitive Na fluxes in nystatin-treated rabbit urinary bladder (Eaton et al. 1984) and frog skin (Ehrenfeld et al. 1992), as well as of isolated Na/K ATPase enzymatic activity (Homareda and Matsui 1985). The

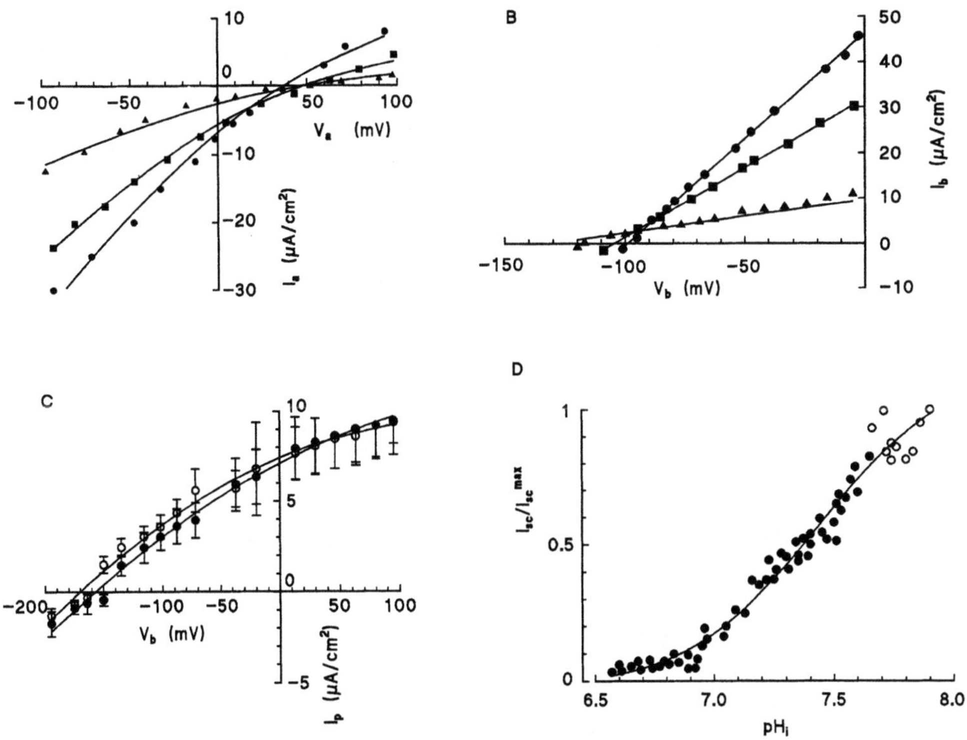

Fig. 3. Effects of intracellular acid load on Na and K transport produced by techniques shown in Fig. 2. Simultaneous recording of **A** amiloride-sensitive Na+ conductance of apical membranes and **B** K^+ conductance of basolateral membranes under control conditions (■) and during an acid load produced by 5% CO (▲), and an alkali load produced by NH_4Cl (●). An intracellular acidification simultaneously decreases the ionic conductance at each membrane, whereas an alkali load increases Na^+ and K^+ conductances. (Adapted from Harvey et al. 1988). **C** Effects of an acid load (●) on Na/K pump I–V relationship. Changing pH_i from 7.3 to 6.9 had little effect on pump current compared with control (○). (Data are shown as means±SEM of 11 cells. (Adapted from Ehrenfeld et al. 1992). **D** The sensitivity of spontaneous transepithelial Na transport rate measured as short-circuit current (●) to intracellular pH. Aldosterone-treated tissues are indicated by (○) (Adapted from Harvey and Ehrenfeld 1988b)

effects of pH_i on short-circuit current (I_{sc} = net transepithelial Na absorption rate) is therefore a result of a reduction in Na uptake across the apical membranes and not due to pH_i effects on Na/K ATPase. The continued operation of the Na/K pump should reduce intracellular Na^+ activity (A^i_{Na}) during an acid load, and this prediction has been experimentally verified (Harvey et al. 1988). With long-term acid loading A^i_{Na} gradually increases, despite inhibition of apical Na entry, and is due to activation of basolateral Na/H exchange (Harvey and Ehrenfeld 1988b). Intracellular pH fulfills the essential properties of a signal of cross-talk by producing rapid, reversible, simultaneous and covariant effects on Na and K transport. In the intact frog-skin epithelium, spontaneous variations in the transepithelial Na transport rate are associated with measured pH_i changes, which agree with the predicted effects on Na^+ and K^+ conductances (Fig. 3A, B). Hormonal stimulation of Na absorption by aldosterone is also associated with raised intracellular pH. The strict relationship between I_{sc} and pH_i raises questions as to whether the pH_i changes are consequent to or the determinant of Na absorption. It now appears likely that both solutions are correct. Modulation of pH_i by Na/H exchange will affect Na absorption (Harvey and Ehrenfeld 1988b,c) and variations in the latter produce changes in pH_i (Harvey and Ehrenfeld 1985; Harvey et al. 1988). Thus, intracellular H^+ acts as servo-control signal that is modulated by, and is a regulator of, Na^+ absorption in high resistance epithelia.

2.4 pH_i Effects on Single Na^+ Channels

The effect of pH_i on transepithelial Na absorption is caused by a reduction in apical Na permeability. The transfer of Na across apical membranes involves a large population of channels and the macroscopic Na current (I_{Na}) across this membrane is a function of the following ion channel parameters:

$$I_{Na} = i_s \cdot N \cdot P_o \quad (3)$$

$$= g_s \cdot (V_a - E_{Na}) \cdot N \cdot P_o, \quad (4)$$

where i_s = single-channel current; N = the total number of channels in the membrane; P_o = the single-channel open probability; g_s = single-channel conductance; V_a = apical membrane potential; E_{Na} = Nernst potential for Na^+. From this relationship it is clear that a decrease in Na uptake could result from a reduction in either single Na^+ channel current or open probability or the total number of active channels in the membrane. We have examined the effects of pH_i on these parameters using the patch-clamp technique. The apical membranes of frog-skin epithelium is covered by a layer of dead cells (stratum corneum), which makes access with a patch pipette impossible. However, the distal renal cell culture line, A6, from *Xenopus laevis* displays a similar Na transport function, and the apical membranes are accessible with patch pipettes. The direct effects of pH_i can be studied in patches of apical membrane that have been excised from intact polarized A6 monolayers. In this configuration the cytosolic side of the patch is exposed to the bath solution, and channel activity can be recorded as function of bath (cytosolic side) pH. Single Na^+ channel activity is extremely sensitive to slight variations in pH_i (Fig. 4), and the probability of the channel being in the open state is greatly reduced over a narrow range of pH_i from 7.4 to 6.9. The decrease in P_o is produced both by a shortening of the mean open time and an increase in mean closed time. In membrane patches that contain more than one Na^+ channel, a decrease in pH_i reduced the number of simultaneously active channels. Because the single-channel conductance (5 pS) is not affected over the physiological pH_i range, it can be concluded that inhibition of Na uptake by hydrogen ions is a result of changes to single-channel open probability and the

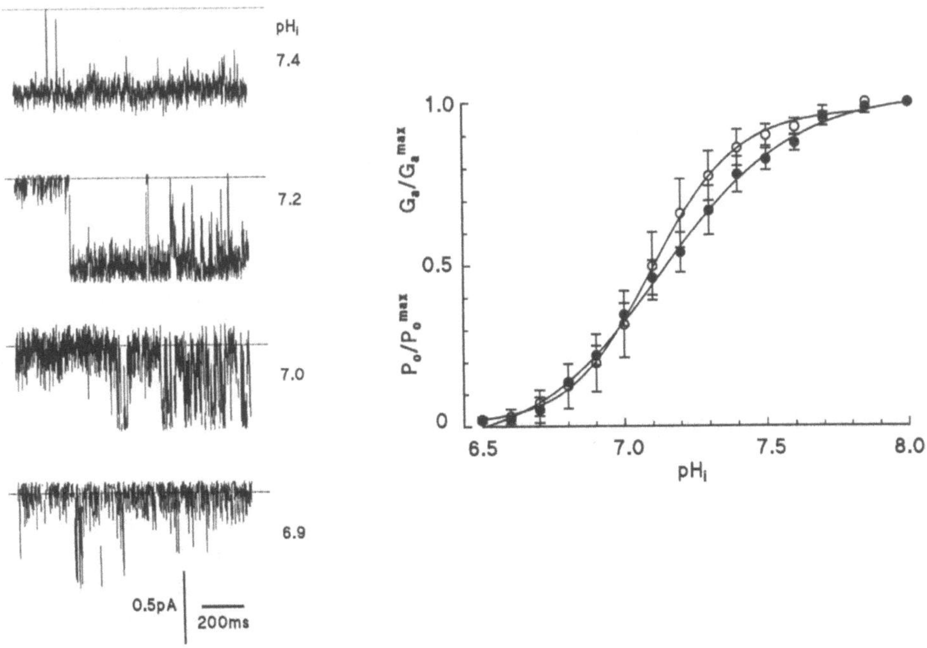

Fig. 4. Effects on single Na^+ channel activity of changing pH at the cytosolic side of inside-out patch excised from the apical membrane of an A6 cell. Patch-clamp recordings of single Na^+-channel inward currents are shown over the range pH_i 7.4–6.9 (Na Ringer solution in patch pipette and K "intracellular-like" solution in bath; Vp = 50 mV; E_{Na} = 50 mV; the *dashed line* denotes the closed state). The single-channel activity is expressed as normalized open-channel probability (○) and is shown as a function of pH_i. The relationship Po/Po_{max} vs pH_i was fitted by a Hill equation (*solid line* through data points):

$$Po/Po_{max} = \frac{(10^{-pK})^n}{(10^{-pK})^n + (H^+)_i^n} \qquad (5)$$

The best fit gave values of pK = 7.25 and a Hill coefficient *n* of 6 (the number of protons interacting with each titratable site). The macroscopic amiloride-sensitive Na^+ conductance (●) of apical membranes determined in principal cells of frog-skin epithelium displays a similar sensitivity to pH_i. (Means ± SEM of 11 cells); (adapted from Harvey et al. 1988, 1992)

number of active Na^+ channels in the membrane. The activity of the Na^+ channel expressed in terms of open probability is very sensitive to slight variations in pH_i between 7.0 and 7.4. Hydrogen ions affect channel closure by interacting with high cooperativity at a titration site with apparent pK = 7.2. Because the normal intracellular pH value is ≈ 7.2, the Na^+ channel activity is optimally sensitive to slight variations in pH_i. The role of pH_i in determining the overall rate of Na absorption can be best appreciated by comparing the pH-dependence of single Na^+ channels and total apical membrane Na^+ conductance (Fig. 4). Both single-channel activity and the macroscopic Na conductance display the same sensitivity to intracellular pH, which is an extremely potent regulator of the rate-limiting step in net transepithelial sodium absorption. The similar response of intact cells and excised membranes to pH_i reinforces the conclusion that hydrogen ions act independently of other possible cytosolic factors. Calcium ions are not involved in this response, because chelation of cytosolic Ca^{2+} does not affect the sensitivity of Na^+ channels to pH_i (Harvey and Thomas 1987). A direct effect of pH on Na channels has also been reported in renal collecting tubule (Palmer and Frindt 1987). An interesting observation for comparative physiology is that the Na^+ channels in apical membranes of distal renal culture A6 cells and of frog-skin granular cells must be similar, because of the close pH-sensitivity of single-channel and macroscopic Na transport in these different tissues (P_o in A6 cells and G_a in frog skin).

2.5 pH_i Effects on Single K^+ Channels

There is convincing evidence that pH_i is a major determinant of K^+ conductance in Na-absorbing epithelia. Intracellular hydrogen ions inhibit the macroscopic K^+ conductance in whole cells and single K^+ channels in renal tubule (Hunter et al. 1988; Hurst and Hunter 1989, 1990; Oberleithner et al. 1988; Schwegler et al. 1990) and total basolateral K^+ conductance in frog skin (Harvey et al. 1988). In contrast to apical Na^+ channel studies, the approach to basolateral membrane channels with patch electrodes is hampered by the presence of cellulose/collagen filters in epithelial cell cultures and by connective tissue in natural epithelia. Recently, we have been successful in recording single K^+ channel currents from basolateral membranes in frog-skin epithelium separated from the corium by collagenase treatment and mounted in an Ussing chamber (Urbach et al. 1994). This setup (Fig. 1) has the advantage of also allowing recording of the macroscopic electrical parameters of Na transport (short-circuit current and membrane potentials and conductances). Using this technique we were able to show that K^+ is recycled across basolateral membranes of frog skin and A6 epithelia via a 25 pS inward-rectifier ATP-sensitive channel (K_{ATP}), which is up-regulated by a membrane-bound G protein. The inward-rectifier K_{ATP} channel is more conductive for K^+ entry than exit from cell (inward rectification). However, the open probability of this channel is increased by membrane depolarization and is maximal under physiological electrochemical gradients that favour K efflux. Single-channel rectification properties reflect the macrosopic inward rectification of basolateral membrane current in frog skin and A6 cells (Nagel 1985; Granitzer et al. 1991; Urbach et al. 1994). The K_{ATP} channel is very sensitive to slight changes in intracellular pH (Fig. 5). In excised inside-out patches the open probability of the channel is maximal for cytosolic-side pH 7.5 and minimal at pH 7. The pH_i effect is voltage-independent, with equal inhibition of single-channel outward and inward currents. Hydrogen ions decrease the mean time the channel spends in the open state and increase the mean closed time by interacting with high cooperativity at a titratable site of apparent pK = 7.2 on the cytosolic side of the membrane outside the channel pore (voltage insensitivity). This is very similar to

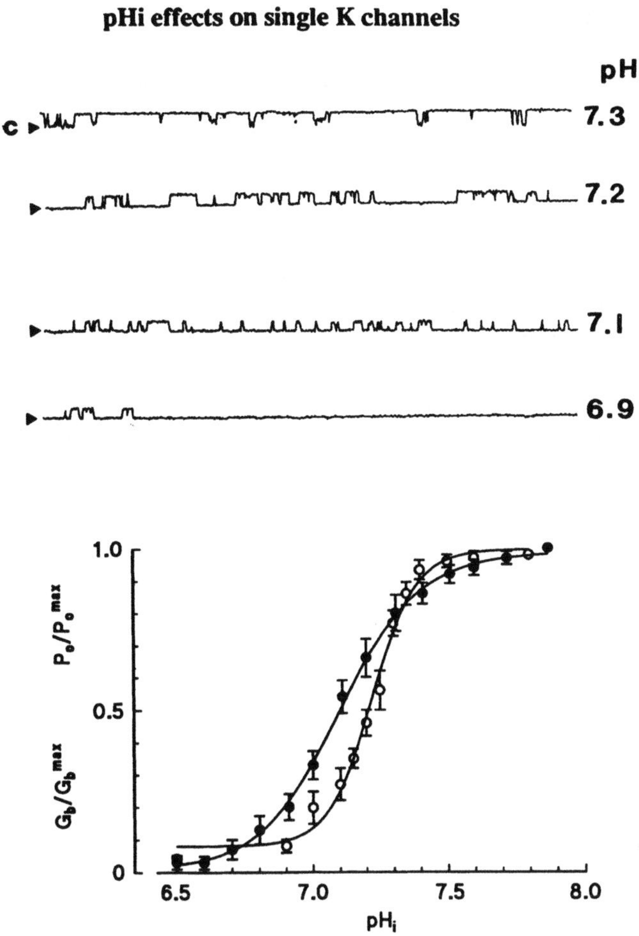

Fig. 5. Effects on single K_{ATP} channel activity of changing pH at the cytosolic side of an inside-out patch excised from basolateral membrane of a principal cell in frog-skin epithelium. Recordings of single K_{ATP} channel outward currents are shown over the pH$_i$ range 7.3–6.9 (Na Ringer solution in patch pipette and K "intracellular-like" solution in bath; Vp = 0 m V; E_K = –95 mV; the *arrow* "*C*" denotes the closed state). The single-channel activity is expressed as normalized open-channel probability (○) and is shown as a function of pH$_i$. The best-fit Hill equation gave values of pK=7.25 and a Hill coefficient of 6, and is similar to the pH$_i$ dependence of basolateral membrane macroscopic K^+ conductance (●) in principal cells of frog-skin epithelium. (Mean ± SEM of 11 cells; adapted from Harvey et al. 1988, 1992)

the mode of action of H^+ on apical Na^+ channels. The macroscopic K^+ conductance of the basolateral membranes displays a pH$_i$ dependence comparable to the single K_{ATP} channel (Fig. 5). Taken together, these results support the conclusion that H^+ can change the activity of the K_{ATP} channel without involving cytosolic factors, and that K_{ATP} channels are the major determinant of the macroscopic conductance of the basolateral membrane.

2.6 Aldosterone and K_{ATP} Channels

One of the advantages of applying the patch-clamp technique to intact epithelia in an Ussing chamber is that hormonal stimulation of Na absorption and single-channel activity can be followed simultaneously. The acute activation of Na/H exchange by aldosterone has major implications in determining the onset of activation of Na^+ and K^+ channels. From the preceding discussion, an intracellular alkalinization would be expected to rapidly activate these channels and stimulate net Na absorption. Using the Ussing chamber patch-clamp setup we have investigated whether the acute effect of aldosterone on the short-circuit current (shown in Fig. 3C) involves activation of basolateral K^+ channels. Aldosterone was found to rapidly activate K_{ATP} channels in cell-attached basolateral membranes by increasing channel open probability, and the number of active channels present simultaneously in the membrane (Fig. 6). Such an

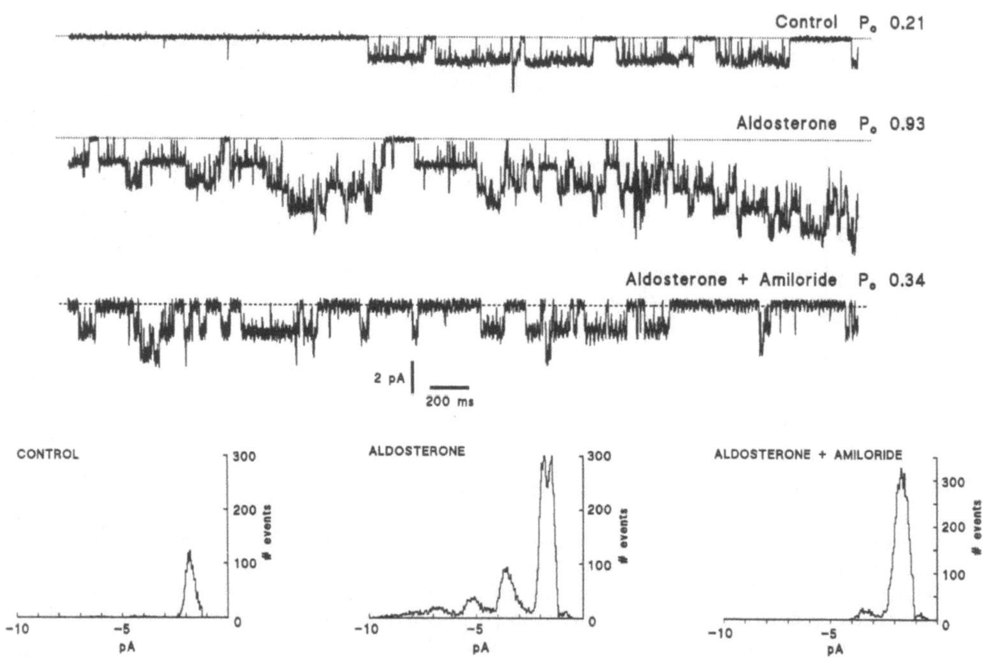

Fig. 6. Aldosterone activation of K_{ATP} channels in cell-attached membranes of frog-skin principal cells in situ in an isolated epithelium mounted in an Ussing chamber as shown in Fig. 1. Single-channel recordings show low spontaneous open probability for inward currents (Vp + 70 mV; 120 mM KCl in pipette; closed-state channel is indicated by *dashed line*). The number of simultaneously open channels is increased following 20-min incubation with aldosterone (10 nM) on basolateral side. The lower trace recorded 60 s after adding amiloride (10 μM basolateral side) shows that the diuretic reduced single-channel activity in the presence of aldosterone. The open-state current-amplitude/frequency histograms clearly show the activation of K_{ATP} channels by aldosterone and their inactivation by amiloride. The inhibitory effect of amiloride is interpreted as being due to closure of K^+ channels by intracellular H^+ following inhibition of Na/H exchange

effect has been reported in renal distal tubule after long-term (overnight) incubation with aldosterone (Wang et al. 1989). The stimulatory effect of the hormone on potassium channel activity can be reduced by adding amiloride (10^{-5} M) to the basolateral side of the epithelium. This concentration of amiloride blocks Na/H exchange and produces a fall in intracellular pH (Harvey and Ehrenfeld 1988b). It is therefore possible that aldosterone activates K_{ATP} channels via stimulation of Na/H exchange and increased pH_i. Such a rapid effect of aldosterone on Na/H exchange has a precedent in frog renal distal tubule within < 20 min (Oberleithner et al. 1987, 1988a, 1989, 1989, 1990, 1991; Oberleithner 1990) and in human lymphocytes < 10 min (Wehling et al. 1991), and may involve a membrane-bound receptor. The stimulatory effect of aldosterone on K_{ATP} channels is extremely rapid considering that there is normally a latency of up to 60 min before stimulation of transepithelial Na transport becomes apparent (Crabbé 1963; Garty 1986; Rossier et al. 1985). Aldosterone exerts pleiotropic effects on ion transport mechanisms. Rapid stimulation of Na absorption and K secretion by this hormone could involve a concerted action of many cytosolic signals. Although cellular ATP is not significantly affected during this period (< 1 h) by aldosterone (Urbach and Harvey 1992), other possible aldosterone-dependent transport regulators, such as exocytosis (Wade 1986; Harvey 1992) may be involved in mediating the acute response to aldosterone. Studies in kidney have demonstrated acute effects of aldosterone with short latency (< 30 min) on Na/K ATPase activity (El Mernissi and Doucet 1983), transepithelial PD (Gross and Kokko 1977), methylation of cytosolic proteins (Minuth et al. 1987), intracellular pH (Oberleithner et al. 1987; Harvey et al. 1988), RNA synthesis (Zwanzig et al. 1990) and exocytosis (Harvey 1992).

2.7 Cell Model of pH_i and Cross-Talk

The cell model and pH_i/ion transport control loop shown in Fig. 7 depicts the role of pH_i as a modulator of Na and K transport at opposite apical and basolateral membranes. Intracellular H^+ control the rate of Na entry and K exit from the cell and can signal changes in apical Na permeability so as to produce corresponding changes in basolateral K^+ permeability. The basolateral Na/H exchanger is the best candidate as transducer of these effects. Its activation by aldosterone (Oberleithner et al. 1987; Harvey et al. 1988; Wehling et al. 1991) and sensitivity to intracellular pH and Na concentration (Harvey and Ehrenfeld 1988b) supports this notion. An intracellular alkalinization exerts a dual effect on Na uptake by directly increasing the number of active Na^+ channels, and indirectly by hyperpolarizing the membrane potential through activation of K^+ channels. These properties, and the equal sensitivity of single apical Na channels and basolateral K^+ channels to cytosolic H^+, provide convincing evidence that pH_i acts as a servo-control signal to couple rates of Na and K transport in tight epithelia (Harvey et al. 1988; Dawson and Richards 1990). The operation of pH_i regulatory mechanisms, ionic exchangers or cotransporters of H^+ (OH^-, HCO_3^-),

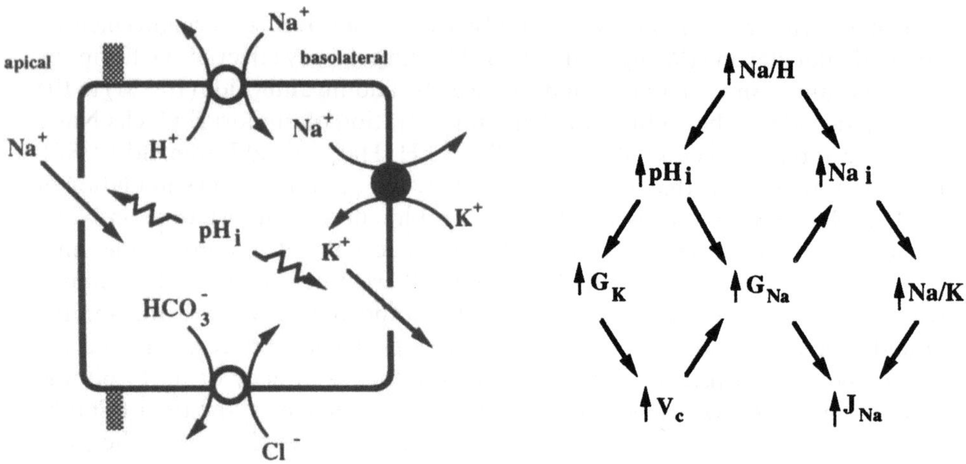

Fig. 7. Cell model of simultaneous regulation of apical Na^+ and basolateral K^+ conductances by cell pH and the effects of stimulating basolateral Na/H exchange on ion channels and the Na/K pump

have been shown to affect transepithelial Na and K transport via pH_i (Harvey and Ehrenfeld 1988c; Oberleithner et al. 1988, 1989, 1990a; Wang et al. 1989; Harvey 1992).

3 Regulation of K^+ Channel Activity by Intracellular ATP: ADP

Adenosine triphosphate (ATP) has been described as an inhibitor (Bleich et al. 1990; Wang et al. 1990) or an activator (Friedrich et al. 1989; Ohnu-Shoshaku et al, 1990) of inward-rectifier K^+ channels in epithelia. In principal cells of frog-skin epithelium, K_{ATP} channel activity is reduced by 50% at cytosolic-side ATP concentrations between 30 and 80 μm (Fig. 8). The normal concentration of cellular ATP in these cells ranges between 1 and 3 mM (Urbach and Harvey 1992). If ATP is the sole regulator, the K_{ATP} channel should obviously be inactivated in the intact cell. Because this is not the case, other factors must be involved in modulating the inhibitory effect of ATP. Patch-clamp recording of single K_{ATP} channels in excised basolateral membranes demonstrate that down-regulation of K_{ATP} channel activity by 1 mmol/1 ATP was largely prevented by 0.1 mM ADP on the cytosolic side (Fig. 8). The ADP is a weak inhibitor of K_{ATP} channel activity in frog-skin epithelium, and can shift the K_i for ATP from the micromolar to the millimolar range. The mechanism of action of ATP on K_{ATP} channels remains a mystery, although it is well established in other cell types that

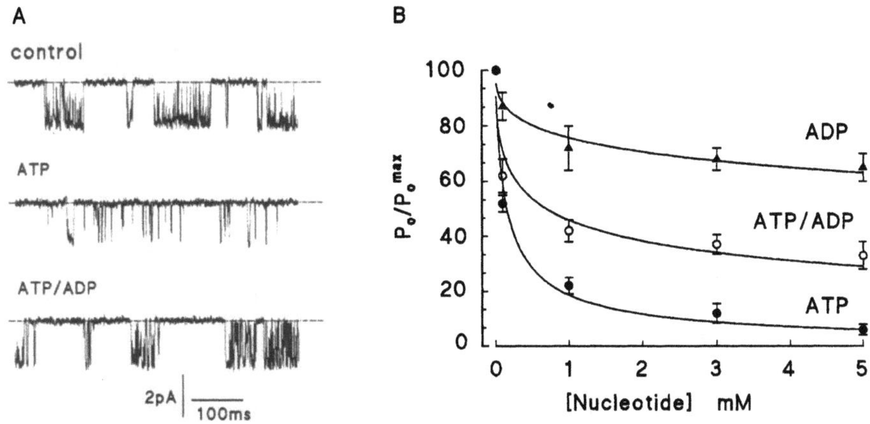

Fig. 8A,B. Effects of ATP and/or ADP on single K_{ATP} channel activity in an inside-out patch excised from a principal cell basolateral membrane of frog skin. Patch-clamp recordings of single-channel inward current are shown in control (K "intracellular-like" solution bathing both sides of patch; Vp + 70 mV) and 60s after adding 1 mM ATP to the cytosolic side of the patch. The channel activity recovered partially when ADP 100 μM was added in the presence of ATP. Single K_{ATP} channel activity is expressed as normalized open probability and is shown as a function of adenosine nucleotide concentration on the cytosolic side of the membrane. The ATP was tested alone (*filled circles*, $n = 11$ patches) or in the presence of a constant background of 100 μM ADP (*open circles*, $n = 11$). The response to ADP alone is shown by the (*triangles*) ($n = 7$)

phosphorylation of the channel is not the mechanism of inhibition by adenosine phosphates (Aschcroft 1988).

3.1 Coupling Between Na/K Pump and K_{ATP} Channels

Cross-talk between Na/K pump and K leak within the basolateral membrane is essential to maintain osmotic balance (Diamond 1982). Basolateral membrane K^+ conductance decreases after inhibition of the Na/K pump (Messner et al. 1985; Harvey et al. 1988; Horisberger and Giebisch 1988), but the underlying cause(s) are unclear. The sensitivity of K_{ATP} channels to the cytosolic ATP : ADP ratio provides one possible mechanism for cross-talk between Na/K pump activity and recycling of K^+ (Beck et al. 1991; Tsuchiya et al. 1991; Urbach and Harvey 1992). This coupling mechanism is illustrated in Fig. 9 with an example of a control loop for cross-talk following an increase in apical Na permeability (e.g. Lapointe and Duplain 1991). An increased entry rate of Na^+ entry across the apical membrane will increase the cytoplasmic Na^+ transport pool (Harvey and Kernan 1984a), which has been shown to stimulate the Na/K pump (Garty 1984; O'Neil and Hayhurst 1985; Haber et al. 1987; Ehrenfeld et al. 1992). Because ≈50% of oxygen consumption is due to Na/K pump activity, (Zerahn 1956; Leaf and Renshaw 1957) its activation will tend to lower the ATP : ADP ratio and

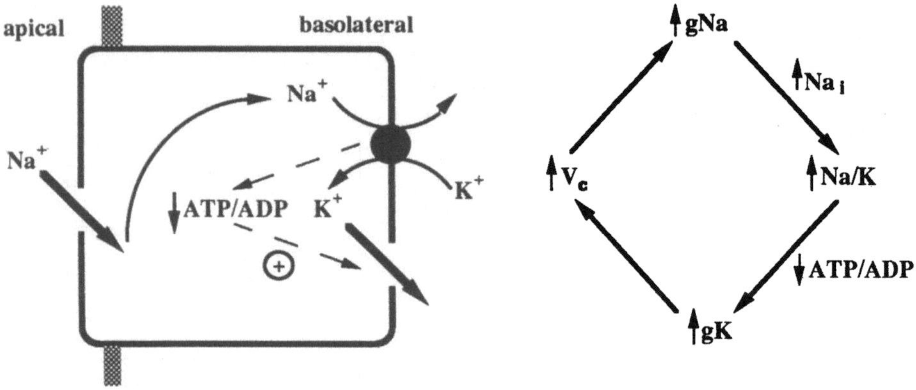

Fig. 9. Metabolic regulation of Na and K transport in frog-skin granular cells. A control loop is shown for cross-talk following an increase in apical Na permeability. The increased rate of Na influx raises $[Na]_i$ and stimulates the Na/K pump, which reduces the ATP : ADP ratio and activates K_{ATP} channels. An increase in basolateral membrane K^+ conductance causes membrane hyperpolarization, which maintains the electrical gradient for apical Na uptake

cause activation of K_{ATP} channels. The increased K^+ efflux across the basolateral membranes will indirectly hyperpolarize the apical cell membrane and limit the depolarization effect of an increased Na^+ conductance. This cross-talk will help to conserve a favourable electrical gradient for Na uptake and maintain equilibrium between Na and K fluxes at opposite cell membranes.

4 Regulation of K^+ Channel Activity by Intracellular Ca^{2+}

Cytosolic Ca^{2+} is considered a prime regulator (usually an activator) of K^+ channels. Intracellular calcium has been variously reported to have strong

Fig. 10. Opposite effects of Ca^{2+}_i on the K_{ATP} channels and K_{Ca} channels in inside-out patches excised from basolateral membranes of isovolume cells and swollen cells, respectively (K "intracellular-like" solution in patch pipette and same solution in bath with pCa modified by Ca^{2+}/EGTA buffer; Chang et al. 1988). The activity of these channels is expressed as normalized single-channel open probability and is presented as a function of cytosolic-side free Ca^{2+} concentration. The data were fitted by a Hill equation (*solid line* through *data points*):

$$Po = \frac{Po_{min} + (Po_{max} - Po_{min})}{1 + ([Ca^{2+}]_{50} / [Ca^{2+}])^n} . \quad (6)$$

The best fit gave values for the concentration of calcium, which produced a 50% reduction in K_{ATP} channel activity ($[Ca^{2}+]_{50}$=183 nmol/1) and a Hill coefficient of 2. For the K_{Ca} channel, the half-maximal activation was produced at Ca^{2+}_i 370 nM, with a Hill coefficient of 2. In the normal expected range of cytosolic free Ca^{2+} (10–100 nmol/l) the K_{ATP} channel is very active. However, when $[Ca^{2+}]_i$ is increased above 100 nmol/l, K_{ATP} channel openings are less frequent and K_{Ca} channels become increasingly activated. (Data are mean ± SEM of 7 patches) inhibitory effects on inward-rectifier K^+ channels in rat thick ascending limb of Henle (Bleich et al. 1990) and in shark rectal gland (Gögelein et al. 1987), or no effect of Ca^{2+} on inward-rectifier K^+ channels in apical membranes of rat cortical collecting tubule (Frindt and Palmer 1989). In frog-skin principal cells the K_{ATP} channel is inactivated by raising Ca^{2+} above 100 nM on the cytosolic side of excised inside-out membranes (Fig. 10). Swelling of epithelial cells is associated with an increased membrane permeability to Ca^{2+} and elevated levels of free intracellular Ca^{2+} (Wong et al. 1990; McCarty and O'Neil 1991). Hypo-osmotic-

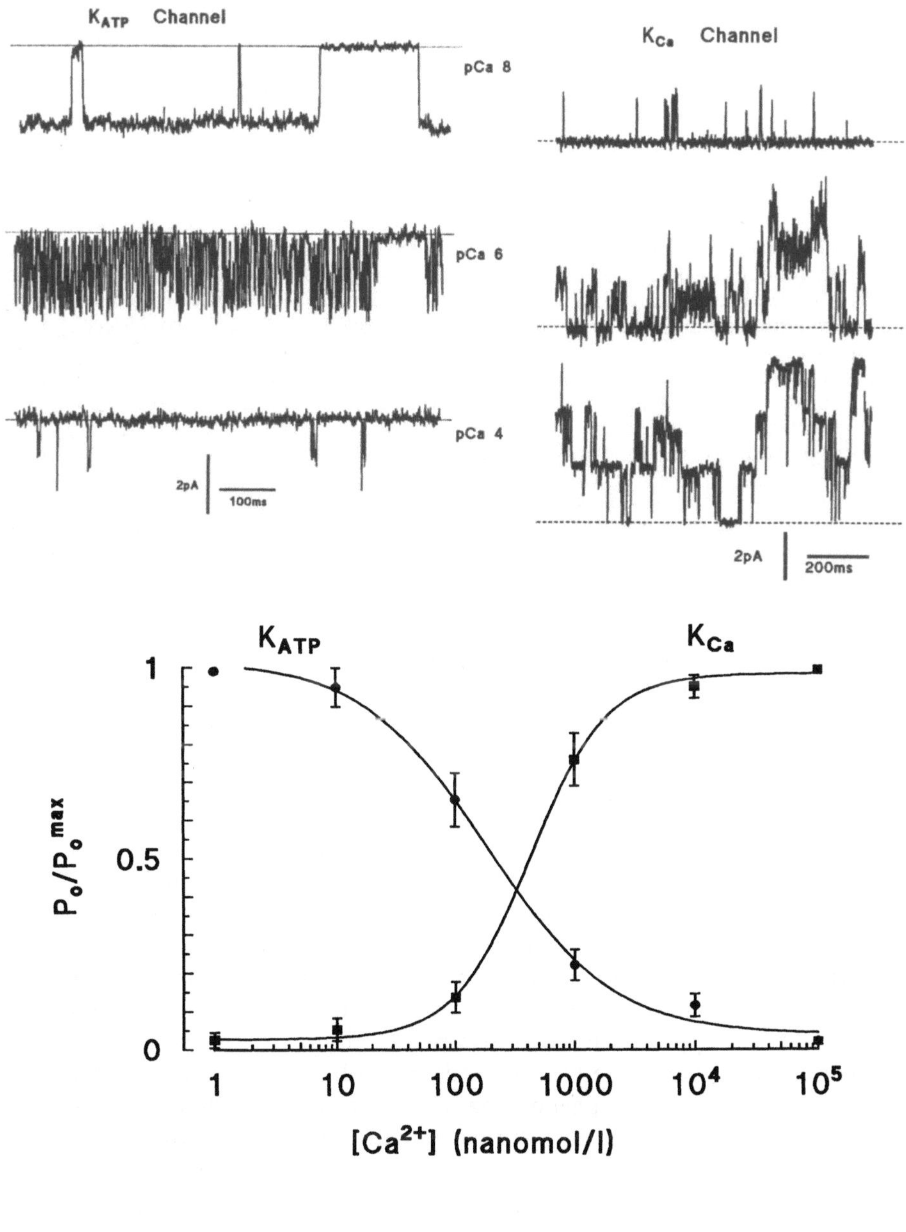

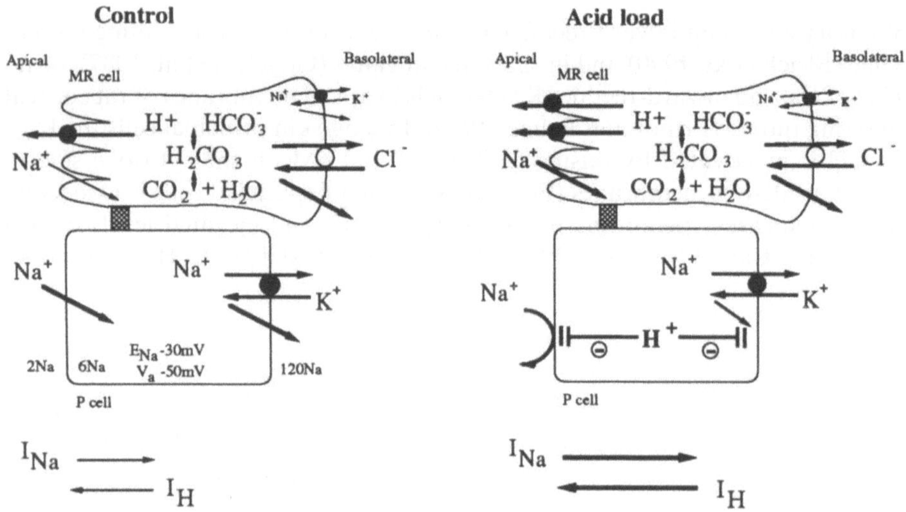

Fig. 11. Ion transport through mitochondria-rich (MR cells) and principal (P) cells of frog-skin epithelium under media conditions comparable with the in vivo state. Net Na uptake in P cells is indirectly electrically coupled to proton secretion in MR cells with 1:1 stoichiometry under open-circuit conditions in the absence of transepithelial anion fluxes. Under control conditions (apical side bathed in Na_2SO_4 or 2 mM NaCl solution buffered to pH 7.4 with 4 mM imidazole; serosal side bathed in NaCl Ringer solution buffered with 5% CO_2 and 24 mM HCO_3^-), proton secretion is driven by a H^+-ATPase pump in the apical membranes of MR cells, and the bulk of Na^+ absorption occurs through amiloride-sensitive channels in P cells (Na chemical gradients are shown in mM). The MR cell and P cell generate equal and opposite H^+ and Na^+ currents, respectively. An acid load stimulates both carbonic anhydrase-dependent H^+ secretion and amiloride-sensitive Na^+ absorption, despite inhibition of Na^+ and K^+ channels by H^+ in P cells. "Proton-coupled" Na^+ absorption in an acid load may occur via MR cells, which can buffer canges in intracellular pH by activation of H^+ pumps and by the dynamic buffering power of carbonic anhydrase

induced swelling of principal cells in frog-skin activates a 35 pS outward-rectifier K^+ channel. The open probability of this channel is increased by cytosolic Ca^{2+} (K_{Ca} channel in Fig. 11) and decreased by quinidine and quaternary ammonium ions. In swollen cells the K_{Ca} channel replaces the K_{ATP} channel in generating the macroscopic K^+ conductance (Harvey et al. 1991b; Urbach and Harvey 1994). Swelling of turtle-colon cells has also been shown to induce a Ca^{2+}-dependent K conductance that is preferentially blocked by lidocaine or quinidine (Germann et al. 1986; Dawson et al. 1988, 1990; Dawson 1991). Increasing internal Ca^{2+} concentration can, therefore, alter the type of K^+ channel present in the membrane. This has important consequences for the interpretation of macroscopic epithelial membrane conductance measurements based on the use of K^+ depolarized tissues and membrane perforation with ionophores (antibiotics or detergents). The appearance of an outwardly rectifying K^+ current in potassium-depolarized and ionophore-treated turtle colon has been ascribed to such osmotic or Ca_i^{2+} -induced phenomena (Dawson et al. 1990). Similar conclusions were also reached in cultured human sweat duct epithelium, from a comparison of the tetraethylammonium sensitivity of normal and metacholine-evoked K^+ conductances (Larsen et al. 1990). Volume-activated and muscarinic agonist-activated K^+ channels do not appear to be responsible for the resting K^+ conductance in nonstimulated turtle-colon epithelium (Dawson et al. 1988). The K_{Ca} channel presents a similar pharmacology and Ca^{2+}-activation profile of other basolateral membrane K^+ channels activated by cell swelling or detergent treatment in turtle colon (Chang and Dawson 1988), in nystatin-treated frog urinary bladder (Van Driessche 1986) and by oxytocin and cyclic adenosine monophosphate (cAMP) in nystatin-treated toad urinary bladder (Van Driessche and Erlij 1988). Exposure of frog-skin principal cells to basolateral hypotonic or high KCl solutions produces cell swelling, which activates both K_{Ca} channels and a 35 pS chloride channel (Harvey et al. 1992). The Ca^{2+}-activated K^+ channels, in parallel with volume-sensitive Cl^- channels, are most likely involved in regulatory volume decrease (Ussing 1987; Okada and Hazama 1989).

5 Proton Permeability in Anti-Diuretic-Hormone-Induced Channels

In the absence of anti-diuretic hormones (ADH), the apical cell membranes of amphibian skin and urinary bladder are relatively impermeable to water and H^+. In frog skin and urinary bladder, oxytocin induces a channel with selective permeability for H^+ and water at the apical cell membranes, which is blocked by inhibitors of F_0 proton channels and by mercury ions (Harvey et al. 1991a). The rapid movement of protons between hydronium ions and neighbouring water molecules can explain the high H^+ mobility in aqueous solutions at 25 °C of 36×10^{-4} cm/s (Glasstone and Lewis 1960; Deawer and Nichols 1989). This value is close to the H^+ permeability we calculated for ADH water channels, and is in agreement with the demonstration by Gluck and Al-Awqati (1980) of vasopressin-induced proton conductance in urinary bladder and high proton conductance of ADH water channel in endocytotic vesicles (Harris et al. 1991).

5.1 Acid Threat of High Proton Permeability in Water Channels

Given the dilute and variable acidic/alkaline nature of external solutions normally in contact with the apical surfaces of amphibian skin and bladder (pond water and urine), it is possible that ADH will produce simultaneous changes in net water flux and pH_i. The proton permeability in water-filled channels may pose an acid threat during anti-diuresis especially at low urine pH. Oxytocin produces changes in intracellular pH in frog-skin (Civan et al. 1988; Harvey et al. 1991a). The ADH-induced intracellular acidification was found to be concurrent with the peak water response in toad bladder (Brem et al. 1986) and in frog-skin (Harvey et al. 1991a). There is extensive evidence that intracellular acidification blocks

ADH-induced water flow at a post-cAMP step (Parisi et al. 1983; Parisi and Bourguet 1984; Parisi and Wietzerbin 1984). Because intracellular pH controls epithelial Na^+ and K^+ conductances that can be stimulated by ADH (Nagel 1978; Erlij et al. 1986), it is possible that hydrogen ions provide a servo-control of transcellular osmolyte and water flux during anti-diuresis.

6 Cellular Transport Pathways for Na^+ Absorption and H^+ Secretion

The frog skin in vivo absorbs sodium and secretes hydrogen ions (Krogh 1939) in an apparent 1:1 stoichiometry (Garcia-Romeu et al. 1969). The amphibian skin and urinary bladder, when isolated in an Ussing chamber, can also actively secrete hydrogen ions under appropriate electrochemical gradients (Frazier and Vanatta 1971; Ludens and Fanestil 1972; Machen and Erlij 1975; Ehrenfeld and Garcia-Romeu 1977; Ramsey 1982). H^+ secretion (JH) is restricted to mitochondria-rich cells (Harvey and Ehrenfeld 1988c; Ehrenfeld et al. 1989) and is a two-step process composed of proton extrusion across the apical membranes via an H^+-ATPase pump, and bicarbonate secretion across the basolateral membranes through a Cl^-/HCO_3^- exchanger (Duranti et al. 1986). Mitochondria-rich cells in amphibian skin, and intercalated cells in turtle urinary bladder and cortical nephron, are the site of acid secretion, whereas Na absorption is confined essentially to principal cells (Durham and Nagel 1986; Koeppen 1987; Steinmetz 1988; Harvey and Ehrenfeld 1988c; Sauer et al. 1990). The proton secretion function and high carbonic anhydrase content of MR cells in amphibian skin (Rosen and Friedley 1973; Ehrenfeld et al. 1985; Katz and Gabbay 1988) indicate a close similarity to the alpha-type intercalated cell of collecting tubule and turtle urinary bladder (Steinmetz 1988; Schwartz et al. 1982). Similar active hydrogen secretion mechanisms have been described in the gill of freshwater fish and crustacea, and in mammalian colon and renal collecting tubule (Larsen 1991; Kirchner 1988; Krippeit-Drews et al. 1989; Avella and Bornancin 1990; Perrone et al. 1990; Frömter 1988; Zeiske et al. 1992).

6.1 Control of Na Absorption by H^+ Pumps in Mitochondria-Rich Cells

Most in vitro studies of Na^+ absorption (JNa) in frog skin use Ringers solution bathing both sides of the tissue, whereas in vivo the Na uptake occurs from more dilute external media. For example, frogs can absorb Na in the absence of a permeant anion from freshwater containing less than 2 mmol/l Na. These observations, first reported in detail by Krogh (1939), have also been described in vitro when isolated skins are exposed to dilute apical saline solutions under open-circuit conditions (OCC; Ehrenfeld and Garcia-Romeu 1977). Krogh suggested that, because Na^+ and Cl^- absorption can be dissociated, a counterion such as H^+ or NH_4^+ must be excreted to maintain electroneutrality. This hypothesis has been tested both in vivo (Garcia-Romeu et al. 1969; Kirchner 1982) and in vitro (Ehrenfeld and Garcia-Romeu 1977; Ehrenfeld et al. 1985), and a model has been developed that uses the proton pump in mitochondria-rich (MR) cells to energize Na^+ absorption in principal cells under "physiological" electrochemical gradients (Harvey and Enrenfeld 1988c; (Fig. 11). The apparent electrical "coupling" between JNa and JH under these conditions results from the equilibrium of circulating cation currents in the absence of transepithelial anionic fluxes.

6.2 Effects of Acid Load on Proton-Coupled Na Absorption

The effects on Na^+ uptake of acid-loading frog-skin epithelium have been shown to be dependent on the relationship existing between Na^+ absorption and H^+ secretion (Harvey and Ehrenfeld 1988c). Under conditions where JNa and JH are "coupled" (open-circuited skins exposed to dilute external saline or impermeable anions), an intracellular acid load produced by CO_2 or NH_4 exposure causes a reversible increase in net Na^+ absorption. The stimulatory effect of CO_2 on JNa presents a paradox in that CO_2-induced intracellular acidification inactivates Na^+ channels in apical membranes of principal cells as discussed in detail here. The difficulty is to explain how JNa is increased when the major route for Na reabsorption through principal cells is inhibited. Because the acid-stimulated JNa is sensitive to amiloride, it must proceed by a cellular route, and this implicates the MR cell. Investigations into sodium absorption by MR cells have lead to conflicting conclusions. An amiloride and ouabain-sensitive swelling of MR cells has been demonstrated by video-image analysis in toad skin (Spring and Ussing 1986; Larsen et al. 1987), which indicates the presence of apical Na^+ channels and basolateral Na^+/K^+ ATPase. In contrast, electronmicroprobe elemental analysis of MR cells in frog skin (Rick et al. 1978; Dörge et al. 1990), failed to show significant changes in cellular Na, K or Rb in response to diuretics. Recent microprobe data, however, have revealed that MR cells that showed changes in Na content following ouabain and amiloride treatment accounted for 50% and 25%, respectively, of total MR cell population (Rick 1992). Microelectrode measurements of membrane potential in turtle urinary bladder and renal nephron intercalated cells do not support the notion that these cells could constitute a major site for transepithelial Na uptake (Durham and Nagel 1986; Koeppen 1987; Bello-Reuss 1991). Recently, this problem was examined by testing for amiloride-sensitive apical membrane Na^+ currents in patch-clamped MR cells (Harvey and Larsen 1992; Harvey 1992). Using this technique, we were able to provide direct evidence for Na^+ channels in apical membranes of frog-skin MR cells. This channel is highly selective for Na^+ relative to K^+. Fitting the GHK equation to single channels I-V relations yielded values for permeability selectivity coefficient PNa/PK=35 and single-channel conductance = 3.6 pS (120 mM Na/3 mM K in the patch pipette and 12 mM Na/120 mM K in the bath perfusate). The low conductance and high Na/K selectivity of Na^+ channels in MR cells compares very closely with the biophysical characteristics of patch-clamped Na^+ channels in principal cells (Palmer and Frindt 1986a; Frings et al. 1988). The excised Na^+ channel from MR cells is very sensitive to cytosolic pH, and its continued activity in these cells during an acid load may indicate that the high intrinsic buffering power of MR cells protect the Na^+ channel from closure by intracellular H^+. A rapid regulation of intracellular pH in MR cells may be achieved by the high carbonic anhydrase activity (Katz and Gabbay 1988) and stimulation of JH directly by H^+ (Cohen and Steinmetz 1980) or by insertion (exocytosis) of H^+ pumps into the apical membranes (Schwartz and Steinmetz 1971; Van Adelsberg and Al-Awqati 1986; Harvey 1992). Under these conditions the MR cell becomes a privileged pathway for Na absorption when uptake via principal cells is inhibited by an acid load.

7 Conclusions

Physiological changes in intracellular pH in principal cells regulate apical and basolateral Na^+ and K^+ channel activity instantaneously, simultaneously and in parallel. Cross-talk between ion channels and the Na/K pump can be mediated by Na/H-induced changes in intracellular pH and [Na]. Cell metabolism also plays an important role in coupling ion flows through pump and channels. The activity of the Na/K pump sets the ATP : ADP ratio in the cell and indirectly regulates K recycling through basolateral K_{ATP} channels.

The proton pump in MR cells appears to be a major determinant of sodium absorption through principal cells of amphibian skin in vivo and in vitro in the absence of transepithelial electrogenic anion flux. When Na^+ channels in principal cells are closed by an acid load, the MR cells may provide an alternative cellular pathway for acid-stimulated, amiloride-sensitive Na^+ uptake.

References

Almers W (1990) Exocytosis. Annu Rev Physiol 52: 607–627

Arruda JAL, Dytko G, Taylor Z (1990) Stimulation of H^+ secretion by CO_2 in turtle bladder: role of intracellular pH, exocytosis and calcium. Am J Physiol 258: R222–R231

Ashcroft FM (1988) Adenosine 5'-triphosphate-sensitive potassium channels. Annu Rev Neurosci 11: 97–118

Avella M, Bornancin M (1990) Ion fluxes in the gills of freshwater and seawater salmonid fish. In: Truchot J-P, Lahlou B (eds) Animal nutrition and transport processes, vol 6. Comparative physiology, Karger, Basel, pp 1–13

Beauwens R, fisher RS, Crabbé J (1990) Sodium transport by "tight" amphibian epithelia. In: Kinne RKH (ed) Basic principles in transport, vol 3. Comp arative physiology. Karger, Basel, pp 1–57

Beck JS, Breton S, Mairbaurl H, Laprade R, Giebisch G (1991) Relationship between sodium transport and intracellular ATP in isolated perfused rabbit proximal convoluted tubule. Am J Physiol 261: F634–F639

Bello-Reuss E (1991) Electrophysiological identification of cell types in cortical collecting duct monolayers. Renal Physiol Biochem 14: 1–11

Bleich M, Schlatter E, Greger R (1990) The luminal K^+ channel of the thick ascending limb of Henle's loop. Pflügers Arch Eur J Physiol 415: 449–460

Brem AS, Pacholski M, Lawler RG (1986) fluctuations in intracellular pH associated with vasopressin stimulation. Am Physiol 251: F897–F903
Cantiello HF, Ausiello DA (1991) Development of epithelial Na channels and regulation by guanine nucleotide regulatory (G) proteins and phospholipids. Current topics in membranes, vol 39. Academic Press, London pp 395–436
Cantiello HF, Patenaude CR, Ausiello DA (1989) G-protein subunit, a_i-3, activates a pertussis toxin-sensitive Na^+ channel from the epithelial cell line, A6. J Bio Chem 264: 20867–20870
Cantiello HF, Stow JF, Prat AG, Ausiello DA (1991a) Actin filaments regulate epithelial Na channel activity. Am J Physiol 261: C882–C888
Cantiello HF, Patenude C, Zaner K (1991b) Osmotically induced electrical signals from actin filaments. Biophys J 59: 1284–1289
Carvounis CP, Levine SD, Hays RM (1979) pH dependence of water and solute transport in toad urinary bladder. Kidney Int 15: 513–519
Chang D, Dawson DC (1988) Digitonin permeabilized colonic cell layers. Gen Physiol 92: 281–306
Chang D, Hsieh PS, Dawson DC (1988) Calcium: a program in BASIC for calculating the composition of solutions with specified free concentrations of calcium, magnesium and other divalent cations. Comput Biol Med 18: 351–366
Chase HS (1984) Does calcium couple the apical and basolateral membrane permeabilities in epithelia? Am J Physiol 247: F869–F876
Chase HS, Al-Awqati Q (1983) Calcium reduces the sodium permeability of luminal membrane vesicles from toad bladder. Studies using a fast reaction apparatus. J Gen Physiol 77: 693–712
Civan M, Cragoe EJ, Peterson-Yantorno K (1988) Intracellular pH in frog skin: effects of Na^+, volume and cAMP. Am J Physiol 255: F126–F134
Civan MM, Oler A, Peterson-Yantorno K, George K, O'Brein TG (1991) Ca^{2+}-independent form of protein kinase C may regulate Na transport across frog skin. J Membr Biol 121: 37–50
Cohen LH, Steinmetz PR (1980) Control of active proton transport in turtle urinary bladder by cell pH. J Gen Physiol 76: 381–394
Crabbé J (1963) Site of action of aldosterone on the bladder of the toad. Nature 200: 787–788
Das S, Garepapaghi M, Palmer LG (1991) Simulation by cGMP of apical Na channels and cation channels in toad urinary bladder. Am J Physiol 260: C234–C241
Davis W, finn AL (1982) Sodium transport inhibition by amiloride reduces basolateral membrane K conductance in tight epithelia. Science 26: 525–527
Dawson DC (1991) Ion channels and colonic salt transport. Annu Rev Physiol 53: 321–339
Dawson DC, Richards NW (1990) Basolateral K conductance: role in regulation of NaCl absorption and secretion. Am J Physiol 28: C181–C195
Dawson DC, Van Driessche W, Helman Sl (1988) Osmotically-induced basolateral K^+ conductance in turtle colon: lidocaine induced K^+ channel noise. Am J Physiol 254: C165–C174
Dawson DC, Wilkinson DJ, Richards NW (1990) Basolateral potassium channel noise: signals from the dark side. In: Helman SI, Van Driessche W (eds) Current topics in membranes and transport, vol 37. Channels and noise in epithetial tissues. Academic Press, New York, pp 191–212
Deawer DW, Nichols JW (1989) Proton flux mechanisms in model and biological membranes J Membr Biol 107: 91–103
Diamond JM (1982) Transcellular cross-talk between epithelial cell membranes. Nature 300: 683–685
Dörge A, Beck FX, Rich R, Nagel W, Thurau K (1990) Effect of amiloride on electrolyte concentrations and rubidium uptake in principal and mitochondria-rich cells of frog skin. Pflügers Arch Eur J Physiol 416: 335–338
Duranti E, Ehrenfeld J, Harvey BJ (1986) Acid secretion through *Rana esculenta* skin: involvement of an anion exchange mechanism at the basolateral membrane. J Physiol 378: 195–211
Durham JH, Nagel W (1986) Evidence for separate cellular origins of sodium and acid–base transport in turtle bladder Am J Physiol 250: C609–C616

Eaton DC, Hamilton KL, Johnson KE (1984) Intracellular acidosis blocks the basolateral Na-K pump in rabbit urinary bladder. Am J Physiol 247: F946–F954
Ehrenfeld J, Garcia-Romeu F (1977) Active hydrogen excretion and sodium absorption through isolated frog skin. Am J Physiol 233: F46–F54
Ehrenfeld J, Garcia-Romeu F, Harvey BJ (1985) Electrogenic active proton pump in *Rana esculenta* skin and its role in sodium transport. J Physiol 359: 331–355
Ehrenfeld J, Garcia-Romeu F, Harvey BJ (1988) Electrogenic active proton pump in *Rana esculenta* skin and its role in sodium ion transport. J Physiol 359: 331–355
Ehrenfeld J, Lacoste I, Harvey BJ (1989) The key role of the mitochondria-rich cell in Na^+ and H^+ transport across frog skin epithelium. Pflügers Arch Eur J Physiol 414: 59–67
Ehrenfeld J, Lacoste I, Garcia-Romeu F, Harvey BJ (1990) Interdependence of Na^+ and H^+ transport in frog skin. In: Truchot J-P, Lahlou B (eds) Animal nutrition and transport processes, comparative physiology, vol 6. Karger, Basel, pp 152–170
Ehrenfeld J, Lacoste I, Harvey BJ (1992) Effects of intracellular signals on Na^+/K^+-ATPase pump activity in the frog skin epithelium. Biochim Biophys Acta 1106: 197–208
El Mernissi G, Doucet A (1983) Short-term effect of aldosterone on renal sodium transport and tubular Na/K ATPase. Pflügers Arch Eur J Physiol 399: 139–146
Els WJ, Helman Sl (1991) Activation of epithelial Na channels by hormonal and autoregulatory mechanisms of action. J Gen Physiol 98: 1197–1220
Erlij D, Van Driessche W, De Wolf I (1986) Oxytocin stimulates the apical K^+ conductance in frog skin. Pflügers Arch Eur J Physiol 407: 602–606
Farquhar MG, Palade GE (1964) Functional organization of amphibian skin. Proc Natl Acad Sci USA 51: 569–577
Farquhar MG, Palade GE (1965) Cell junctions in amphibian skin. J Cell Biol 26: 263–291
Frazier LW, Vanatta JC (1971) Excretion of H^+ and NH_4^+ by the urinary bladder of the acidotic toad and the effect of short-circuit current on the excretion. Biochim Biophys Acta 241: 20–29
Friedrich F, Weiss H, Paulmichi M, Lang F (1989) Activation of potassium channels in renal epithelioid cells (MDCK) by extracellular ATP. Am J Physiol 256: C1016–C1021
Frindt G, Lee CO, Yang JM, Windhager EE (1988) Potential role of cytoplasmic calcium ions in the regulation of sodium transport in renal tubules. Miner Electrolyte Metab 14: 40–47
Frings S, Purves RD, Macknight ADC (1988) Single-channel recordings from the apical membrane of the toad urinary bladder epithelial cell. J Membr Biol 106: 157–172
Frömter E (1988) Mechanisms and regulation of ion transport in the renal collecting duct. Comp Biochem Physiol A 90: 701–777
Funder TH, Ussing HH, Wieth TO (1967) The effects of CO_2 and hydrogen ions on active Na transport in the isolated frog skin. Acta Physiol Scand 71: 65–76
Garcia-Romeu F, Ehrenfeld J (1975) Effect of acetazolamide on Na^+ transport through the isolated frog skin at low and high external sodium concentrations. Life Sci 17: 579–582
Garcia-Romeu F, Salibian A, Pezzani-Hernandez S (1969) The nature of the in vivo sodium and chloride uptake mechanisms through the epithelium of the Chilean frog *Calyptocephalella gayi* (Dum. et Bibr. 1841). Exchanges of hydrogen against sodium and bicarbonate against chloride. J Gen Physiol 53: 816–835
Garty H (1984a) Current-voltage relations of basolateral membrane in tight amphibian epithelia: use of nystatin to depolarize the apical membrane. J Membr Biol 77: 213–222
Garty H (1984b) Amiloride blockable sodium fluxes in toad bladder membrane vesicles. J Membr Biol 82: 268–279
Garty H (1986) Mechanisms of aldosterone action in tight epithelia. J Membr Biol 90: 193–205
Garty H, Benos D (1988) Characteristics and regulatory mechanisms of the amiloride-blockable Na^+ channel. Physiol Rev 68: 309–373
Garty H, Civan ED, Civan MM (1985) Effects of internal and external pH on amiloride-blockable Na transport across toad urinary bladder vesicules J Membr Biol 87(1): 67–75
Garty H, Asher C, Yeger O (1987) Direct inhibition of epithelial sodium channels by a pH-dependent interaction with calcium and by other divalent ions. J Membr Biol 95(2): 151–162

Garty H, Yeger O, Yanovsky A, Asher C (1989) Guanosine nucleotide dependent activation of the amiloride-blockable Na^+ channel. Am J Physiol 256: F965–F969

Germann WJ, Ernst SA, Dawson DC (1986) Resting and osmotically induced basolateral K conductances in turtle colon. J Gen Physiol 88: 253–274

Glasstone S, Lewis D (1960) Elements of physical chemistry, 2nd ed. Macmillan, London, pp 446

Gluck S, Al-Awqati Q (1980) Vasopressin increases water permeability by inducing pores. Nature 284: 631–632

Gluck S, Cannon C, Al-Awqati Q (1982) Exocytosis regulates urinary acidification by rapid insertion of H^+ pumps into the luminal membrane. Proc Natl Acad Sci USA 79: 4327–4331

Gögelein H, Greger R, Schlatter E (1987) Potassium channels in basolateral membranes of the rectal gland of Squalus acanthias. Pflügers Arch Eur J Physiol 409: 107–113

Granitzer M, Nagel W, Crabbé J (1991) Voltage dependent membrane conductances in cultured renal distal cells. Biochim Biophys Acta 1069: 87–93

Grinstein S, Erlij D (1978) Intracellular calcium and the regulation of sodium transport in the frog skin. Proc Roy Soc Lond 202: 353–360

Gross JB, Kokko JP (1977) Effect of aldosterone and potassium-sparing diuretics on electrical potential differences across the distal nephron. J Clin Invest 59: 82–89

Haber RS, Pressley TA, Loeb JN, Ismail-Beigi F (1987) Ionic dependence of active Na-K transport: "clamping" of cellular Na^+ with monensin. Am J Physiol 253: F26–F33

Harris HW Jr, Kikeri D, Janoshazi A, Solomon AK, Zeidel ML (1990) High proton flux through membranes containing antidiuretic hormone water channels. Am J Physiol 259: F366–F371

Harvey BJ (1992) Energization of sodium absorption by the proton-ATPase pump in mitochondria-rich cells of frog skin. J Exp Biol 172: 289–309

Harvey BJ, Ehrenfeld J (1985) Electrogenic active proton pump in *Rana esculenta* skin and its role in sodium ion transport. J Physiol 359: 331–355

Harvey BJ, Ehrenfeld J (1986) Regulation of intracellular sodium and pH by the electrogenic H^+ pump in frog skin. Pflügers Arch Eur J Physiol 406: 362–366

Harvey BJ, Ehrenfeld J (1988a) Dual role for intracellular protons in the control of sodium transport in frog skin epithelium. Pflügers Arch 411(Suppl 1): R66

Harvey BJ, Ehrenfeld J (1988b) Role of Na^+/H^+ exchange in the control of intracellular pH and cell membrane conductances in frog skin epithelium. J Gen Physiol 92: 793–810

Harvey BJ, Ehrenfeld J (1988c) Epithelial pH and ion transport regulation by proton pumps and exchangers. In: Proton passage across cell membranes. Ciba Foundation Symposium 139, Wiley, Chichester, pp 39–164

Harvey BJ, Kernan RP (1984a) Intracellular ionic activities in relation to external sodium and effects of amiloride and/or ouabain. J Physiol 349: 501–517

Harvey BJ, Larsen BJ (1993) Sodium and chloride currents in single mitochondria-rich cells of amphibian skin. J Physiol (Lond) 459: 241

Harvey BJ, Thomas RC (1987) Intracellular pH and calcium effects on sodium conductance and transport in isolated frog skin epithelium. J Physiol 353: C87

Harvey BJ, Urbach V (1992) Cellular regulation of ATP- and calcium-sensitive K^+ channels in frog skin and distal renal cell culture (A6) epithelia. 46th Ann Symp Soc Gen Physiol J Gen Physiol 100: p19a

Harvey BJ, Thomas SR, Ehrenfeld J (1988) Intracellular pH controls cell membrane Na^+ and K^+ conductances and transport in frog skin epithelium. J Gen Physiol 92: 767–791

Harvey BJ, Lacoste I, Ehrenfeld J (1991a) Common channels for water and protons at apical and basolateral cell membranes of frog skin and urinary bladder epithelia. J Gen Physiol 97: 749–776

Harvey BJ, Urbach V, van Kerkhove E (1991b) Inward rectifier K^+ channels in principal cells of isolated frog skin epithelium and in cultured A6 cells. J Physiol (Lond) 438: 264

Homareda H, Matsui H (1985) Effect of pH on Na^+ and K^+ binding to NaK ATPase. In: Glynn J, Ellroy C (eds) The sodium pump. 4th Annu Int Conf on Na, K ATPase? The company of biologists. Charlesworth & Co, Huddersfield, pp 251–254

Horisberger J-D, Giebisch G (1988) Voltage dependence of basolateral membrane in the Amphiuma collecting tubule. J Membr Biol 105: 257–263
Hunter M, Oberleithner H, Henderson RM, Giebisch G (1988) Whole-cell potassium currents in single early distal tubule cells. Am J Physiol 255: F699–F703
Hurst AM, Hunter M (1989) Apical K channels of frog diluting segment: inhibition by acidification. Pflügers Arch Eur J Physiol 415(1): 115–117
Hurst AM, Hunter M (1990) Acute changes in channel density of amphibian diluting segment. Am J Physiol 259: C1005–C1009
Katz U, Gabbay S (1988) Mitochondria-rich cells and carbonic anhydrase content of toad skin epithelium. Cell Tissue Res 251: 425–431
Khadouri C, Marsy S, Barlet-base C, Doucet A (1989) Short-term effect of aldosterone on NEM-sensitive ATPase in rat collecting tubule. Am J Physiol 257: F177–F181
Kirschner LB (1982) NaCl transport in intact frogs and isolated frog skins: a reconciliation. In: Corradino RA (ed) Functional regulation at the cellular and molecular levels. Elsevier North Holland, Amsterdam, pp 199–215
Kirschner LB (1988) Basis for apparent saturation kinetics of Na^+ influx in freshwater hyperregulators. Am J Physiol 254: R984–R988
Koefoed-Johnsen V, Ussing HH (1958) The nature of the frog skin potential. Acta Physiol Scand 42: 293–308
Koeppen BM (1987) Electrophysiological identification of principal and intercalated cells in the rabbit outer medullary collecting duct. Pflügers Arch Eur J Physiol 409: 138–141
Krippeit-Drews P, Drews G, Graszynski K (1989) Effects of ion substitution on the transepithelial potential difference of the gills of the fiddler crab *Uca-tangeri*: evidence for a proton pump in the apical membrane. J Comp Physiol (B) 159 (1): 43–50
Kristensen P, Ussing HH (1985) Epithelial organization. In Seldin GW, Giebisch G (eds) The kidney: physiology and pathophysiology. Raven Press, New York
Krogh A (1939) Osmotic regulation in aquatic animals. Cambridge Univ Press, Cambridge
Kubota TB, Biagi BA, Giebisch G (1983) Effects of acid–base disturbances on basolateral membrane potential and intracellular potassium activity in the proximal tubule of necturus. J Membr Biol 73: 61–68
Lacoste I, Harvey BJ, Ehrenfeld J (1991) Cl permeability of the basolateral membrane of the *Rana esculenta* epithelium: activation of Cl/HCO3 exchange by alcaline intracellular pH. Biochem Biophys Acta 1063(1): 103–110
Lang F, Rehwald W (1992) Potassium channels in renal epithelial transport regulation. Physiol Rev 72 (1): 1–32
Lapointe J-Y, Duplain M (1991) Regulation of basolateral membrane potential after stimulation of Na transport in proximal tubules. J Membr Biol 120: 165–172
Larsen E H (1991) Chloride transport by high-resistance heterocellular epithelia. Physiol Rev 71 (1): 235–283
Larsen EH, Ussing HH, Spring KR (1987) Ion transport by mitochondria-rich cells in toad skin. J Membr Biol 99: 25–40
Larsen EH, Novak I, Pedersen PS (1990) Cation transport by sweat ducts in primary culture, ionic mechanism of cholinergically evoked current oscillations. J Physiol 424: 109–131
Larsen EH, Willumsen NJ, Christoffersen BC (1992) Role of proton pump of mitochondria-rich cells for active transport of chloride ions in toad skin epithelium. J Physiol 450: 203–216
Latorre R, Miller C (1983) Conduction and selectivity in potassium channels. J Membr Biol 71: 11–30
Leaf A, Renshaw A (1957) Ion transport and respiration of isolated frog skin. Biochem J 65: 82–90
Leaf A, Keller A, Dempsey EF (1964) Stimulation of sodium transport in toad bladder by acidification of mucosal medium. Am J Physiol 207: 547–552
Lewis SA, DeMoura JLC (1984) Apical membrane area of rabbit urinary bladder increases by fusion of intracellular vesicles: an electrophysiological study. J Membr Biol 82: 123–136
Ling BN, Eaton DC (1989) Effects of luminal Na on single Na channels in A6 cells, a regulatory role for protein kinase. C. Am J Physiol 256: F1094–F1103

Loo DDF, Lewis SA, Ifshin MS, Diamond JM (1983) Turnover, membrane insertion, and degradation of sodium channels in rabbit urinary bladder. Science (Wash DC) 221: 1288–1290

Ludens JH, Fanestil DD (1972) Acidification of urine by the isolated urinary bladder of the toad. Am J Physiol 223: 1338–1344

Machen T, Erlij D (1975) Some features of hydrogen (ion) secretion by the frog skin. Biochim Biophys Acta 406: 120–130

Macknight ADC, DiBona DR, Leaf A (1980) Sodium transport across toad urinary bladder: a model "tight" epithelium. Physiol Rev 60: 615–715

MacRobbie EAC, Ussing HH (1961) Osmotic behaviour of the epithelial cells of frog skin. Acta Physiol Scand 53: 348–365

Mandel LJ (1978) Effects of pH, Ca, ADH and theophylline on kinetics of Na entry in frog skin. Am J Physiol 235: C35–C48

McCarty NA, O'Neil RG (1991) Calcium-dependent control of volume regulation in renal proximal tubule cells. 1. Swelling-activated Ca^{2+} entry and release. J Membr Biol 123: 149–160

Messner G, Wang W, Paulmichl M, Oberleithner H, Lang F (1985) Ouabain decreases apparent potassium conductance in proximal tubules of the amphibian kidney. Pflügers Arch Eur J Physiol 405: 131–137

Mills JW, Ernst SA, Dibona DR (1977) Localization of Na^+ pump sites in the frog skin. J Cell Biol 73: 88–110

Minuth WW, Steckelings U, Gross P (1987) Complex physiological and biochemical action of aldosterone in toad urinary bladder and mammalian renal collecting duct cells. Renal Physiol 10: 297–310

Nagel W (1976) The intracellular electrical potential profile of the frog skin epithelium. Pflügers Arch Eur J Physiol 365: 135–143

Nagel W (1978) Effects of antidiuretic homone upon electrical potential and resistance of apical and basolateral membranes of frog skin. J Membr Biol 42: 99–122

Nagel W (1985) Basolateral membrane ionic conductance in frog skin. Pflügers Arch 405 (Suppl 1): S39–S43

Nagel W, Garcia-Diaz JF, Armstrong WM (1981) Intracellular ionic activities in frog skin. J Membr Biol 61: 127–134

Oberleithner H (1990) Aldosterone-regulated ion transporter in the kidney. Klin Wochenschr 68: 1087–1090

Oberleithner H (1991) Acute aldosterone action in renal target cells. Cell Physiol Biochim 1: 2–12

Oberleithner H, Weigt M, Westphale HJ, Wang W (1987) Aldosterone activates Na/H exchange and raises cytoplasmic pH in target cells of the amphibian kidney. Proc Natl Acad USA 84: 1464–1468

Oberleithner H, Kersting U, Gassner B (1988a) Aldosterone-controlled linkage between Na/H exchange and K channels in fused renal epithelial cells. Ciba Found Symp 139: 201–219

Oberleithner H, Kersting U, Hunter M (1988b) Cytoplasmic pH determines K^+ conductance in fused renal epithelial cells. Proc Nat Acad Sci USA 85: 8345–8349

Oberleithner H, Kersting U, Silbernagl S, Steigner W, Vogel U (1989) Fusion of cultured dog kidney (MDCK) cells II. Relationship between cell pH and K conductance in response to aldosterone. J Membr Bio 111: 49–56

Oberleithner H, Wogel U, Kersting U, Steigner W (1990) Aldosterone stimulated sodium ion hydrogen ion and chloride bicarbonate exchange. Pflügers Arch Eur J Physiol 416(5): 533–539

Ohnu-Shoshaku T, Kubota T, Yamaguchi J, Fujimoto M (1990) Regulation of inwardly rectifying K^+ channels by intracellular pH in opossum kidney cells. Pflügers Arch Eur J Physiol 416: 138–143

Okada Y, Hazama A (1989) Volume-regulatory ion channels in epithelial cells NIPS 4: 238–242

O'Neil RG, Hayhurst RA (1985) Sodium dependent modulation of renal Na/K ATPase: influence of mineralocorticoids on the cortical collecting duct. J Membr Biol 85: 169–179

Onken H, Zeiske W, Harvey B (1990) Effect of mucosal hydrogen ion and chemical modification on transcellular potassium ion current in frog skin. Biochim Biophys Acta 1042 (1): 95–102
Palmer LG (1984) Voltage-dependent block by amiloride and other monovalent cations of apical Na channels in the toad urinary bladder. J Membr Biol 80 (2): 153–165
Palmer LG (1985) Modulation of apical Na permeability of the toad urinary bladder by intracellular Na, Ca, and H. J Membr Biol 83: 57–69
Palmer LG, Frindt G (1986a) Amiloride-sensitive Na channels from the apical membrane of rat cortical collecting tubule. Proc Natl Acad Sci USA 83: 2767–2770
Palmer LG, Frindt G (1986b) Epithelial sodium channels: characterisation by using the patch-clamp technique. Fed Proc 45(12): 2708–2712
Palmer LG, Frindt G (1987) Effects of cell calcium and pH on sodium channels from rat cortical collecting tubule. Am J Physiol 253: F333–F339
Parisi M, Bourguet J (1984) Effects of cellular acidification on ADH-induced intramembrane particle aggregates. Am J Physiol 246: C157–C159
Parisi M, Wietzerbin J (1984) Cellular pH and the ADH-induced hydroosmotic response in different ADH target epithelia. Pflügers Arch 402: 211–215
Parisi M, Wietzerbin J, Bourguet J (1983) Intracellular pH, transepithelial pH gradients and ADH induced water channels. Am J Physiol 244: F712–F718
Perrone RD, Nye JP, McBride DE, Zahniser DJ, Geyer JJ, Ucci AA (1990) Aldosterone induces apical vesicles in rat distal colon. Am J Physiol 259: F65–F71
Ramsay AG (1982) Membrane potentials, resistances and conductances of toad *Bufo marinus* bladder during sodium and hydrogen ion transport. Proc Soc Exp Biol Med 170 (1): 94–102
Rick R (1992) Intracellular ion concentrations in the isolated frog skin epithelium: evidence for different types of mitochondria-rich cells. J Membr Biol 127: 227–236
Rick R, Dörge A, von Arnim E, Thurau K (1978) Electron microprobe analysis of frog skin epithelium: evidence for a syncytial sodium transport compartment. J Membr Biol 39: 313–331
Rosen S, Friedley NJ (1973) Carbonic anhydrase activity in *Rana pipiens* skin: biochemical and histochemical analysis. Histochimie 36: 1–4
Rossier BC, Paccolat MP, Verrey F, Kraehenbühl JP, Geering K (1985) Mechanism of action of aldosterone: a pleiotropic response. Horm Cell Regul 9: 209–225
Sauer M, Flemmer A, Thurau K, Beck F-Z (1990) Sodium entry routes in principal and intercalated cells of the isolated perfused cortical collecting duct. Pflügers Arch Eur J Physiol 416: 88–93
Schaerer E, Neutra MR, Kraehenbuhl J-P (1991) Molecular and cellular mechanisms involved in transepithelial transport. J Membr Biol 123: 93–103
Schafer JA, Hawk CT (1992) Regulation of Na^+ channels in the corical collecting duct by AVP and mineralocorticoids. Kidney Int 41: 255–268
Schwegler JS, Steigner W, Heuner A, Silbernagl S (1990) pHi dependent membrane conductance of proximal tubule cells in culture OK differential effects on potassium ion conductive and sodium ion conductive channels. J Membr Biol 117(3): 243–252
Schoen H, Erlij D (1985) Current-voltage relations of the apical and basolateral membranes of the frog skin. J Gen Physiol 86: 257–287
Schultz SG (1992) Membrane cross-talk between epithelial cells. In: Seldin DW, Giebish G (eds) The kidney: physiology and pathophysiology, 2nd edn. Raven Press, New York, pp 287–298
Schwartz GJ, Al-Awqati Q (1985) Carbon dioxide causes exocytosis of vesicles containing hydrogen ion pumps in isolated perfused proximal and collecting tubules. J Clin Invest 75 (5): 1638–1644
Schwartz JH, Steinmetz PR (1971) CO_2 requirements for H^+ secretion by the isolated badder. Am J Physiol 220: 2051–2057
Schwartz J, Bethencourt D, Rosen S (1982) Specialized function of carbonic anhydrase-rich and granular cells in turtle bladder Am J Physiol 242: F627–F633
Sharin SH, Blankemeyer JT (1989) Demonstration of gap junctions in frog skin epithelium. Am Phyisol 257: C658–C664
Shearman MS, Sekiguchi K, Nishizuka Y (1989) Modulation of ion channel activity: a key function of the protein kinase C enzyme family. Pharmacol Rev 41 (2): 211–237

Smith PR, Benos DJ (1991) Epithelial Na^+ channels. Annu Rev Physiol 53: 509–530
Spring KR, Ussing HH (1986) The volume of mitochondria-rich cells of frog skin epithelium. J Membr Biol 92: 21–26
Steinmetz PR (1988) Electrogenic proton transport by intercalated cells of tight urinary epithelia. Ciba Foundation Symp 139. Proton passage across cell membranes. Wiley, Chichester, pp 122–138
Stetson DL, Steinmetz PR (1983) Role of membrane fusion in CO_2 stimulation of proton secretion by turtle bladder. Am J Physiol 245: C113–C120
Taylor A, Windhager EE (1979) Possible role of cytosolic calcium and Na–Ca exchange in regulation of transepithelial sodium transport. Am J Physiol 236: F505–F512
Taylor A, Windhager EE (1985) Cytosolic calcium and its role in regulation of transepithelial ion and water transport. In: Seldin DW, Giebish G (eds) The kidney, vol II. Raven Press, New York, pp 1297–1322
Thomas RC (1984) Experimental displacement of intracellular pH and the mechanism of its subsequent recovery. J Physiol (Lond) 354: 3–22
Tsuchiya K, Wang W, Giebisch G, Welling PA (1991) ATP is a coupling modulator of parallel Na/K ATPase-K channel activity in the rabbit proximal tubule. J Am Soc Nephrol 2: 753
Turneim K (1991) Intrinsic regulation of apical sodium entry in epithelia. Physiol Rev 71 (2): 429–445
Urbach V, Harvey BJ (1994) Potassium channels and ATP in epithelial cells of frog skin. J Physiol (Lond) 459: 283
Urbach V, Van Kerkhove E, Harvey BJ (1994) Inward-rectifier potassium channels in basolateral membranes of frog skin epithelium. J Gen Physiol 103: 583–604
Ussing HH (1987) Role of potassium in epithelial transport illustrated by experiments on frog skin epithelium. In: Arnostkleinzeller et al. (eds) Current topics in membrane transport, vol 31. Academic Press, New York
Van Adelsberg J, Al-Awqati Q (1986) Regulation of cell pH by Ca^{2+}-mediated exocytotic insertion of H^+-ATPases. J Cell Biol 102: 1638–1645
Van Driessche W, Erlij D (1988) Activation of K^+ conductance in basolateral membrane of toad urinary bladder by oxytocin and cAMP. Am J Physiol 254: C816–C821
Wade JB (1986) Role of membrane fusion in hormonal regulation of epithelial transport. Annu Rev Physiol 48: 213–223
Wang W, Giebisch G (1991) Dual effect of ATP on the apical small conductance potassium channel of the rat cortical collecting duct? J Gen Physiol 98(1): 35–62
Wang W, Schab A, Giebisch G (1990) Regulation of small conductance K channel in apical membrane of rat cortical collecting tubule. Am J Physiol 259: F494–F502
Wang W, Dietl P, Silbernagl S, Oberleithner H (1987) Cell membrane potential: a signal to control intracellular pH and transepithelial hydrogen secretion in frog kidney. Pflügers Arch Eur J Physiol 409: 289–295
Wang W, Wang Y, Silbernagl S, Oberleithner H (1988) Fused cells of frog proximal tubule: voltage-dependent intracellular pH. J Membr Biol 101(3): 259–265
Wang W, Henderson RM, Geibel J, White S, Giebisch G (1989) Mechanism of aldosterone-induced increase of K^+ conductance in early distal renal tubule cells of the frog. J Member Biol 111: 277–289
Wang W, White S, Geibel J, Giebisch G (1990) A potassium channel in the apical membrane of rabbit thick ascending limb of Henle's loop. Am J Physiol 258: F244–F253
Wehling M, Käsmayr J, Theisen K (1991) Rapid effects of mineralocorticoids on sodium-proton exchanger: genomic or nongenomic pathway? Am J Physiol 260: E719–E726
Wong SME, DeBell MC, Chase HS Jr (1990) Cell swelling increases intracellular free [Ca] in cultured toad bladder cells. Am J Physiol 258: F292–F296
Zeiske W (1992) Invertebrate epithelial Na^+ channels: amiloride-induced current-noise in crab gill. Biochim Biophys Acta 1105: 245–252
Zerahn K (1956) Oxygen consumption and active sodium transport in the isolated and short-circuited frog skin. Acta Physiol Scand 36: 300–318
Zwanzig M, Minuth WW, Gross P (1990) Action of aldosterone on cultured renal collecting duct cells during the latent period. Renal Physiol Biochem 13: 285–294

Chapter 8

Intracellular Signals Controlling Ionic and Acid–Base Regulation in Avian Nasal Gland Cells

T.J. Shuttleworth

Contents

1 Introduction

The elimination of excess salt represents a significant problem for the vast majority of vertebrate species whose kidneys are generally not capable of producing a urine that is more concentrated than the body fluids. Only in mammalian species, with the full development of the loop of Henle and the counter-current multiplier system of the renal tubule, is the kidney routinely able to concentrate the urine sufficiently to handle most excess salt loads. Consequently, wherever nonmammalian species are exposed to the problem of excess salt, extrarenal mechanisms for its elimination have evolved. The tissues concerned are diverse and include the branchial epithelium of teleosts, the rectal gland of elasmobranchs, and various cranial glands (lacrimal, nasal, salivary, and lingual) in reptiles, and nasal glands in birds. Despite this diversity in the tissues utilized, current evidence suggests a certain consistency in overall function

Department of Physiology, University of Rochester School of Medicine and Dentistry, Rochester, NY 14642, USA

Advances in Comparative and Environmental Physiology, Vol. 22

and mechanism of action, at least at the cellular level (Shuttleworth 1987). Of these different tissues, the avian salt-secreting nasal gland, in addition to its continued inherent interest for comparative physiologists, has attracted the attention of a variety of physiologists interested in certain unique properties and particular advantages that this tissue present to investigate more general phenomena. For example, together with the shark rectal gland, the avian nasal gland played an important role in elucidating the basic mechanisms of transepithelial chloride secretion (Ernst and Mills 1977). More recently, the avian nasal gland has become increasingly used in the study of intracellular signaling pathways responsible for the control of exocrine ion secretion, providing an important model system for investigating the general features offluid secretion as performed by such tissues as salivary and lacrimal glands, and pancreatic acini. It is this aspect of their physiology that is the subject of this brief review. In fact, the avain nasal gland has a particularly unique place in the history of the study of intracellular signals in that some of the very earliest investigations of the increased turnover of membrane phosphoinositides following muscarinic receptor activation were performed by Lowell and Mabel Hokin in the 1950s using this tissue. Unfortunately, the all-important link between the metabolism of these membrane phospholipids and the initiation and stimulation of cellular activity was not realized until some 25 years later.

2 Secretory Mechanism

The avian nasal gland is a paired structure usually lying in a supraorbital position with ducts that drain the secretion into the nasal cavity. Histologically, each gland is a multilobular organ with each secretory lobule consisting of a mass of scretory cells surrounding a secretory duct. A detailed description of the anatomy of the gland and its variations in different species, is beyond the scope of this review, and interested readers are referred to the excellent article by Butler et al. (1991). There is considerable evidence suggesting that the primary mechanism of ion secretion by the avian nasal gland generally conforms to similar transport processes described in many other tissues, namely the secondary active transport of chloride (Frizzell et al. 1979). In this process (Fig. 1), the activity of a basolateral Na–K pump generates a gradient for the downhill movement of sodium into the cell across the basolateral membrane. This downhill movement of sodium is directly coupled to the uphill movement of chloride into the cell across the same membrane by means of a Na–2Cl–K cotransporter. This results in an intracellular accumulation of chloride to a level sufficient to allow the passive exit of this ion across the apical membrane, via apical chloride channels. Potassium ions brought into the cell on the cotransporter, as well as by the Na–K pump, exit via basolateral potassium channels. The resulting transpithelial movement of chloride ions generates a lumen-negative electrical potential gradient that induces the passive movement of sodium into the secretory lumen via a

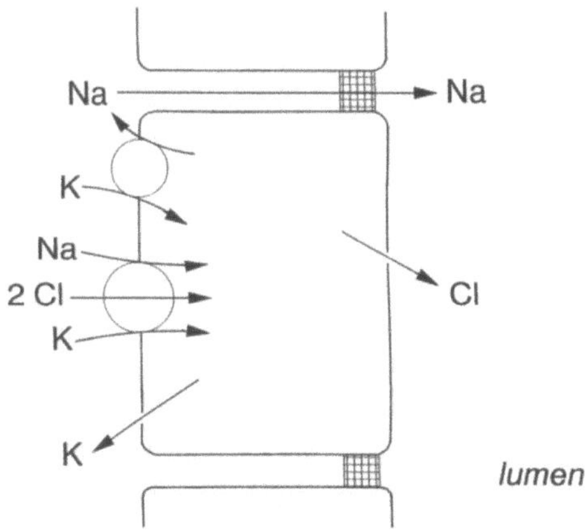

Fig. 1. A model for secretion in the avian nasal gland involving the secondary active transport of chloride

paracellular route. The net result is a transepithelial secretion of sodium and chloride. As mentioned previously, the evidence in support of this mechanism in the avian gland is fairly comprehensive and includes the sensitivity of secretory activity to inhibition by ouabain and by the loop diuretics furosemide and bumetanide, which block the contransporter (Hootman and Ernst 1980, 1981; Lowy et al. 1985; Shuttleworth and Thompson 1987), the localization of the Na–K pump to the basolateral membrane (Ernst and Mills 1977), the ability of basolateral barium to inhibit secretion presumably by blocking K channels (Lowy et al. 1989), the dependence of secretory activity on the presence of both sodium and chloride ions in the extracellular fluid (Hootman and Ernst 1981; Lowy et al. 1989) and the absence of any net secretion of sodium under short-circuit conditions (Lowy et al. 1989). Despite all this, other mechanisms of secretion have been proposed, particularly a model involving the primary section of an iso-osmotic fluid, which is then modified by water reabsorption during passage along the secretory ducts (Marshall et al. 1985, 1987). Also, alternative mechanisms for the basolateral uptake of sodium and chloride involving separate Na^+/H^+ and Cl^-/HCO_3^- exchangers have been suggested (Peaker 1971). Although such exchangers may contribute to the overall uptake of sodium and chloride ions (particularly as they may be active in the regulation of acid–base balance following stimulation of secretion; see below), and a limited modification of the secreted fluid may indeed occur in the secretory ducts, the weight of evidence that the principal mechanism of primary secretion is via a process similar to the secondary active chloride transport described previously, seems overwhelming.

3 Signals for Secretion

It is clear that the principal mechanism controlling secretion rate in vivo is stimulation via the secretory nerve (see Peaker and Linzell 1975). Various attempts to describe or assign an assortment of hormonal influences on the functioning of the gland have generally been inconclusive. Thus, at various times angiotensin (Gerstberger et al. 1984), substance P (Cheeseman et al. 1975), and atrial natriuretic peptide (Schütz and Gerstberger 1990), as well as adrenocorticotropic hormone (Holmes et al. 1972), corticosterone (Phillips et al. 1961), prolactin (Phillips and Ensor 1972) and thyroid hormones (Ensor et al. 1970) have all been implicated in the control of nasal gland secretion. However, subsequent investigations have generally revealed that the observed effects were either the result of methodological errors or were due to indirect effects on salt gland function, particularly via changes in blood flow, rather than direct effects on the secretory cells themselves (see Butler et al. 1989; Wilson 1987a). A possible exception to this is the reported response to atrial natriuretic peptide (ANP). Schütz and Gerstberger (1990) have shown the presence of receptors for this peptide in nasal gland tissue. Their studies indicate that the ANP-induced response is only transient, which may explain the failure of earlier studies where longer collection periods were used to show any direct effect of this hormone on salt gland function (Wilson 1987b). The fact that in other tissues this peptide acts intracellularly via elevations in cyclic guanosine monophosphate (GMP) is also of interest in view of the reported stimulatory effect of exogenous cyclic GMP on secretory activity in nasal gland cells (Stewart et al. 1979; Stewart and Sen 1981). This is clearly an area that deserves further study.

Despite all this, the overall conclusion is that the principal, although not exclusive, control mechanism of secretory cell activity in the avian nasal gland is via the release of acetylcholine from the postsynaptic parasympathetic fibers of the secretory nerve. The action of the released acetylcholine is mediated via muscarinic receptors located on the basolateral membrane of the secretory cells to activate secretion. The details of the intracellular effects of this receptor activation are described in the sections that follow.

3.1 Phosphoinositides and Inositol Phosphates

The essential result of muscarinic receptor activation in the avian nasal gland, as in most exocrine cells, is the enhanced breakdown of a specific pool of certain membrane phospholipids, the phosphoinositides (for review, see Bansal and Majerus 1990; Downes and Macphee 1990; Majerus et al. 1990; Rana and Hokin 1990; Catt et al. 1991). Generally, these phosphoinositides are quantitatively only a minor component (2–8%) of the total membrane phospholipids and consist principally of phosphatidylinositol (PI), phosphatidylinositol 4-phosphate (PIP), and phosphatidylinositol 4, 5-bisphosphate (PIP_2) existing in

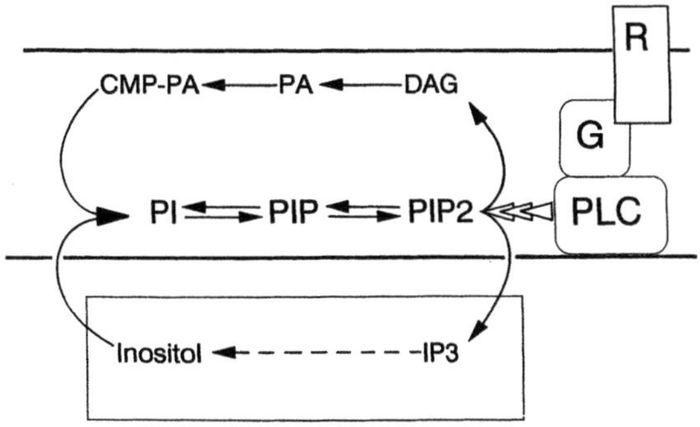

Fig. 2. The "phosphoinositide cycle" and the role of receptor-activated phospholipase C. The detailed metabolism of IP3 to give inositol eventually (see *box*) is shown in Fig. 3. *R* Receptor; *G* guanine nucleotide binding protein (G protein); *PLC* phospholipase C; *PI* phosphatidylinositol; *PIP* phosphatidylinositol 4-phosphate; *PIP2* phosphatidylinositol 4,5-bisphosphate; *IP3* inositol 1,4,5-trisphosphate; *DAG* diacylglycerol; *PA* phosphatidic acid; *CMP-PA* (≡ CDP-DG, cytidine-5' -diphospho-1,2-diacyl-*sn*-glycerol)

an equilibrium so that approximately 80% of the total phosphoinositide pool is in the form of PI. The receptor-enhanced break-down of these phosphoinositides results from the activation of a membrane phosphoinositide-specific phospholipase C (PLC). Consistent with other similar signaling pathways, coupling between the surface membrane receptor and the intracellular effector enzyme is via a heterotrimeric guanine nucleotide binding protein (G protein). The specific nature of the G protein involved in such PLC-coupled responses has only recently been resolved (Hildebrandt and Shuttleworth 1993) and shown to be a member of the G_q/G_{11} family (Pang and Sternweis 1990; Strathmann and Simon 1990). In reality, PLC is a group of at least five related enzymes (Crooke and Bennett 1989; Rhee et al. 1989) whose action on phosphoinositides results in the generation of a variety of inositol phosphates, inositol cyclic phosphates and diacylglycerol. Of these, the inositol phosphates and cyclic phosphates are metabolized through a series of steps to eventually produce inositol, and the diacylglycerol is converted, at least in part, first to phosphatidic acid and then CMP phosphatidate before recombining with inositol to reform PI (Fig. 2). It was just such a "PI cycle", involving the alternate turnover of the phosphatidic acid and phosphoinositide pools on agonist addition and subsequent removal, that was first described by Hokin and Hokin in the mid-fifties partly as a result of their studies on the avian nasal gland (see Hokin 1987). Unfortunately, it was not until Michell in 1975 proposed the link between this PI cycle response and signals that influenced a variety of cell activities via changes in the intracellular calcium concentration that its true significance began to be realized.

It is now clear that the primary target for agonist-stimulated PLC action is not PI, as was first supposed, but PIP_2 with the resulting generation of diacylglycerol and inositol 1,4,5–trisphosphate [Ins $(1,4,5)P_3$] (Fig. 2). These both have clearly defined actions as intracellular second messengers with diacylglycerol activating protein kinase C, and Ins $(1,4,5)P_3$ being involved in

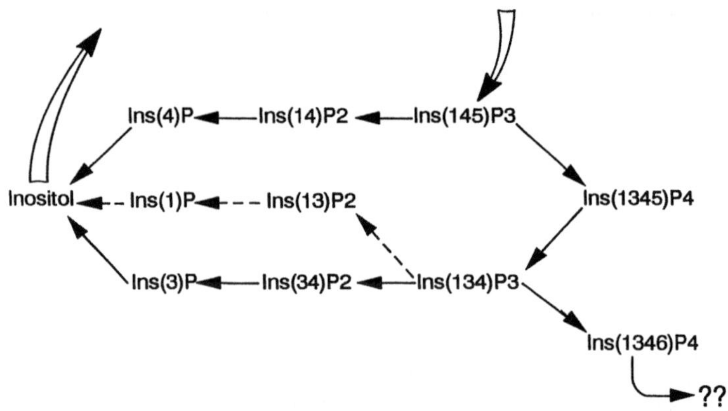

Fig. 3. Principal pathways of inositol phosphate metabolism in the avian nasal gland

generating changes in intracellular calcium ion concentrations ($[Ca^{2+}]_i$). The agonist-stimulated breakdown of phosphoinositides and the generation of Ins(1,4,5)P_3 have been well documented in the avian nasal gland (Fisher et al. 1983; Snider et al. 1986; Shuttleworth and Thompson 1989) and the complex subsequent metabolism of the generated inositol phosphates described in some detail (Shuttleworth and Thompson 1989; Hildebrandt and Shuttleworth 1992). Essentially, the pattern conforms to that seen in other cell types (Fig. 3), with the Ins(1,4,5)P_3 being either dephosphorylated by a 5-phosphatase to give inositol 1,4-bisphosphate, or phosphorylated by a 3-kinase to produce inositol 1,3,4,5-tetrakisphosphate [Ins(1,3,4,5)P_4]. This latter moiety is then dephosphorylated by the 5-phosphatase to give another inositol trisphosphate, namely inositol 1,3,4-trisphosphate [Ins(1,3,4)P_3]. Subsequent metabolism generally involves further dephosphorylation steps to give, mainly, inositol 3,4-bisphosphate and inositol 3-monophosphate, together with inositol 4-monophosphate [derived from the Ins(1,4)P_2], and finally inositol. Of all these various inositol phosphates formed as a result of receptor activation only one, Ins(1,4,5)P_3, has a clearly identified role as an intracellular signal. This involves its action in releasing calcium from a specific intracellular store (Berridge and Irvine 1984), as discussed below. Much attention has been paid to a possible role for Ins(1,3,4,5)P_4 in the cell, with some groups claiming an essential role in the activation of Ca^{2+} entry (see below), and others proposing roles (some positive, others negative) in the mobilization of intracellular Ca^{2+} stores (see Joseph et al. 1989; Hill and Boynton 1990; Gawler et al. 1990; Parker and Ivorra 1991). This is currently an area of much controversy, and a conclusive picture has yet to be resolved. The specific interest in Ins(1,3,4,5)P_4 arises largely from the fact that the cell has a perfectly adequate way of removing the Ins(1,4,5)P_3 signal by dephosphorylation via the 5-phosphatase to Ins(1,4)P_2, yet at least a portion of the generated Ins(1,4,5)P_3 is phosphorylated to produce Ins(1,3,4,5)P_4. This phosphorylation is, of course, at the expense of adenosine triphosphate (ATP). The generation of Ins(1,3,4,5)P_4 following receptor activation is rapid in onset and, in in vitro studies, the 3-kinase responsible for its generation can be shown to be activated by Ca^{2+} in a calmodulin-dependent manner (Biden and Wollheim 1986; Daniel et al. 1988; Li et al. 1989; Takazawa et al. 1990). However, recent studies on the avian gland have shown that the Ca^{2+}-dependent activation of the 3-kinase is probably of limited physiological relevance in the intact cell, and the major parameter determining the generation of Ins(1,3,4,5)P_4 following receptor activation is simply the corresponding increase in Ins(1,4,5)P_3 levels (Hildebrandt and Shuttleworth 1992). Interestingly, examination of the kinetic properties of the relevant enzymes involved in Ins(1,4,5)P_3 and Ins(1,3,4,5)P_4 metabolism (i.e. the 5-phosphatase and the 3-kinase) indicates that, although the generation of Ins(1,4,5)P_3 and its dephosphorylation by the 5-phosphatase are fairly directly linked in a quantitative manner to the degree of receptor stimulation, maximum levels of Ins(1,3,4,5)P_4 generation are achieved at very low substrate [i.e. Ins(1,4,5)P_3] concentrations (Hildebrandt and Shuttleworth 1992). The interesting corollary raised by this is that intracellular Ins(1,3,4,5)P_4, rather than showing a graded response to different agonist concentrations, may act much more like a simple "on/off" switch in the cell. Unfortunately, the physiological meaning of this remains unclear in the absence of any definitive identification of the role, if any, of Ins(1,3,4,5)P_4 in the cell.

The overall metabolism of inositol phosphates in cells is certainly more complex than the simple outline given previously. In particular, various cell types have been shown to contain a range of additional inositol phosphates, especially certain more highly phosphorylated moieties including other $InsP_4$s [e.g. Ins(1,3,4,6)P_4 and Ins(3,4,5,6)P_4], as well as Ins(1,3,4,5,6)P_5 and $InsP_6$. Although commonly ignored, these moieties are frequently present at very high concentrations in the cells and, although their concentrations or turnovers are not rapidly modulated as a result of receptor activation, certain long-term changes in their metabolism have been noted, and possible roles in growth and differentiation responses have been suggested (Michell et al. 1990). However, other than the putative identification of Ins(1,3,4,6)P_4 in salt gland homogenates (Hildebrandt and Shuttleworth 1992; see Fig. 3), these compounds have not been systematically studied in the avian gland.

3.2 Intracellular Calcium

As noted previously, Ins(1,4,5)P_3 generated as a result of muscarinic receptor activation acts within the cell to release calcium from certain intracellular stores, thereby generating at least the initial part of the rise in intracellular free Ca^{2+} ion concentration ($[Ca^{2+}]_i$). The identity and location of the specific Ins(1,4,5,)P_3-sensitive store within cells is not entirely clear, but it apparently represents a distinct component (usually less than half) of the vesicular nonmitochondrial store within the cell. Ins(1,4,5,)P_3 releases Ca^{2+} by acting at a specific receptor on this store and opening a Ca^{2+} channel in the membrane surrounding it. Recently, the Ins(1,4,5)P_3 receptor in mammalian cerebellum has been isolated and its

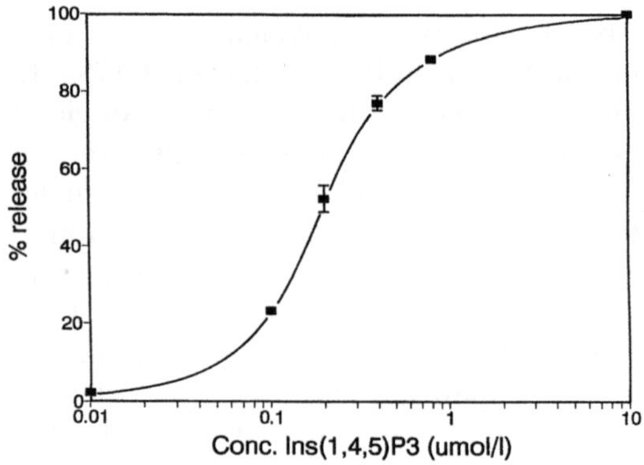

Fig. 4. Concentration response curve for the release of calcium from intracellular stores by exogenous inositol 1,4,5-trisphosphate [Ins(1,4,5)P_3]. Cell preparations were permeabilized by addition of saponin and the release of calcium into the medium determined using indo-1 (free acid). (See Shuttleworth 1992 for technical details)

sequence determined (Furuichi et al. 1989; Mignery et al. 1990), although similar studies in nonmammalian species have not been performed. Incorporation of this protein into liposomes or into lipid bilayers results in an Ins(1,4,5)P_3-gated Ca^{2+} conductance indicating that the Ins(1,4,5)P_3 receptor and the Ca^{2+} channel activities are located in the same protein (Ferris et al. 1989; Bezprozvanny et al. 1991). In the avian gland, Ins(1,4,5)P_3-induced Ca^{2+} release from intracellular stores has been studied in cells whose plasma membranes have been permeabilized by low concentrations of saponin (Shuttleworth 1992). In such preparations exogenous Ins(1,4,5)P_3 releases Ca^{2+} from the store with an EC_{50} of approximately 0.2 μmol l^{-1} (Fig. 4). This high sensitivity to Ins(1,4,5)P_3 raises a problem however, as estimates of the intracellular content of Ins(1,4,5)P_3 in resting cells, based on specific protein-binding assays, give values of at least 1–2 μmol l^{-1} (Shuttleworth 1990a,b; Hildebrandt and Shuttleworth 1991). Experiments show that such concentrations would result in the virtual complete depletion of Ca^{2+} from the Ins(1,4,5)P_3-sensitive stores. Similar findings have been reported for a variety of other cell types (Challiss et al. 1990; Nogimori et al. 1990; Bird et al. 1991), and have given rise to the suggestion of either some kind of intracellular "natural inhibitor" that is lost on permeabilization, or that intracellular Ins(1,4,5)P_3 is compartmentalized with a significant proportion unavailable for reaction with its receptor on the intracellular stores.

Whatever the specific details, it is clear that the generation of Ins(1,4,5)P_3 following receptor activation results in the rapid release of intracellular Ca^{2+}, and that this is responsible for the initial rise in $[Ca^{2+}]_i$. Thus, measurements of $[Ca^{2+}]_i$ using the intracellular fluorescent probe indo-1 reveal that $[Ca^{2+}]_i$ increases from resting values of approximately 100 nmol l^{-1} to 450–500 nmol l^{-1} within 1–3 s of the addition of an agonist such as carbachol (Fig. 5); (Shuttleworth and Thompson 1989; Shuttleworth 1990a). Under normal conditions, $[Ca^{2+}]_i$ subsequently declines to a sustained plateau value of approximately 300–350 nmol l^{-1} (or more than 3 times resting values) that is maintained as long as receptor activation continues. Addition of an antagonist, such as atropine, rapidly returns $[Ca^{2+}]_i$ to resting levels of around 100 nmol l^{-1}. Experiments show that the value of the sustained plateau in $[Ca^{2+}]_i$ is directly related to medium Ca^{2+}

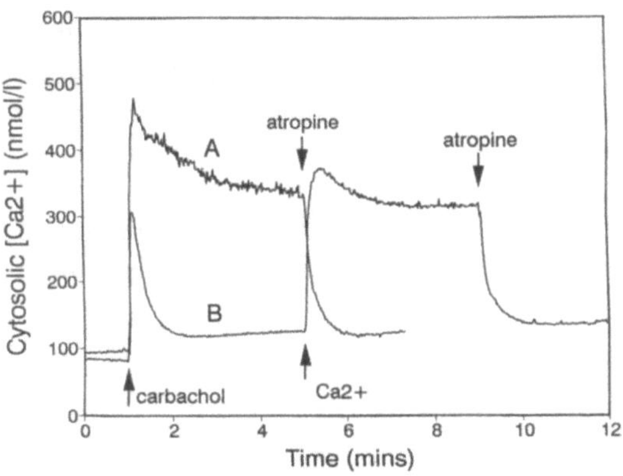

Fig. 5. Effect of muscarinic receptor activation on intracellular Ca^{2+} concentrations ($[Ca^{2+}]_i$) in avian nasal gland cells. Suspensions of isolated cells were loaded with the fluorescent probe indo-1 and $[Ca^{2+}]_i$ measured (see Shuttleworth and Thompson 1989 for details). Trace *A* represents cells in normal medium ($[Ca^{2+}]$ = 1.3 mmol l^{-1}). Trace *B* represents cells suspended in a low-Ca^{2+} medium ($[Ca^{2+}]_i$ = 0.04 mmol l^{-1}). At the point indicated (Ca2+), extracellular Ca^{2+} was restored to 1.3 mmol l^{-1}. Representative traces are shown

concentration, whereas the initial rise in $[Ca^{2+}]_i$ is only slightly affected. Thus, in cells suspended in a low-Ca^{2+} medium ($[Ca^{2+}]_0$ = < 0.04 mmol l^{-1}), addition of carbachol produces a somewhat attenuated initial rise in $[Ca^{2+}]_i$, which then spontaneously declines to resting values within 1 min, despite the continued presence of agonist (Fig. 5). Restoration of normal extracellular Ca^{2+} to such cells results in a rapid increase in $[Ca^{2+}]_i$ to reach a sustained plateau similar to that seen in the cells stimulated in the normal-Ca^{2+} medium (i.e., ~300–350 nmol l^{-1}). Further experiments have revealed that the rate of the increase in $[Ca^{2+}]_i$ observed in such "restoration" experiments is related to the value to which extracellular Ca^{2+} is raised, but in a manner that shows saturation phenomena (Shuttleworth and Thompson 1989). Such experiments support the idea that the initial rise in $[Ca^{2+}]_i$ is caused by intracellular release, presumably as a result of the action of Ins(1,4,5)P_3, but further show that the sustained $[Ca^{2+}]_i$ signal is due to the entry of Ca^{2+} from the extracellular medium. Because it is now clear that it is specifically this *sustained* component of the overall elevation in $[Ca^{2+}]_i$ that is the physiologically relevant signal for the activation of secretion (Shuttleworth and Thompson 1989), much attention has been paid to the nature and control of this receptor-activated Ca^{2+} entry process in exocrine cells in recent years. Despite this effort, in which studies using the avian nasal gland as a model have played an integral part, the picture is still far from clear. From various pharmacological studies, as well as the effects of membrane depolarization, it is clear that the Ca^{2+} entry pathway is distinct from the fairly well-characterized, voltage-activated Ca^{2+} channels of excitable cells (e.g., see Stuenkel and Ernst 1990). However, the precise mechanism of activation, either in the avian nasal gland or in nonexcitable cells in general, is still unresolved and is an area of considerable current controversy. One model, based largely on studies in lacrimal cells, proposes that a combined action of Ins(1,4,5)P_3, depleting intracellular Ca^{2+} stores, together with Ins(1,3,4,5)P_4 results in the activation of a plasma-membrane Ca^{2+} entry channel (Petersen1989; Irvine 1990). Another model, the so-called 'capacitative' model, suggests that Ca^{2+} entry is independent of any direct action of inositol phosphates and that it is specifically the emptying of the intracellular Ca^{2+} stores per se, by whatever means, that activates the Ca^{2+} entry channel (Putney et al. 1989; Takemura and Putney 1989; Putney 1990) Detailed examination of both of these models reveals that they have a critical feature in common, namely that Ca^{2+} entry is dependent on, and strictly subsequent to, prior emptying of the Ins(1,4,5)P_3-sensitive intracellular stores. However, directly contradicting this are certain results obtained in studies using the avian gland in which $[Ca^{2+}]_i$ signals were initiated by direct activation of the G protein

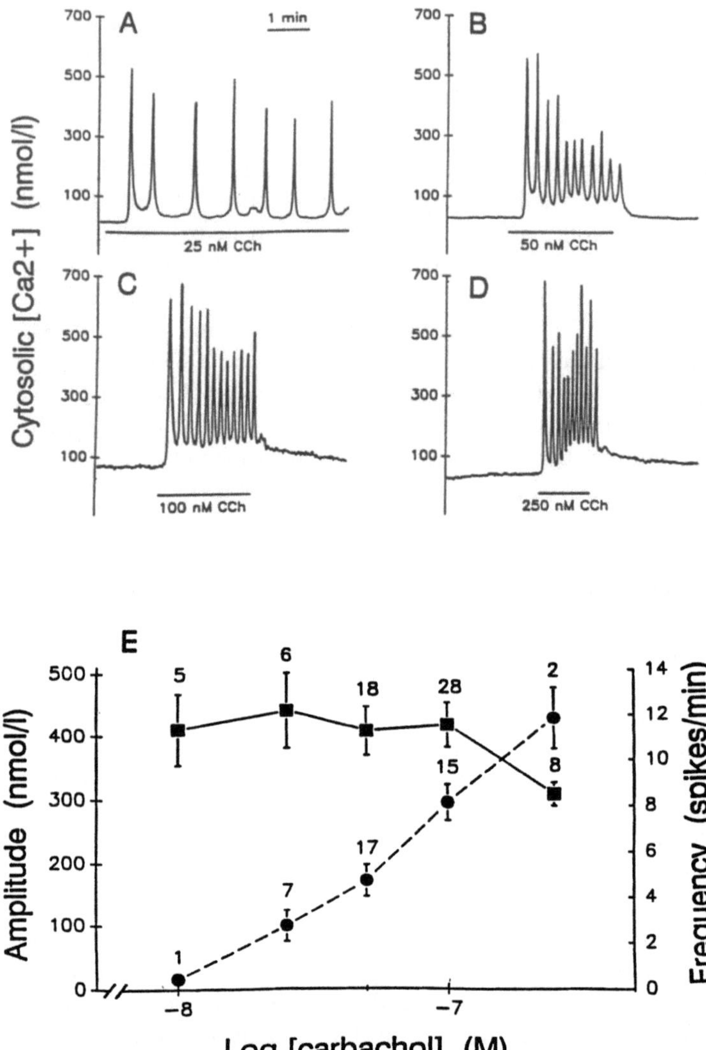

Fig. 6. Oscillatory $[Ca^{2+}]_i$ responses measured in individual avian nasal gland cells and the effect of agonist concentration. Traces **A–D** are representative $[Ca^{2+}]_i$ responses from individual cells to increasing concentrations of carbachol. Trace **E** illustrates the effect of agonist (carbachol) concentration on oscillation amplitude (*squares*) and oscillation frequency (*circles*). Data are means (± SE) with sample size given above each point. (Data from Crawford et al. 1991)

using fluoroaluminate (Shuttleworth 1990b). In these studies fluoroaluminate activated the same Ca^{2+} entry pathway as carbachol, but did so with a much slower time course. This slower time course permitted the detailed analysis of the overall response and revealed that under these conditions the Ca^{2+} entry component was initiated significantly before any detectable release of intracellular Ca^{2+}. These data suggest the possibility of more direct link between receptor activation and the initiation of Ca^{2+} entry, one that is independent of the mobilization of Ca^{2+} from intracellular stores (Shuttleworth 1990b). In summary, it seems possible that there are multiple factors controlling Ca^{2+} entry, and it is clear that this most important aspect of the cellular response to muscarinic receptor activation is far from being fully understood.

The previous description of the overall $[Ca^{2+}]_i$ signal in the avian gland generally refers to the response seen in cell suspensions. When single individual cells are studied using microfluorimetric techniques, marked oscillations in $[Ca^{2+}]_i$ are frequently seen (Crawford et al. 1991; Martin and Shuttleworth 1994a), particularly at low agonist concentrations or at subphysiological temperatures (Fig. 6). This feature of the response is obscured in experiments using cell suspensions, because individual cells do not oscillate in synchrony, at least when isolated. The oscillations observed in single cells are of a fairly constant amplitude, but their frequency is related to the concentration of agonist reaching a maximum rate of around 12 spikes per min at carbachol concentrations of 250 nmol l^{-1} (Fig. 6E). At higher agonist concentrations a sustained elevation in $[Ca^{2+}]_i$ is seen essentially the same as that in cell suspensions. Similar oscillatory $[Ca^{2+}]_i$ responses have been observed and described in many different cell types, and various models have been proposed to explain their generation (Woods et al. 1987; Gray 1988; Jacob 1990; Berridge 1991; Harootunian et al. 1991; Martin and Shuttleworth 1994a). Nevertheless, their precise physiological relevance remains unclear. One suggestion is that the frequency-coding of signals via these oscillations may be physiologically more important than the amplitude-coding implied by sustained elevations of $[Ca^{2+}]_i$ (Berridge 1991).

3.3 Protein Kinase C

As stated previously, the other signaling moiety generated as a result of receptor-enhanced phospholipase C activity acting on membrane phosphoinositides is diacylglycerol. It should also be noted that the generation of the two second messengers, diacylglycerol and Ins(1,4,5)P_3, is not always directly linked because, at least in some systems, the breakdown of PIP (and possibly PI) can result in the generation of diacylglycerol without the simultaneous generation of Ins(1,4,5)P_3. However, no direct evidence for this exists in the avian gland, and the precise physiological implications of this phenomenon are currently unclear. The signaling properties of diacylglycerol lie in its ability to activate protein kinase C (Nishizuka 1988). In the presence of diacylglycerol and certain phospholipids (particularly phosphatidylserine), the calcium sensitivity of this enzyme is increased such that it essentially becomes fully active at resting levels of $[Ca^{2+}]_i$, an increase in sensitivity that is associated with a translocation of the enzyme from the cytosol to the membrane. In common with many other systems, the precise consequences of protein kinase C activation are far from clear, and although it presumably results in the phosphorylation of certain proteins, the identity of those proteins and their physiological roles are unknown. In an early study Fisher et al. (1983) failed to detect any change in protein phosphorylation as a result of carbachol action in the nasal gland, but recently Torchia et al. (1991) have shown that stimulation with carbachol results in a specific protein-kinase-C-dependent phosphorylation of a 170-kDa protein. This protein was apparently localized to the plasma membrane. Interestingly, (and unusually) its phosphorylation required an elevation in $[Ca^{2+}]_i$ in addition to protein kinase C activation. This unusual requirement for a simultaneous elevation in $[Ca^{2+}]_i$ for phosphorylation to occur is, however, consistent with the fact that unlike many other systems, protein kinase C activation alone does not appear to directly affect intracellular Ca^{2+} signals in the avian nasal gland. For example, addition of the diacylglycerol analog OAG (1-oleoyl–2acetyl-glycerol) failed to affect either the Ins(1,4,5)P_3-induced initial $[Ca^{2+}]_i$ transient or the sustained elevation in $[Ca^{2+}]_i$ resulting from enhanced Ca^{2+} entry (Shuttleworth, unpubl. data). Similar results were obtained in experiments using the phorbol ester PMA (phorbol 12-myristate 13-acetate) to directly activate protein kinase C. The definitive identification of the 170-kDa protein described by Torchia et al. (1991) will obviously be important in determining possible roles of the diacylglycerol/protein kinase C pathway in response of the gland to muscarinic receptor activation. For example, evidence suggests that protein kinase C may be involved in the activation of the putative Na^+/H^+ exchanger that appears to play a major role in the regulation of intracellular pH (pH_i) following the activation of secretion (Shuttleworth and Wood 1992, see below). In addition, and consistent with current ideas of possible roles for protein kinase C in cell growth, it is possible that the generation of diacylglycerol and the consequent activation of protein kinase C may well be important in the profound adaptive differentiation and proliferation response seen in this tissue.

3.4 Cyclic Adenosine Monophosphate

As noted previously, it is clear that the major physiological control of secretory activity is via release of acetylcholine from the secretory nerve acting on muscarinic cholinergic receptors to raise $[Ca^{2+}]_i$ in salt gland cells. However, it is also clear that other potential pathways for the stimulation and control of secretory activity exist in these cells. Thus, despite previous reports to the contrary (Peaker 1971; Peaker and Linzell 1975; Stewart et al. 1979), elevation of intracellular levels of cyclic adenosine monophosphate (AMP), either via addition of exogenous cyclic AMP analogs or by the activation of endogenous adenylate cyclase by forskolin, results in the initiation of secretion (Lowy and

Ernst 1987; Shuttleworth and Thompson 1987). However, the identity of the primary messenger normally responsible for activation of this pathway in these cells and the precise role that it plays in the overall control and modulation of secretion in vivo has not been determined. Stimulatory responses via β-adrenergic receptors have been described (Lowy et al. 1985; Lowy and Ernst 1987), but it appears that these receptors may only be expressed in birds of a certain age or following specific salt-loading regimes. Secretory responses to vasoactive intestinal peptide (VIP), which operates via an adenylate cyclase pathway, have also been observed (Matin and Shuttleworth 1994b), and VIP has been identified in the avian nasal gland tissue (Lowy et al. 1987; Gerstberger 1988; Gerstberger et al. 1988). Similarly, addition of adenosine to freshly isolated cells, apparently acting at A_2-type receptors (Van Calker et al. 1979), results in an increase in intracellular cyclic AMP and marked stimulation of secretory activity (Shuttleworth, unpubl. data). However, in all cases the physiological relevance of the observed response is unknown, and no clearly defined role for any of these possible primary messengers that couple to the cyclic-AMP-mediated pathway has been identified in the intact animal.

3.5 Sites of Action in the Secretory Mechanism

The previous discussion of the generation of various intracellular second messengers in response to the activation of appropriate receptors does not tell us how these signals actually act to initiate and control the secretion of ions by the cells. Although the increased secretory activity resulting from receptor activation can be shown to be sensitive to inhibitors, such as ouabain and loop diuretics (e.g., furosemide or bumetanide), this does not mean that either the Na-K pump or the Na–2Cl-K cotransporter are the primary sites of action for these intracellular signals. As noted previously, as far as the cholinergic/$[Ca^{2+}]_i$-mediated system is concerned, it is clear that it is specifically the *sustained* component of the elevation in $[Ca^{2+}]_i$ that is the physiologically relevant signal for secretion. For example, addition of carbachol to cells in a low-Ca^{2+} medium, where the generation of inositol phosphates and the transient rise in $[Ca^{2+}]_i$ are both largely unaffected, is not sufficient to activate secretion (Shuttleworth and Thompson 1989). However, this does not tell us where the sustained elevation in $[Ca^{2+}]_i$ is having its effects. Only recently has direct evidence relating to this question been obtained for the avian gland (Martin and Shuttleworth 1994c; Martin et al. 1994). This evidence has revealed that the mechanisms involved are basically similar to those known to occur in other exocrine secretory tissues (particularly the parotid gland, lacrimal gland, and pancreatic acini) in that the primary sites of action for the intracellular signals are now known to be the ion channels of the basolateral and apical membranes (see Fig. 1). For example, elevations in $[Ca^{2+}]_i$ result in the activation of basolateral K-channels and apical Cl-channels (Martin and Shuttleworth 1994c), resulting in an increased efflux of these ions from the cytosol. At agonist concentrations that produce oscillatory

$[Ca^{2+}]_i$ signals, the induced Cl^- and K^+ currents are also oscillatory (Martin and Shuttleworth 1994c). The consequent decline in intracellular levels of both ions presumably increases the activity of both the basolateral Na-2Cl-K cotransporter and the Na-K pump, resulting in the sustained secretory activity of the cells. As to the other stimulatory pathway, that involving cyclic AMP, it appears that this specifically acts to increase an apical Cl^- conductance (Martin et al. 1994). This again corresponds to findings from other chloride-secreting tissues (Frizzell et al. 1979). The specific nature of the channels activated is, as yet, unclear but the kinetic properties of the Cl^- conductance activated by cyclic AMP are clearly distinct from that activated by Ca^{2+} (Martin et al. 1994). In addition, Richards et al. (1989) reported finding two types of K^+-channels in cells from the avian nasal gland, one of which was identified as being Ca^{2+}-dependent and specifically activated as a result of muscarinic agonist action.

4 Metabolism and Intracellular pH

In the avian nasal gland the stimulation of secretory activity results in a greatly increased rate of metabolism in the secretory cells. For example, addition of a maximal concentration of carbachol (500 μmol l^{-1}) to a suspension of isolated nasal gland cells increases oxygen consumption approximately 2.5-fold (Shuttleworth and Thompson 1989; Hildebrandt and Shuttleworth 1991). That this increase in oxygen consumption is directly related to secretory activity is

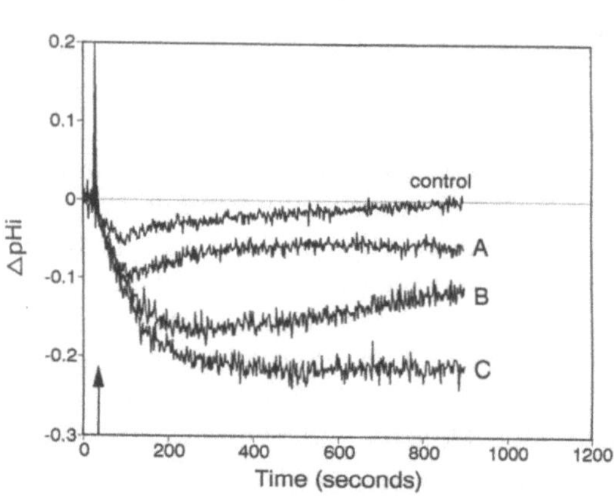

Fig. 7. Effect of muscarinic receptor activation on intracellular pH (pH_i) in avian nasal gland cells. Suspensions of isolated cells were loaded with the fluorescent probe BCECF and the pH_i was measured (see Shuttleworth and Wood 1992 for details). At the point indicated by the *arrow*, carbachol was added and the change in pH_i from resting levels was determined (ΔpH_i). The trace marked *control* represents the response seen in cells in normal bicarbonate-containing medium. Traces '*A–C*' represent cells suspended in bicarbonate-free media. Trace '*B*' was from cells exposed to amiloride, and trace '*C*' from cells exposed to both amiloride and DIDS. Traces are means of a minimum of five separate experiments

shown by the fact that in the presence of bumetanide, oxygen consumption only increases by approximately 20% following addition of carbachol. As expected, the profound increase in metabolic activity that this increase in oxygen consumption represents, also has important implications for acid–base balance in the secretory cells, because acid equivalents, in the form of carbon dioxide, are generated. Investigation of the changes in pH_i associated with the activation of secretion, using the intracellular fluorescent pH-sensitive probe BCECF (Shuttleworth and Wood 1992), have shown that addition of carbachol is indeed followed by a rapid fall in pH_i, but that this subsequently recovers to regain resting levels within 5–6 min (Fig. 7). More detailed analysis of the response showed that the initial fall in pH_i was largely a result of cell activation per se, but that the subsequent recovery involved a combination of bicarbonate-dependent, DIDS-sensitive and amiloride-sensitive components. Indeed, when each of these components were eliminated, addition of carbachol revealed a profound underlying intracellular acidification that presumably reflects the acid–base consequences of the increased metabolic activity following the stimulation of secretion (Fig. 7). The principal component restoring pH_i was the amiloride-sensitive process, and its activation always resulted in an alkalinization of the cytosol. Physiologically, this activation appeared to be closely coupled to muscarinic receptor occupation, but it appeared to be independent of any sustained elevation in $[Ca^{2+}]_i$, suggesting a possible involvement of diacylglycerol and protein kinase C. The DIDS-sensitive component(s), on the other hand, appeared to be activated principally by displacement of the pH_i away from resting levels and acted in either direction in a manner appropriate to restore resting pH_i. Although it was not possible to precisely identify each of these compensatory mechanisms, amiloride-sensitive Na^+/H^+ exchange, bicarbonate-dependent increases in intracellular buffering capacity and DIDS-sensitive (and possibly bicarbonate-sensitive) anion exchangers (e.g. Cl^-/HCO_3^-) were clear candidates. As noted, through the activation of these compensatory mechanisms, stimulation of secretion does not result in any sustained change in pH_i. This suggests that pH_i itself is unlikely to play any significant role in the intracellular signaling of secretory activity. However, it is important to remember that the increased activity of these various acid–base regulatory mechanisms responsible for restoring pH_i may well have important additional functions in the cell. For example, it can be anticipated that this increased activity of acid–base regulatory mechanisms will likely be important in enhancing the ability of the cell to maintain its pH_i in the face of changing metabolic activities. In addition, the activation of Na^+/H^+ and Cl^-/HCO_3^- exchangers in the basolateral membrane could, under appropriate circumstances, contribute to the uptake of sodium and chloride ions across this membrane independently of any increase in the activity of the Na–2Cl–K cotransporter. Finally, the increased activity of these various exchangers may play additional roles in the overall functioning of the cell during secretion, most particularly in cell-volume regulation and possibly as part of the more long-term adaptive proliferation and differentiation responses shown by these cells (see below).

5 Muscarinic Receptor Characterization

As discussed, the action of acetylcholine released from the secretory nerve is via muscarinic receptors on the secretory cells. Studies from a variety of different tissues have shown that muscarinic receptors in fact represent a family of receptor subtypes. Genetically, five distinct subtypes (ml–m5) have been identified. These different subtypes have been shown to demonstrate certain specificity both in tissue distribution and in the intracellular signaling pathways to which they are coupled (Kubo et al. 1986; Bonner et al. 1987; Peralta et al. 1987, 1988; Neher et al. 1988). Thus, the ml, m3, and m5 subtypes are generally tightly coupled to the phosphoinositide system, whereas the m2 and m4 subtypes are usually associated with the inhibition of adenylate cyclase. The pharmacological characterzation of muscarinic receptors, however, has generally revealed only three subtypes (Ml–M3) which show only partial correlation with the aforementioned molecular classification (Hammer et al. 1980). Concerning the avian nasal gland, pharmacological studies of the receptors on the secretory cells have revealed a somewhat ambiguous characterization (Hootman and Ernst 1982; Snider et al. 1986; Hildebrandt and Shuttleworth 1994). For example, the order of potency for the displacement of ^{3}H-QNB binding in these cells was found to be: atropine ≈ 4-DAMP > pirenzipine > carbachol (Hildebrandt and Shuttleworth 1994). This potency order is characteristic of M3-type receptors and is broadly consistent with data reported for mammalian exocrine tissues. However, the muscarinic ligand McN-A-343, which shows high selectivity for M1-type receptors, but not M3-type receptors, was found to be an effective partial agonist in nasal gland cells (Hildebrandt and Shuttleworth 1994). Attempts to resolve this contradiction by studies using oligonucleotide probes specific to the different molecular subtypes revealed positive hybridization reactions in Northern blots with only the m1-specific probe (Hildebrandt and Shuttleworth 1994). The conclusion was that the receptor on the avian nasal gland has some pharmacological characteristics of the mammalian M3 subtype but, genetically, may be related to the mammalian m1 subtype.

6 Proliferation and Differentiation Responses

In the previous discussion, we described the immediate response of nasal gland cells to secretory stimuli. There is, however, an extremely important more long-term response shown by this tissue to chronic stimulation. This response involves a profound adaptive proliferation and differentiation of the secretory cells and is, for example, most readily seen as an approximate doubling of total gland weight within 48 h of the animals being presented with a saline load in the form of 1% NaCl as drinking water. This dramatic adaptive response is associated with large increases in Na-K-ATPase, plasma membrane amplification, RNA content, and in vivo secretory capacity (Fletcher et al. 1967; Ernst et al. 1967; Ernst and Ellis 1969; Stewart et al. 1976; Hossler et al. 1978; Lingham et al. 1980; Merchant et al. 1985). Evidence demonstrating increases in ^{3}H-thymidine incorporation and total DNA contents (Holmes and Stewart 1968; Knight and Peaker 1979; Hossler 1982) indicates that the overall response involves hyperplasia as well as hypertrophy, although the relative individual contribution of each of these processes and their precise timing is somewhat variable, perhaps depending on the age of the animal, the nature of the salt-loading protocol, and whether the salt load is being experienced for the first time.

An important question that obviously has implications far beyond the avian nasal gland is the nature of the regulatory signals responsible for this proliferative effect. It has been known for some time that the adaptive response is directly associated with the nervous stimulation of secretion. For example, unilateral postganglionic transection of the nerve to the gland prior to the administration of a salt load prevents the normally observed adaptive increase in gland weight, RNA concentration and RNA : DNA ratios only in the denervated gland whilst the intact contralateral gland responds normally (Pittard and Hally 1973; Hanwell and Peaker 1975). Furthermore, in vivo administration of the muscarinic receptor antagonist atropine blocks the adaptive response (Hanwell and Peaker 1975). The interesting conclusion from these studies is that the normal pathways responsible for the activation of secretion via muscarinic receptors are apparently those responsible for, or at least a necessary component of, the initiation of these adaptive proliferation and differentiation responses in the gland too. This immediately limits the number of possible signals involved and makes the avian gland a particularly suitable model system for the study of this important general phenomenon. Comparison between the intracellular signals generated following receptor activation in cells before and after the adaptive response have revealed profound changes, particularly in the magnitude of inositol phosphate turnover and Ins(1,4,5)P_3 generation (Hildebrandt and Shuttleworth 1991). However, these differences could be entirely ascribed to corresponding changes in receptor number, and molecular analysis of the receptor subtype present in the cells indicates that the adaptive growth response is not associated with any change in the nature of the muscarinic receptors present (Hildebrandt and Shuttleworth 1994). The conclusion is that some facet of the signals generated as a result of normal muscarinic activation of secretion (i.e., phosphoinositide turnover, generation of inositol phosphates and diacylglycerol, increases in $[Ca^{2+}]_i$ and activation of protein kinase C) are intimately involved in some way with long-term adaptive proliferation and differentiation responses. Based on studies from other tissues that are admittedly very different, the most likely candidates would seem to be protein kinase C and possibly certain of the higher phosphorylated inositol phosphates [e.g., Ins(1,3,4,6)P_4 and Ins(1,3,4,5,6)P_5].

7 Conclusions

As already noted, the avian nasal gland has been intimately involved in investigations on the general physiology of cellular signals, particularly in exocrine cells, from the very earliest studies. In this brief review I have tried to highlight the important role that the avian gland continues to play, particularly in current investigations of the nature of receptor-activated Ca^{2+} entry, and the signals for adaptive cell proliferation and differentiation. I hope that these examples, and the many other important questions that remain, illustrate that this involvement shows every indication of continuing for some time to come.

Acknowledgments. I wish to thank Jill Thompson for her invaluable technical assistance and Jan-Peter Hildebrandt for helpful comments on the manuscript. Much of the research described in this chapter from the author's laboratory was supported by NIH grant GM-40457.

References

Bansal VS, Majerus PW (1990) Phosphatidylinositol-derived precursors and signals. Annu Rev Cell Biol 6: 41–67
Berridge MJ (1991) Cytoplasmic calcium oscillations: a two pool model. Cell Calcium 12: 63–72
Berridge MJ, Irvine RF (1984) Inositol trisphosphate, a novel second messenger in cellular signal transduction. Nature 312: 315–321
Bezprozvanny I, Watras J, Ehrlich BE (1991) Bell-shaped calcium-response curves of Ins(1,4,5)P_3^- and calcium-gated channels from endoplasmic reticulum of cerebellum. Nature 351: 751–754
Biden TJ, Wollheim CB (1986) Ca^{2+} regulates the inositol tris/tetrakisphosphate pathway in intact and broken preparations of insulin-secreting RINm5F cells. J Biol Chem 261: 11931–11934
Bird GSJ, Oliver KG, Horstman DA, Obie J, Putney JW Jr (1991) Relationship between the calcium-mobilizing action of inositol 1,4,5-trisphosphate in permeable AR4-2J cells and the estimated levels of inositol 1,4,5-trisphosphate in intact AR4-2J cells. Biochem J 273: 541–546
Bonner TI, Buckley NJ, Young AC, Brann MR (1987) Identification of a family of muscarinic acetylcholine receptor genes. Science 237: 527–532
Butler DG, Siwanowicz H, Puskas D (1989) A re-evaluation of experimental evidence for the hormonal control of avian nasal salt glands. In: Hughes MR, Chadwick A (eds) Progress in avian osmoregulation. Leeds Philosophical and Literary Soc, Leeds, pp 127–141
Butler DG, Youson JH, Campolin E (1991) Configuration of the medial and lateral segments of duck (*Anas platyrhynchos*) salt glands. J Morphol 207: 201–210
Catt KJ, Hunyady L, Balla T (1991) Second messengers derived from inositol lipids. J Bioenerg Biomembr 23: 7–27
Challiss RAJ, Chilvers ER, Willcocks AL, Nahorski SR (1990) Heterogeneity of [^{3}H]inositol 1,4,5-trisphosphate binding sites in adrenal-cortical membranes. Characterization and validation of a radioreceptor assay. Biochem J 265: 421–427
Cheeseman J, Cheeseman P, Phillips JG (1975) Effect of substance P on nasal gland secretion in the domestic duck (*Anas platyrhynchos*). Gen Comp Endocrinol 27: 527–530
Crawford KM, Stuenkel EL, Ernst SA (1991) Agonist-induced frequency modulation of Ca^{2+} oscillations in salt gland secretory cells. Am J Physiol Cell Physiol 261: C177–C184

Crooke ST, Bennett CF (1989) Mammalian phosphoinositide-specific phospholipase C isoenzymes. Cell Calcium 10: 309–323

Daniel JL, Dangelmaier CA, Smith JB (1988) Calcium modulates the generation of inositol 1,3,4-trisphosphate in human platelets by the activation of inositol 1,4,5-trisphosphate 3-kinase. Biochem J 253: 789–794

Downes CP, Macphee CH (1990) *myo*-Inositol metabolites as cellular signals. Eur J Biochem 193: 1–18

Ensor DM, Thomas DH, Phillips JG (1970) The possible role of the thyroid in extrarenal secretion following a hypertonic saline load in the duck (*Anas platyrhynchos*). J Endo 46: X

Ernst SA, Ellis RA (1969) The development of surface specialization in the secretory epithelium of the avian salt gland in response to osmotic stress. J Cell Biol 40: 305–321

Ernst SA, Mills JW (1977) Basolateral plasma membrane localization of ouabain-sensitive transport sites in the secretory epithelium of the avian salt gland. J Cell Biol 75: 74–94

Ernst SA, Goertemiller CC, Ellis RA (1967) The effect of salt regimens on the development of (Na^+–K^+)-dependent ATPase activity during the growth of salt glands of ducklings. Biochim Biophys Acta 135: 682–692

Ferris CD, Huganir RL, Supattapone S, Snyder SH (1989) Purified inositol 1,4,5-trisphosphate receptor mediates calcium flux in reconstituted lipid vesicles. Nature 342: 87–89

Fisher SK, Hootman SR, Heacock AM, Ernst SA, Agranoff BW (1983) Muscarinic stimulation of phospholipid turnover in dissociated avian salt gland cells. FEBS Lett 155: 43–46

Fletcher GL, Stainer IM, Holmes WN (1967) Sequential changes in the adenosinetriphosphatase activity and the electrolyte capacity of the nasal glands of the duck (*Anas platyrhynchos*) during the period of adaptation to hypertonic saline. J Exp Biol 47: 375–391

Frizzell RA, fleld M, Schultz SG (1979) Sodium-coupled chloride transport by epithelial tissues. Am J Physiol 236: F1–F8

Furuichi T, Yoshikawa S, Miyawaki A, Wada K, Maeda N, Mikoshiba K (1989) Primary structure and functional expression of the inositol 1,4,5-trisphosphate-binding protein $P_{400.}$ Nature 342: 32–38

Gawler DJ, Potter BVL, Nahorski SR (1990) Inositol 1,3,5-tetrakisphosphate-induced release of intracellular Ca^{2+} in SH-SY5Y neuroblastoma cells. Biochem J 272: 519–524

Gerstberger R (1988) Functional vasoactive intestinal polypeptide (VIP)-system in salt glands of the Pekin duck. Cell Tissue Res 252: 39–48

Gerstberger R, Gray DA, Simon E (1984) Circulatory and osmoregulatory effects of angiotensin II perfusion of the third ventricle in a bird with salt glands. J Physiol 349: 167–182

Gerstberger R, Sann H, Simon E (1988) Vasoactive intestinal peptide stimulates bloodflow and secretion of avian salt glands. Am J Physiol 255: R575–R582

Gray PTA (1988) Oscillations of free cytosolic calcium evoked by cholinergic and catecholaminergic agonists in rat parotid acinar cells. J Physiol (Lond) 406: 35–53

Hammer R, Berrie CP, Birdsall NJM, Burgen ASA, Hulme EC (1980) Pirenzipine distinguishes between different subclasses of muscarinic receptors. Nature 283: 90–92

Hanwell A, Peaker M (1975) The control of adaptive hypertrophy in the salt glands of geese and ducks. J Physiol 248: 193–205

Harootunian AT, Kao JPY, Paranjape S, Adams SR, Potter BVL, Tsien RY (1991) Cytosolic Ca^{2+} oscillations in REF52 fibroblasts: Ca^{2+}-stimulated IP_3 production or voltage-dependent Ca^{2+} channels as key positive feedback elements. Cell Calcium 12: 153–164

Hildebrandt J-P, Shuttleworth TJ (1991) Inositol phosphates and $[Ca^{2+}]_i$ signals in a differentiating exocrine cell. Am J Physiol 261: C210–C217

Hildebrandt J-P, Shuttleworth TJ (1992) Calcium-sensitivity of inositol 1,4,5-trisphosphate metabolism in exocrine cells from the avian salt gland. Biochem J 282: 703–710

Hildebrandt J-P, Shuttleworth TJ (1993) A G_q-type G protein couples muscarinic receptors to inositol phosphate and calcium signaling in exocrine cells from the avian salt gland. J Memb Biol 133: 183–190

Hildebrandt J-P, Shuttleworth TJ (1994) Muscarinic receptor characterization in differentiating avian exocrine cells. Am J Physiol 266: R674–R681

Hill TD, Boynton AL (1990) Inositol tetrakisphosphate-induced sequestration of Ca^{2+} replenishes an intracellular pool sensitive to inositol trisphosphate. J Cell Physiol 142: 163–169
Hokin LE (1987) The road to the phosphoinositide-generated second messengers. TIPS 8: 53–56
Holmes WN, Lockwood LN, Bradley EL (1972) Adenohypophyseal control of extrarenal excretion in the duck (*Anas platyrhynchos*). Gen Comp Endocrinol 18: 59–68
Holmes WN, Stewart DJ (1968) Changes in the nucleic acid and protein composition of the nasal glands from the duck (*Anas platyrhynchos*) during the period of adaptation to hypertonic saline. J Exp Biol 48: 509–519
Hootman SR, Ernst SA (1980) Dissociation of avian salt gland: separation procedures and characterization of dissociated cells. Am J Physiol 238: C184–C195
Hootman SR, Ernst SA (1981) Effect of methacholine on Na^+ pump activity and ion content of dispersed avian salt gland cells. Am J Physiol 241: R77–R86
Hootman SR, Ernst SA (1982) [^{3}H]QNB binding to muscarinic receptors in intact avian salt gland cells. Am J Physiol 243: C254–C261
Hossler FE (1982) On the mechanism of plasma membrane turnover in the salt gland of ducklings. Cell Tissue Res 226: 531–540
Hossler FE, Sarras MP, Barrnett RJ (1978) Ouabain binding during plasma membrane biogenesis in duck salt gland. J Cell Sci 31: 179–197
Irvine RF (1990) "Quantal" Ca^{2+} release and the control of Ca^{2+} entry by inositol phosphates—a possible mechanism. FEBS Lett 263: 5–9
Jacob R (1990) Calcium oscillations in electrically non-excitable cells. Biochim Biophys Acta Mol Cell Res 1052: 427–438
Joseph SK, Hansen CA, Williamson JR (1989) Inositol tetrakisphosphate mobilizes calcium from cerebellum microsomes, Mol Pharmacol 36: 391–397
Knight CH, Peaker M (1979) Adaptive hyperplasia and compensatory growth in the salt glands of ducks and geese. J Physiol 294: 145–151
Kubo T, Fukuda K, Mikami A, Maeda A, Takahashi H, Mishina M, Haga T, Haga K, Ichiyama A, Kangawa K, Kojima M, Matsuo H, Hirose T, Numa S (1986) Cloning, sequencing and expression of complimentary DNA encoding the muscarinic acetylcholine receptor. Nature 323: 411–416
Li G, Comte M, Wollheim CB, Cox JA (1989) Mode of activation of bovine brain inositol 1,4, 5-trisphosphate 3-kinase by calmodulin and calcium. Biochem J 260: 771–775
Lingham RB, Stewart DJ, Sen AK (1980) Induction of (Na^++K^+) ATPase in the salt gland of the duck. Biochim Biophys Acta 601: 229–234
Lowy RJ, Ernst SA (1987) β-Adrenergic stimulation of ion transport in primary cultures of avian salt glands. Am J Physiol 252: C670–C676
Lowy RJ, Dawson DC, Ernst SA (1985) Primary culture of duck salt gland II. Neurohormonal stimulation of active transport. Am J Physiol 249: C41–C47
Lowy RJ, Schreiber JH, Ernst SA (1987) Vasoactive intestinal peptide stimulates ion transport in avian salt gland. Am J Physiol 253: R801–R808
Lowy RJ, Dawson DC, Ernst SA (1989) Mechanism of ion transport by avian salt gland primary cell cultures. Am J Physiol 256: R1184–R1191
Majerus PW, Ross TS, Cunningham TW, Caldwell KK, Jefferson AB, Bansal VS (1990) Recent insights in phosphatidylinositol signaling. Cell 63: 459–465
Marshall AT, Hyatt AD, Phillips JG, Condron RJ (1985) Isoosmotic secretion in the avian nasal salt gland: X-ray microanalysis of luminal and intracellular ion distributions. J Comp Physiol B 156: 213–227
Marshall AT, King P, Condron RJ, Phillips JG (1987) The duct system of the avian salt gland as a transporting epithelium: structure and morphometry in the duck *Anas platyrhynchos*. Cell Tissue Res 248: 179–188
Martin SC, Shuttleworth TJ (1994a) Ca^{2+} influx drives agonist-activated $[Ca^{2+}]_i$ oscillations in an exocrine cell. FEBS Lett 352: 32–36
Martin SC, Shuttleworth TJ (1994b) Vasoactive intestinal peptide stimulates a cAMP-mediated Cl^- current in avian salt gland cells. Regul Peptides 52: 205–214

Martin SC, Shuttleworth TJ (1994c) Muscarinic-receptor activation stimulates oscillations in K^+ and Cl^- currents which are acutely dependent on extracellular Ca^{2+} in avian salt gland cells. Pflügers Archiv Eur J Physiol 426: 231–238
Martin SC, Thompson J, Shuttleworth TJ (1994) Potentiation of Ca^{2+}–activated secretory activity by a cAMP-mediated mechanism in avian salt gland cells. Am J Physiol 267: C255–C265
Merchant JL, Papermaster DS, Barrnett RJ (1985) Correlation of Na^+, K^+-ATPase content and plasma membrane surface area in adapted and de-adapted salt glands of ducklings. J Cell Sci 78: 233–246
Michell RH (1975) Inositol phospholipids and cell surface receptor function. Biochim Biophys Acta 415: 81–147
Michell RH, Connroy LA, Flnney M, French PJ, Brown G, Creba JA, Bunce CM, Lord JM (1990) Inositol lipids and phosphates in the regulation of the growth and differentiation of haemopoietic and other cells. Philos Trans R Soc Lond B 327: 193–207
Mignery GA, Newton CL, Archer BT III, Südhof TC (1990) Structure and expression of the rat inositol 1,4,5-trisphosphate receptor. J Biol Chem 265: 12679–12685
Neher E, Marty A, Fukuda K, Kubo T, Numa S (1988) Intracellular calcium release mediated by two muscarinic receptor subtypes. FEBS Lett 240: 88–94
Nishizuka Y (1988) The molecular heterogeneity of protein kinase C and its implications for cellular regulation. Nature 334: 661–665
Nogimori K, Menniti FS, Putney JW Jr (1990) Identification in extracts from AR4-2J cells of inositol 1,4,5-trisphosphate by its susceptibility to inositol 1, 4,5-trisphosphate 3-kinase and 5-phosphatase. Biochem J 269: 195–200
Pang I-H, Sternweis PC (1990) Purification of unique α subunits of GTP-binding regulatory proteins (G proteins) by affinity chromatography with immobilized $\beta\gamma$ subunits. J Biol Chem 265: 18707–18712
Parker I, Ivorra I (1991) Inositol tetrakisphosphate liberates stored Ca^{2+} in *Xenopus* oocytes and facilitates responses to inositol trisphosphate. J Physiol (Lond) 433: 207–227
Peaker M (1971) Avian salt glands. Philos Trains R Soc Lond B 262: 289–300
Peaker M, Linzell JL (1975) Salt glands in birds and reptiles. Cambridge Univ Press, London
Peralta EG, Ashkenazi A, Winslow JW, Smith DH, Ramachandran J, Capon DJ (1987) Distinct primary structures, ligand-binding properties and tissue-specific expression of four human muscarinic acetylcholine receptors. EMBO J 6: 3923–3929
Peralta EG, Ashkenazi A, Winslow JW, Ramachandran J, Capon DJ (1988) Differential regulation of PI hydrolysis and adenylyl cyclase by muscarinic receptor subtypes. Nature 334: 434–437
Petersen OH (1989) Does inositol tetrakisphosphate play a role in the receptor-mediated control of calcium mobilization? Cell Calcium 10: 375–383
Phillips JG, Ensor DM (1972) The significance of environmental factors in the hormone mediated changes in nasal (salt) gland activity in birds. Gen Comp Endocrinol (Suppl)3: 393–404
Phillips JG, Holmes WN, Butler DG (1961) The effect of total and subtotal adrenalectomy on the renal and extrarenal responses of the domestic duck (*Anas platyrhynchos*) to saline loading. Endocrinology 69: 958–969
Pittart JB, Hally AD (1973) The effect of denervation on genotypic and compensatory growth of the immature avian salt gland. J Anat 114: 303
Putney JW Jr (1990) Capacitative calcium entry revisited. Cell Calcium 11: 611–624
Putney JW Jr Takemura H, Hughes AR, Horstman DA, Thastrup O (1989) How do inositol phosphates regulate calcium signaling. FASEB J 3: 1899–1905
Rana RS, Hokin LE (1990) Role of phosphoinositides in transmembrane signaling. Physiol Rev 70: 115–164
Rhee SG, Suh P-G, Ryu S-H, Lee SY (1989) Studies of inositol phospholipid-specific phospholipase C. Science 244: 546–550
Richards NW, Lowy RJ, Ernst SA, Dawson DC (1989) Two K^+ channel types, muscarinic agonist-activated and inwardly rectifying, in a Cl^- secretory epithelium: the avian salt gland. J Gen Physiol 93: 1171–1194

Schütz H, Gerstberger R (1990) Atrial natriuretic factor controls salt gland secretion in the Pekin duck (*Anas platyrhynchos*) through interaction with high affinity receptors. Endocrinology 127: 1718–1726

Shuttleworth TJ (1987) Salt gland function in osmoregulation in terrestrial and aquatic environments. In: Dejours P, Bolis L, Taylor CR, Weibel ER (eds) Comparative physiology: life in water and on land. Fidia Research Series IX, Livinia Press, Padova, pp 537–548

Shuttleworth TJ (1990a) Receptor-activated calcium entry in exocrine cells does not occur via agoinst-sensitive intracellular pools. Biochem J 266: 719–726

Shuttleworth TJ (1990b) Fluoroaluminate activation of different components of the calcium signal in an exocrine cell. Biochem J 269: 417–422

Shuttleworth TJ (1992) Ca^{2+} release from inositol trisphosphate-sensitive stores is not modulated by intraluminal [Ca^{2+}]. J Biol Chem 267: 3573–3576

Shuttleworth TJ, Thompson JL (1987) Secretory activity in salt glands of birds and turtles: stimulation via cyclic AMP. Am J Physiol 252: R428–R432

Shuttleworth TJ, Thompson JL (1989) Intracellular [Ca^{2+}] and inositol phosphates in avian nasal gland cells. Am J Physiol 257: C1020–C1029

Shuttleworth TJ, Wood CM (1992) Changes in pH_i associated with activation of ion secretion in avian nasal salt gland cells. Am J Physiol 262: C271–C228

Snider RM, Roland RM, Lowy RJ, Agranoff BW, Ernst SA (1986) Muscarinic receptor-stimulated Ca^{2+} signaling and inositol lipid metabolism in avian salt gland cells. Biochim Biophys Acta 889: 216–224

Stewart DJ, Sen AK (1981) Role of cyclic GMP in cholinergic activation of Na-K pump in duck salt gland. Am J Physiol 240: C207–C214

Stewart DJ, Semple EW, Stewart GT, Sen AK (1976) Induction of the catalytic protein of (Na^++K^+)-ATPase in the salt gland of the duck. Biochim Biophys Acta 419: 150–163

Stewart DJ, Sax J, Funk R, Sen AK (1979) Possible role of cyclic GMP in stimulus-secretion coupling by salt gland of the duck. Am J Physiol 237: C200–C204

Strathmann M, Simon MI (1990) G protein diversity: a distinct class of α subunits is present in vertebrates and invertebrates. Proc Natl Acad Sci USA 87: 9113–9117

Stuenkel EL, Ernst SA (1990) Multiple calcium mobilization pathways in single avian salt gland cells. Am J Physiol 258: C289–C298

Takazawa K, Lemos M, Delvaux A, Lejeune C, Dumont JE, Erneux C (1990) Rat brain inositol 1,4,5-trisphosphate 3-kinase, Ca^{2+}-sensitivity, purification and antibody production. Biochem J 268: 213–217

Takemura H, Putney JW Jr (1989) Capacitative calcium entry in parotid acinar cells. Biochem J 258: 409–412

Torchia J, Qu Y, Francis J, Pon DJ, Sen AK (1991) Carbachol-stimulated phosphorylation of a 170-kDa endogenous protein in avian salt gland cells. Am J Physiol Cell Physiol 261: C543–C549

Van Calker D, Müller M, Hamprecht B (1979) Adenosine regulates via two different types of receptors, the accumulation of cyclic AMP in cultured brain cells. J Neurochem 33: 999–1005

Wilson JX (1987a) Adrenergic and osmotic responses to atrial natriuretic peptide and angiotensin II in the duck (*Anas platyrhynchos*). Can J Physiol Pharmacol 65: 1995–1999

Wilson JX (1987b) Reevaluation of the effect of substance P on nasal salt gland secretion in the duck (*Anas platyrhynchos*). Gen Comp Endocrinol 67: 256–262

Woods NM, Cuthbertson KSR, Cobbold PH (1987) Agonist-induced oscillations in cytoplasmic free calcium concentration in single rat hepatocytes. Cell Calcium 8: 79–100

Hypoxia and Metabolic Adaptation

Chapter 9

Interrelationships Between Hypoxia and Thermoregulation in Vertebrates[1]

S.C. Wood

Contents

1 Introduction

All vertebrates have a region of the brain used for thermoregulation. The sensory and integrative circuits are similar for both ectothermic and endothermic vertebrates (Crawshaw and Hammel 1974). Most or all thermoregulatory functions can be interpreted by single or dual set point models (Hammel et al.

[1] This chapter is dedicated to Randall Keith Dupré, whose suicide at age 33 years in September 1991 stunned and grieved all who knew him. Keith was a terrific friend and colleague who contributed greatly to the area of thermoregulatory physiology and to many of the concepts cited herein.

Department of Physiology, East Carolina University School of Medicine, Greenville, NC 27858, USA

Advances in Comparative and Environmental Physiology, Vol. 22

1973; Barber and Crawford 1977). The outputs of the thermoregulatory center, behavioral and physiological, provide feedback control of core body temperature (T_b). Perhaps more surprising is the fact that animals without a brain also show temperature preferences and regulation. The protozoan *Paramecium caudatum* is a good example (see below).

Ectotherms must use behavioral mechanisms to control their T_b. Some physiological mechanisms are available to ectotherms, but they are limited to color changes and autonomic control of cutaneous blood flow. Endotherms also use behavior for T_b control, but need physiological mechanisms unavailable to ectotherms (shivering and nonshivering thermogenesis, sweating, panting, etc.) when ambient temperatures are outside of the thermoneutral zone the range of environmental temperatures within which metabolic rate is minimal and independent of environmental temperatures. The distinction between ectotherms and endotherms is often blurrred by examples of facultative endothermy of ectotherms (e.g. brooding pythons, honey bee colonies, warm-muscled tunas, etc.) and facultative ectothermy of endotherms (e.g. torpor, hypoxia-induced hypothermia, and hypoglycemia-induced hypothermia in mammals).

Most ectotherms display a T_b preference when they are in a heterothermal environment. A variety of factors can modify the T_b selected by ectotherms (Huey 1982). Selection of a higher-than-normal T_b in response to pyrogen injection was first demonstrated in lizards (Kluger et al. 1975). This proved to be an adaptive response to induced infection, i.e. it enhanced survival. Behavioral fever was subsequently described for amphibians and fishes (reviewed by Reynolds and Casterlin 1982). It is now known that fever enhances survival via increased production of T cells and increased effectiveness of interferon, an antiviral agent. On the other hand, a number of factors, including hypoxia, elicit the occurrence or selection of lower-than-normal T_b.

2 Hypoxia and Thermoregulation

Vertebrate animals attempt to cope with oxygen deprivation in a variety of ways. Physiological responses are well known, with the most potent ones being increased ventilation and cardiac output. These responses increase tissue O_2 supply. Behavioral responses fall into two categories: exploration to find higher levels of ambient O_2 and exploration to find lower ambient temperatures. The latter, behavioral hypothermia, decreases tissue O_2 demand. Altered thermoregulatory bahavior is involved in hypoxia-induced hypothermia in poikilothermic, ectothermic, and homeothermic vertebrates. This section discusses the interactions between hypoxia and thermoregulation of vertebrates. Similar patterns of hypoxia-induced hypothermia have been reported for invertebrates (Wood et al. 1985). These are not discussed in this chapter, but results for the protozoan *Paramecium caudatum* are presented. Later, some related stresses that induce hypothermia are also described.

Body temperature (T_b) has a marked effect on oxygen uptake (VO_2) of resting animals (Krogh 1914). For most animals the temperature coefficient (Q_{10}) is ~ 2.5, so resting VO_2 changes about 11% per °C change in T_b. Consequently, hyperthermia is deleterious for hypoxic animals. Conversely, hypothermia could be beneficial, particularly for O_2 sensitive organs, e.g. heart and brain. Indeed, most studies shown that hypothermia is a normal and adaptive response to hypoxia in both ectotherms and endotherms.

2.1 Physiological Hypothermia

It is well established that heat loss occurs in hypoxic mammals from peripheral vasodilation (Lintzel 1931; Gellhorn and Janus 1936; Adolph 1951; Miller and Miller 1966). The recent finding of behavioral hypothermia in hypoxic mammals (see below) is significant, because selection of a lower ambient temperature will accelerate the rate of heat loss (Fig. 1). There is also decreased heat production in hypoxic mammals. This is due to redistribution of blood away from brown fat (Szelenyi and Donhoffer 1968; Schubring 1986).

Within a range of ambient temperatures (thermoneutral zone) homeotherms maintain T_b without supplementary heat production or sweat secretion. At ambeint temperatures below thermoneutrality T_b of homeotherms is maintained by thermogenic responses. Therefore, it is possible that hypothermia could interact with hypoxia by eliciting a thermogenic response. This would be deleterious by increasing O_2 demand of tissues involved in shivering and nonshivering thermogenesis. However, in hypoxic rats the normal thermogenic response to reduced temperature is depressed (Dupré et al. 1988; Gautier et al. 1991). A reduced thermogenic response to cold also occurs in hypoxic and hypercapnic hamsters (Kuhnen et al. 1987). Other evidence that the drop in T_b results from a change in central regulation includes a decrease impulse frequency of cold receptors during hypoxia (Dodt 1956) and a change in the set point of preoptic neurons during hypoxia (Tamaki and Nakayama, 1987). Thus, hypothermia during hypoxia can be interpreted using the set point model (Hammel et al. 1973) as a widening of the thermoneutral zone and downward shift of the threshold temperature for shivering thermogenesis.

In ectotherms, the potential offsets of shivering and nonshivering thermogenesis do not occur during behavioral hypothermia. As with mammals, physiological and behavioral mechanisms act in concert to lower T_b. However, in ectotherms hypothermia is primarily behavioral but the rate of drop in T_b may be augmented by control of cutaneous blood flow. Under normoxic conditions lizards cool more slowly than they heat, indicating decreased peripheral blood flow. This increases the rate of cooling during hypoxia, shortening the time required for T_b to drop after hypoxic animals select a lower ambient temperature. Hypoxia abolishes this difference in the iguana (Hicks and Wood 1985).

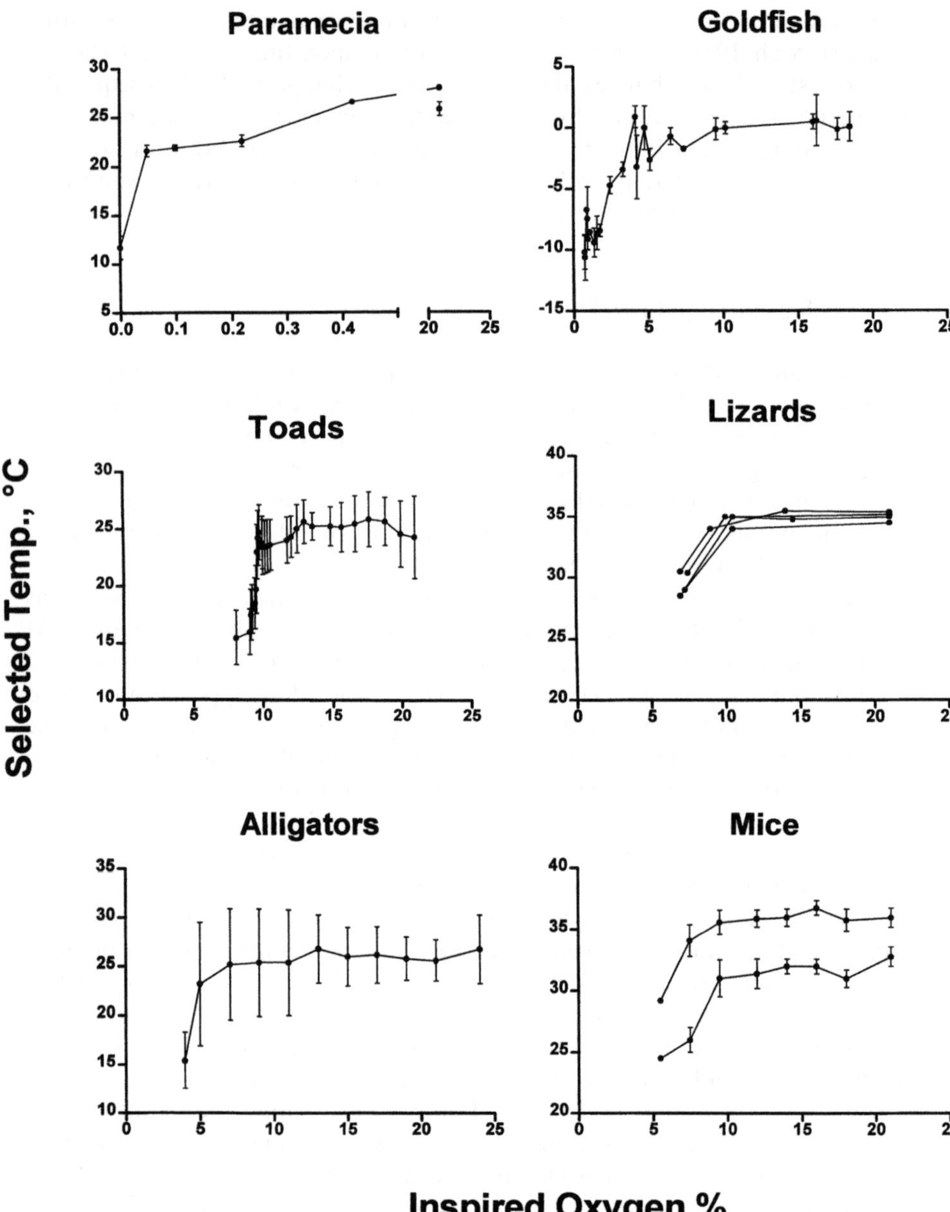

Fig. 1. Effect of ambient PO_2 on temperature selection of various species during exposure to graded hypoxia in a thermal gradient. *Paramecium caudatum* data show recovery of normal selected temperture (+) (Malvin and Wood 199b), goldfish (Rausch and Crawshaw 1990), toads (*Bufo marinus;* Wood and Malvin 1991), lizards (four species; Hicks and Wood 1989), alligators (*Alligator mississippiensis;* Fincher et al. 1992), and mice showing both selected temperature (T_S) and body temperature (Gordon, pers. comm.)

2.2 Behavioral Hypothermia

Most vertebrates, including mammals and birds, utilize behavior to regulate T_b in heterothermal environments. In nature or laboratory thermal gradients, most species show definite thermal preferences. Some so-called cold-blooded animals actually prefer a T_b of 35°C or higher (e.g. lizards). As shown in Fig.1, when exposed to graded hypoxia all types of vertebrates (as well as a protozoan) show a similar pattern of behavioral hypothermia where normal selected temperature is maintained as ambient O_2 is reduced until some threshold value is reached. At lower values of ambient O_2, the animals move to a colder region of the thermal gradient and maintain a lower body temperature. The threshold ranges from 0.05% in *Paramecium caudatum* to ~ 10% ambient O_2 in toads, lizards, and mice. In goldfish and alligators, two species very tolerant of hypoxia, the threshold is ~5% ambient O_2.

An interaction between hypoxia and thermoregulation of ectotherms was first predicted from a model of O_2 transport (Wood 1984). The premise of this model was that animals having venous admixture due to shunts will have a normal arterial PO_2 near the shoulder of the oxygen dissociation curve rather than on the upper flat portion as in normoix mammals. Lacking this normal reserve for hypoxia, ectotherms were predicted to select a lower T_b. By doing so, their arterial saturation should increase (left shift of the oxygen dissociation curve as the same time as their oxygen demand declines (Q_{10} effect). This hypothesis was confirmed in lizards (Hicks and Wood 1985) and other ectotherms including crayfish, amphibian larvae, and teleost fish (Dupré and Wood 1988; Bryan et al. 1984; Rausch and Crawshaw 1990; Wood et al. 1985; Wood 1991).

How widespread among animals is this behavioral response to hypoxia? The answer is not known yet because relatively few species of each vertebrate taxa and a few invertebrate species have been studied so far. A recent search for the simplest animal to show this response studied a protozoan (Malvin and Wood 1991b). *Paramecium caudatum* were placed in an aquatic thermal gradient, exposed first to air and then to nitrogen. The mean temperature selected during air exposure was 26 °C, as shown by Mendelssohn (1895). During extreme hypoxia (Po_2 ~ 0.3 mm Hg) the selected temperature was reduced to ~16 °C (Fig. 1). This hypothermia was completely reversed when the gradient was returned to normoxia. These results are consistent with membrane depolarization at a lower-than-normal temperature during hypoxia. One explanation for this phenomenon is based on the work of Hennessey et al. (1983), in which temperature selection results from the following processes:

1. Heat at the selected temperature is transduced by a mechanism that increases ion conductance of the body membrane presumably by opening membrane ion channels.
2. Ions flowing through channels cause depolarization if resting membrane voltage is less than –20 mV (Flores and Frieden 1968), and depolarization

triggers opening of calcium-dependent channel in ciliary membranes, and initiating an action potential.

3. The action potential(Ca influx) causes a change in the beat direction of the ciliary axoneme and backward swimming. Under severe hypoxia this sequence occurs at a lower temperature, because the resting membrane potential at a given temperature will be decreased. *Paramecium caudatum* may provide a useful model for studies of mechanisms and significance of thermoregulation at the single cell level.

Behavioral hypothermia also occurs in hypoxic mammals (Brauer et al. 1986; Dupré and Owen 1989; Gordon, pers comm). These findings further blur the distinction between ectotherms and homeotherms. Data for mice (Fig. 1) show a degree of behavioral hypothermia during hypoxia that closely parallels the pattern seen in ectotherms. The threshold for a behavioral response in both mice and lizards is at an ambient O_2 of ~10%.

2.3 Mechanisms and Mediators

2.3.1 Neural Mechanisms

Behavioral themoregulation of terrestrial ectotherms has been described using minimum and maximum temperatures based on the early work of Coweles and Bogert (1944). Experiments with shuttle boxes, in which animals are forced to choose between boxes either colder or hotter than their preferred T_b, revealed oscillations of T_b between reproducible limits (Heath 1970). These limits, or exit temperatures, were shown to be normally distributed, suggesting a dual threshold control system (Heath 1970; Barber and Crawford 1977).

Experimental evidence suggests that behavioral hypothermia is a regulated response to hypoxia (vs random movement to a cold area where animals become immobile). Firstly, the response is rapidly reversible. Secondly, the hypoxic ectotherms show a frequency distribution of selected temperature similar to normoxic animals, but shifted to lower temperatures. Thirdly, there is a resetting of exit temperatures in shuttle-box experiments (Hicks and Wood 1985). Behavioral hypothermia of the iguana lizard, shown in Fig. 2, illustrates the exit-temperature mechanism of regulated hypothermia. When iguanas were exposed to 10% O_2 after 24 h of normoxia, they quickly moved to the cold end of the thermal gradient. In a separate shuttle-box experiment the upper (hot) and lower (cold) exit temperatures were determined under normoxic and hypoxic conditions. These are drawn as horizontal lines in Fig. 2. Under hypoxic conditions the upper exit temperature is reduced from ~37 to 31 °C, below the normal preferred T_b. Therefore, the iguanas exit the warm end of the chamber as they would exit a hot shuttle box. The new selected T_b during hypoxia is ~22 °C above the lower exit temperature during hypoxia. When normoxia is restored in the chamber, the selected T_b during hypoxia is below the lower exit temperature.

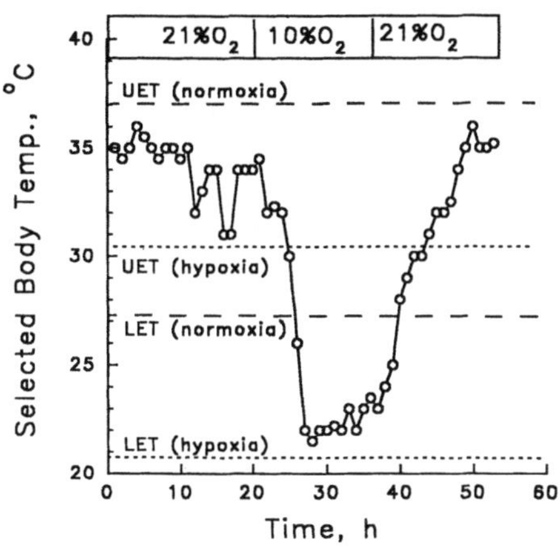

Fig. 2. Selected body temperature (*open circles*) of an iguana lizard in a 10–45 °C thermal gradient chamber during air exposure and exposure to 10% inspired O_2. *Dashed lines* show upper (*UET*) and lower (*LET*) exit temperatures determined in a separate shuttle-box experiment. Selected body temperature during both normoxia and hypoxia lies between the UET and LET. Hypoxic hypothermia is explained by the decrease in UET and LET. (Data from Hicks and Wood 1985)

Therefore, the animal exits the cold end of the chamber. Similar results were obtained with three other species of lizards (Hicks and Wood 1985).

2.3.2 Hormonal Mechanisms

Exogenous and endogenous stimuli that elicit hypothermia are diverse, e.g. ethanol, urine, morphine, prostaglandins, histamine, hypoxia, dry air, anemia, pesticides, food deprivation, and heavy metals (Gordon 1988). There are also diverse pyrogens that, without crossing the blood-brain barrier, alter central thermoregulation (Dascombe 1985). There are a host of potential mediators of hypothermia including endogenous opioid peptides (Kavaliers et al. 1984). For hypothermia induced by hypoxic hypoxia, anemic hypoxia, hypercapnia, and dry air, one mediator seems most likely, i.e. arginine vasopressin (AVP) [arginine vasotocin (AVT) in ectotherms]. Although the evidence is convincing for involvement of AVP (or AVT), this has yet to be demonstrated.

There are several lines of evidence favoring the AVP hypothesis:

1. Hypoxia stimulates the release of AVP in rats (Walker 1986), in fetal sheep (Stegner et al. 1984), and in human infants born after hypoxic distress (Ruth et al. 1988). Hypoxic hypoxia in dogs caused a 250% increase in blood flow to the neurohypophysis and a concurrent rise in plasma AVP (Wilson et al. 1987).
2. AVP is an endogenous antipyretic peptide in mammals. centrally administered AVP elicits hypothermia (Naylor et al. 1986). A vasotocin neurosecretory system has been identified in lizards (Bons 1983).
3. There is enhanced AVP release in response to hypercapnic acidosis in rats (Walker 1987). Hypercapnia induces behavioral hypothermia in toads (Riedel and Wood 1988).
4. Water deprivation is known to increase the release of AVP in response to increased extracellular tonicity and diminished plasma volume in mammals (Wade et al. 1983) and amphibians (Shoemaker et al. 1989). The toad *Bufo*

marinus selects significantly lower T_bs when exposed to dry air (Malvin and Wood 1991a).

5. AVT injections alter behavior in amphibians (reproductive behavior), and the altered behavior can be blocked by AVP antagonists (Moore and Miller 1983). Two approaches should be useful in testing the AVP/AVT hypothesis: (1) administering AVP and AVT antagonist to animals exposed to conditions known to elicit hypothermia, and (2) using the Brattleboro rat, a strain that is genetically AVP-deficient.

2.4 Physiological Significance

Known or putative effects of hypothermia that are pertinent to hypoxia are : (1) left shift of the oxyhemoglobin dissociation curve and improvement of O_2 loading in the lungs, (2) decreased VO_2 according to the Q_{10} effect, e.g., VO_2 would decrease ~11% per °C, and (3) energetically costly responses to hypoxia, e.g. increased cardiac output and ventilation, may be avoided. Known benefits of hypothermia in hypoxic animals encompass several categories:

2.4.1 Survival

Neonatal mammals (Miller and Miller 1966) and anemic rabbits (Gollan and Aono 1973) have increased survival rates when they are hypothermic, Asphyxiated neonated treated by hypothermia have a mortality rate of 11.5% compared with 48% in 288 cases not using hypothermia (Dunn and Miller 1969). Mice exposed to 5% O_2 had increased survival times when T_b was reduced from 37 to 35 °C, and decreased survival when T_b was elevated to 40 °C (Artru and Michenfielder 1981). Hypothermia also improves survival of rats following acute carbon monoxide poisoning (Sutariya et al. 1989). There is 100% survival of lizards allowed to cool during hypoxia vs 100% mortality of animals prevented from cooling down (Hicks and Wood 1985). This benefit of behavioral hypothermia was also demonstrated in the protozoan, *Paramecium caudatum* (Malvin and Wood 199b). As shown in Fig. 1, this organism has an extremely low threshold (~ 0.05% ambient O_2) before a behavioral response is elicited. As shown in Fig. 3, the distribution of *Paramecium caudatum* in a thermal gradient shifts from a mean selected temperature of 26 °C during normoxia to a mean of 16 °C under severely hypoxic conditions. This reduction of selected temperature greatly enhances survival, increasing the survival time for 50% of the population from 110 to 370 h.

2.4.2 Control of Breathing

In reptiles and amphibians lowering T_b markedly reduces the ventilatory response to hypoxia (Glass et al. 1983; Kruhøffer et al. 1987). The arterial PO_2

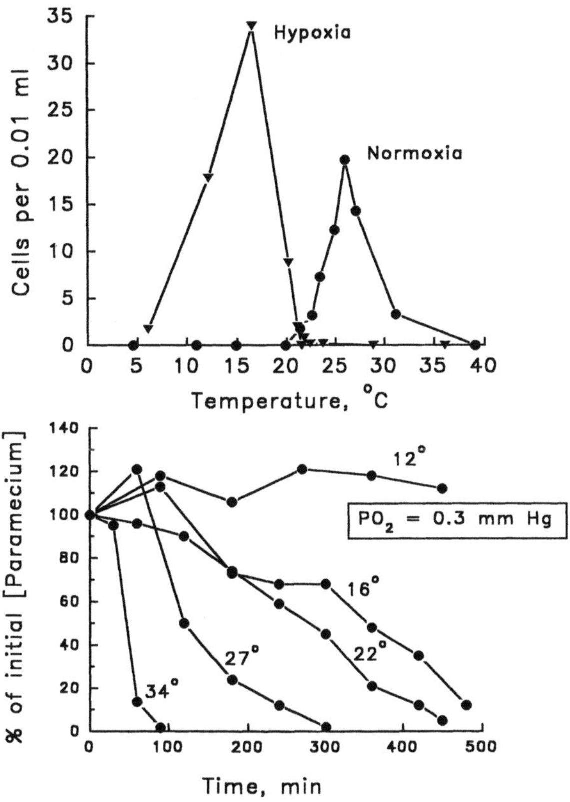

Fig. 3. *Upper panel* Distribution of selected temperature of the protozoan *Paramecium caudatum* in an aquatic thermal gradient under normoxic (ambient PO_2 = 123 mm Hg) and hypoxic (ambeint PO_2 = 0.3 mm Hg) conditions. *Lower panel* Survival curves of *Paramecium caudatum* exposed to ambeint PO_2 of 0.3 mm Hg at temperatures varying from 12 to 34 °C. Increasing numbers (above 100%) indicate population growth during the experiment. (Data from Malvin and Wood 1991b)

needed to elicit a ventilatory response of turtles becomes progressively less as T_b is lowered (Glass et al. 1983). Despite varying arterial PO_2 values at the hypoxic threshold, the arterial $[O_2]$ is relatively constant (Glass et al. 1983; Hicks and Wood 1985). This is due to the effect of temperature on the position of the oxygen dissociation curve of blood. As shown in Fig. 4 for *Bufo paracnemis* and *Bufo marinus,* the reduction of selected T_b during hypoxia reduces or eliminates the ventilatory response to hypoxia. At first glance this seems adaptive by reducing the work of breathing. However, the metabolic cost of breathing in amphibians and reptiles is several orders of magnitude less than in mammals (Perry 1989). As shown in Fig. 5, the metabolic rate of *Bufo marinus* is independent of inspired O_2 down to 4% O_2, despite the dramatic increase in breathing effort shown in Fig. 4.

2.4.3 O_2 Consumption and Transport

The effect of normoxic hypothermia is predictable for $\dot{V}O_2$ (Q_{10} effect). However, during hypoxia the influence of mental alertness, activity, and catecholamine

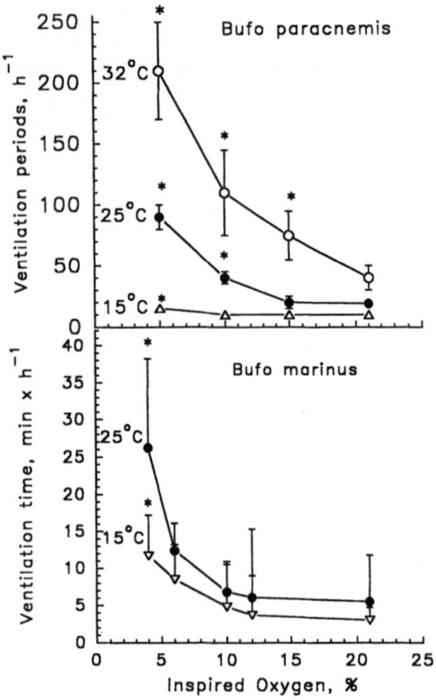

Fig. 4. Effect of body temperature on ventilatory response to hypoxia in the toads *Bufo paracnemius* and *Bufo marinus.* Ventilation periods are the number of breathing cycles per hour; ventilation time is minutes per hour spent breathing. (Data from Kruhøffer et al. 1987; Wood and Malvin 1991a)

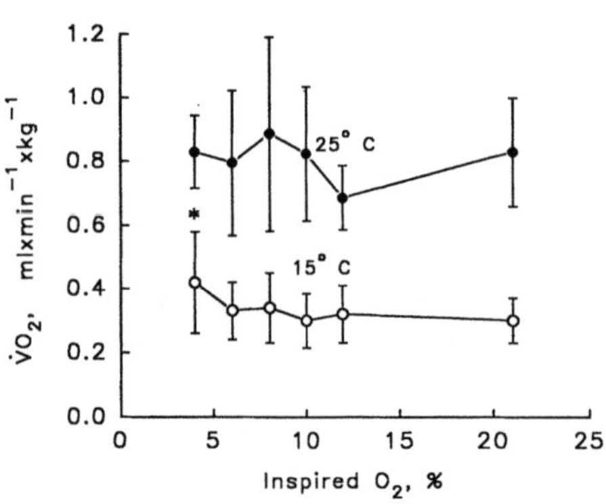

Fig. 5. Metabolic rate of the toad *Bufo marinus* during exposure to graded hypoxia (1h at each level) while kept at a body temperature of 15 of 25 °C. (Data from Wood and Malvin 1991a)

release may offset the Q_{10} effect. In the ectotherm *Bufo marinus,* Fig. 5 shows that $\dot{V}O_2$ seems to follow a predictable Q_{10} effect during graded hypoxia at 25 and 15 °C. At 15 °C there was a significant increase in $\dot{V}O_2$ with 4% inspired O_2, which was related to increased animal activity. A similar trend at 25 °C was not significant. Similar results were obtained for *Bufo marinus* by Pörtner et al. (1991), who suggested that the increase in $\dot{V}O_2$, instead of the expected decrease at the critical PO_2, may reflect a stress response mediated by catecholamine release (Boutilier and Lantz 1989). For the mammal *Rattus norvegicus*, Fig. 6 shows that the hypoxia-induced hypothermia also reported for mice (Fig. 1), results in a considerable reduction in oxygen uptake at ambient temperature below the thermoneutral zone, i.e. a reduced thermogenic response in hypoxia compared with normoxia. As shown in the upper panel of Fig. 6, hypoxia does not abolish thermoregulation in rats; if so, the rat's body temperature would lie in on the isothermal line. However, the rat regulates at a lower-than-normal temperature at each ambient temperature. Although the metabolic rats. More importantly, hypoxia is known to elicit a reduction in brain temperature, and, presumably, brain metabolic rate (see below).

It is also not possible to predict the effect of hypothermia on systemic O_2 transport (SOT), the product of cardiac output (CO) x arterial O_2 content (C_aO_2). Cardiac output may or may not decrease during hypothermia depending on sympathetic reflexes or possible alterations in cardiac contractility. C_aO_2 may or may not change depending on degree of hypoxia and left shift of the oxyhemoglobin dissociation curve (Lister 1984) and whether hemoconcentration occurs (see below). Both cardiac output and C_aO_2 hinge on acid–base status, which may change due to metabolic, respiratory, or temperature effects.

Under normal conditions $\dot{V}O_2$ is a measure of metabolic demand and is not limited by SOT. However, if SOT decreases, $\dot{V}O_2$ may fall below resting levels and become transport-limited. This has been demonstrated in numerous studies of newborn and adult mammals (Lister 1984) and recently confirmed in hypoxic infants and children (Berman et al. 1987). Newborn mammals are more susceptible to transport-limited $\dot{V}O_2$, because of limited cardiac reserve and limited ability to increase O_2 extraction, due to high hemoglobin–O_2 affinity (Lister et al. 1984).

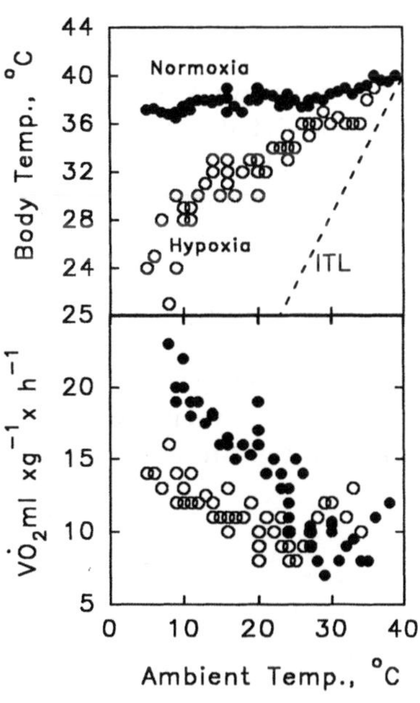

Fig. 6. Effect of ambient temperature on body temperature (*upper panel*) and metabolic rate (*lower panel*) of rats under normoxic and hypoxic (10% inspired O_2) conditions. *ITL* Isothermal line.

From the relationships that SOT = CO × C_aO_2 and $\dot{V}O_2$ = CO × (C_aO_2 – $C_{\bar{v}}O_2$ an equation may be derived showing that $\dot{V}O_2$ = SOT × EO_2, where EO_2 is O_2 extraction ($C_aO_2 - C_{\bar{v}}O_2$)/ C_aO_2. There are three possible outcomes of the measurements of $\dot{V}O_2$ and SOT in animals during hypothermia and hypoxia. Firstly, if hypothermia reduces $\dot{V}O_2$ more than SOT, EO_2 will decrease. This could be beneficial by increasing venous and tissue PO_2. Secondly, SOT and $\dot{V}O_2$ could decrease by the same amount (same Q_{10}) with no net effect on EO_2. Thirdly, SOT might decrease more than $\dot{V}O_2$ with a potentially deleterious increase in EO_2. Experimental data are needed for any species studied in order to determine what happens to these variables during hypoxia-induced hypothermia.

Some aspects of the physiological significance of hypoxia-induced hypothermia have been examined for the toad *Bufo marinus,* exposed to graded hypoxia at 25 °C (its normal selected temperature) and 15 °C (its selected temperature at inspired [O_2] below 10%. As shown in Fig. 7, the selection of a 15 °C body temperature by hypoxic toads has several important advantages, e.g. the ventilatory response is greatly reduced and the animals need not resort to anaerobic metabolism and produce lactic acid. Seen in the top panel of Fig. 7, there is a pronounced hemoconcentration when toads are breathing 4% inspired O_2 at 25 °C, but not at 15 °C. The hemoconcentration is rapidly reversible upon return to normoxic conditions, suggesting shifts of plasma volume and not splenic involvement. Previous studies of exercising toads have also reported hemoconcentration associated with plasma loss (Boutilier et al. 1986). Hemoconcentration has also been reported during hypoxia in splenectomized toads (Malvin et al. 1992). The mechanism of the hemoconcentration is not known with certainty. The onset of hemoconcentration at 25 °C is correlated with appearance of lactate in the plasma (Fig. 7), and an osmotic flux mechanism has been suggested (Boutilier et al. 1986). However, the study of Malvin et al. (1992) on *Bufo woodhouseii* and recent data for *Bufo marinus* (Wood, unpubl.) show that the hemoconcentration in hypoxic toads is correlated very closely with frequency of lymph heart contractions. There is a normal loss of plasma to the lymph system (Middler et al. 1968). Under hypoxic conditions at about 4% inspired O_2, lymph hearts cease to function, and the fluid is not returned to plasma resulting in the observed, rapidly reversible, hemoconcentration.

One of the most beneficial effects of hypoxic toads selecting a lower body temperature is a marked increase in arterial saturation. As shown in Fig. 8, arterial saturation at 15 °C is higher than at 25 °C at each level of graded hypoxia. The mechanism of this is not clear, but the left shift of the oxygen dissociation curve would clearly contribute, as could a reduction of right-to-left shunt (see Hicks and Malvin 1995).

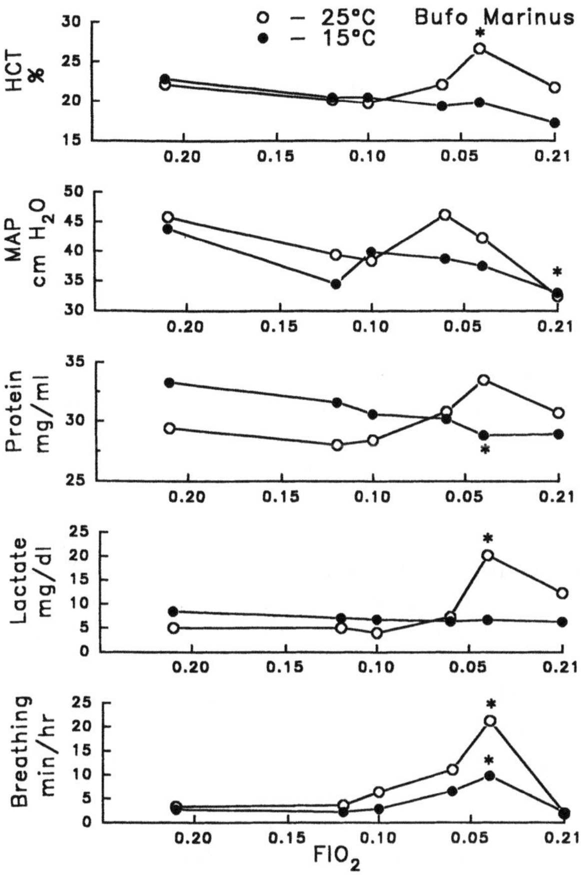

Fig. 7. Effect of graded hypoxia on *Bufo marinus* at 25 °C (normal preferred temperature) and 15°C (preferred temperature when hypoxic). (Data from Wood and Malvin 1991a).

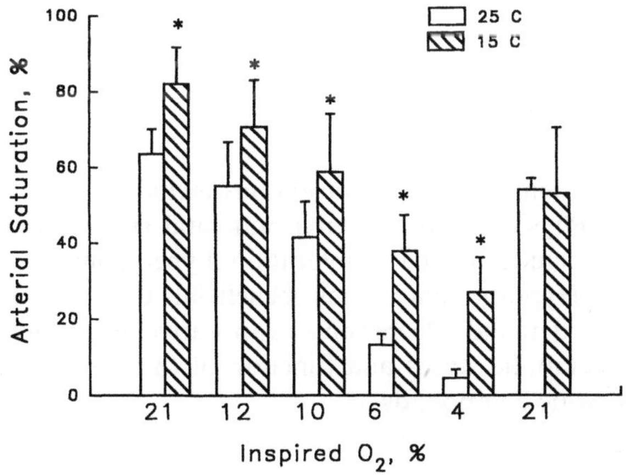

Fig. 8. Arterial saturation of unanesthetized toads (*Bufo marinus*) during graded hypoxia at 25 and 15 °C. (Data from Wood and Malvin 1991a)

2.4.4 Brain Metabolism and Function

Hypoxia presents two problems for neuronal function: acidosis due to lactic acid and energy depletion due to O_2 lack. A growing body of literature suggests that hypothermia is very beneficial to hypoxic brains. Hypothermia (27 °C) has a protective effect on brain metabolism of hypoxic rats (Carlsson and Siesjo 1976). This was attributed to the left shift of the oxyhemoglobin dissociation curve and the reduction in brain metabolic rate. Similar results with hypoxic–hypothermic rats also showed that hypothermia offers better cerebral protection than a dose of phenobarbital producing the same reduction of cerebral metabolic rate (Hagerdal and Keykhah 1978).

The neuropathological consequences of hypoxia or ischemia depend on brain temperature, and hypothermia tends to limit the damage. Brain temperature of rats decreases spontaneously from 36 to 30–31 °C during cerebral ischemia (Busto et al. 1987). This cooling of the brain prevents cell damage, whereas animals whose brain temperature was maintained by heat lamps showed significant histopathology in the hippocampus following ischemia. Deep hypothermia dramatically increases tolerance of the brain to ischemia. White et al. (1973) showed that circulation to primate brain could be stopped for 60 min without adverse effects if the brain were cooled to 5–8 °C. Even mild hypothermia (1–2 °C drop in brain temperature) has been shown to favorably alter the outcome of brain ischemia (Hirsh and Müller 1962; Minamisawa et al. 1990). Ikonomidou et al. (1989) showed that hypothermia (core temperature of 34 °C) was as effective as pretreatment with the neuroprotective drug MK-801 in preventing hypoxic/ischemic damage to rat-pup brains.

The hippocampus is important for spatial learning (Olton et al. 1979) and is particularly sensitive to hypoxia (Misgeld and Frotscher 1982; Aitken and Schiff 1986). Therefore, a protective effect of hypothermia on brain function can be examined by determining spatial learning after hypoxia exposure in control (temperature-clamped) and hypothermic animals.

3 Other Stresses and Thermoregulation

Many factors affect the preferred temperature of ectotherms including season, gender, and diet (Huey 1982). Physiological and environmental stresses that induce behavioral hypothermia include anemia, dehydration, hypercapnia, and pesticides (references below). Hypothermia seems beneficial under these conditions by reducing the O_2 demand, decreasing evaporative water loss (amphibians), protecting the metabolic status and function of the brain, and decreasing the toxicity of xenobiotic compounds.

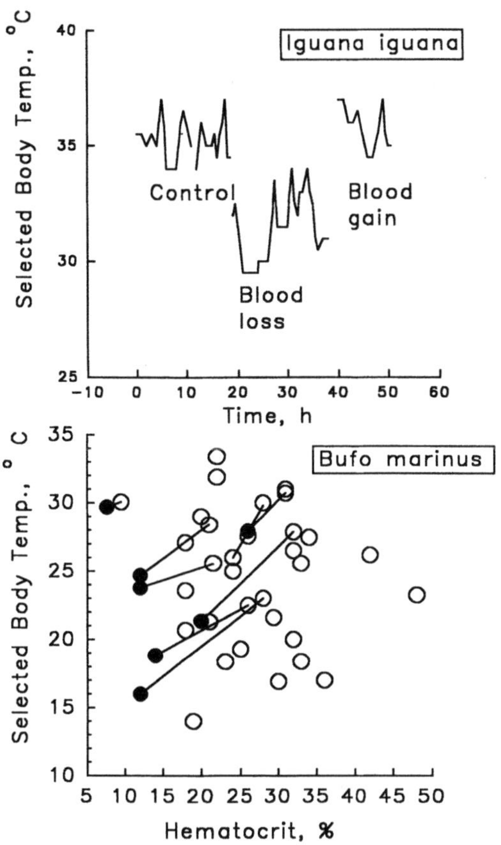

Fig. 9. *Upper panel:* Effect of blood withdrawal and reinfusion on selected body temperature of an iguana lizard in a 10–40 °C thermal gradient. (Data from Hicks and Wood 1985). *Lower panel:* Cross-sectional and longitudinal analysis of the influence of hematocrit on selected body temperature of toads (*Bufo marinus*) in a 10–40 °C thermal gradient. (Data from Wood 1990)

3.1 Anemia

Whatever an animal's preferred T_b is, it must have the O_2 transport capacity to satisfy the resting VO_2 for that temperature. This premise led to the hypothesis that ectothermic animals faced with a reduced O_2 transport capacity, e.g. hematocrit reduction, would select a lower than normal T_b. In the iguana (Fig. 9) and three other species of lizards there was a rapid behavioral hypothermia of ~8 °C in response to hematocrit reduction and a return to normal T_b selection after spontaneous recovery of hematocrit (Hicks and Wood 1985). Similar results were obtained in the amphibian, *Bufo marinus* (Wood 1990). In cross-sectional analysis (Fig. 9), there was no correlation between selected T_b and hematocrit of toads. *Bufo marinus displayed* a wide variation in both hematocrit and temperature selection. In homeotherms the hematocirt normally shows a small individual variation. For example, the coefficient of variation (CV = SD/×) is about 6% for humans. In ectotherms hematocrit generally has much more individual variation. For example, the CVs are ~15% for some amphibians and reptiles, and was 37% in *Bufo marinus*. Poor regulation of hematocrit probably reflects low rates of aerobic metabolism and high tolerance of anaerobic metabolism. For example, amphibians survive with hematocrits of 0% following injection of the hemolytic agent phenylhydrazine (Flores and Frieden 1968).

Mean selected T_b of *Bufo marinus* also shows considerable individual variation, with a CV of 19.5% (Wood 1990). Therefore, the lack of correlation in cross-sectional analyses between selected T_b and hematocrit is not surprising. Even if there were a correlation, it would not suggest a causal relationship. Longitudinal studies provide a more rigorous test of causal realtionships. As shown in Fig. 9, when hematocrit was lowered by removal of blood, all toads selected a significantly lower temperature. For example, a fall in hematocrit from 35 to 25% elicited the selection of a 4.9 °C cooler T_b. This reduced $\dot{V}O_2$ by ~51%, assuming a Q_{10} of ~2.5. When the hematocrit of individual lizards or toads is reduced, it is clear that selected T_b is reduced. Furthermore, for *Bufo marinus*, the amount of T_b selection is "tracking" hematocrit. Once again, however, it is speculative to infer cause and effect from this correlation.

3.2 Hypercapnia

Physiological thermoregulation of endotherms becomes impaired not only under hypoxic conditions, but also under hypercapnic conditions. In humans hypercapnia reduces heat production and increases heat loss by augmenting sweating and inhibiting shivering (Schaefer et at. 1975). Little is known about behavioral thermoregulation of ectotherms or endotherms in response to hypercapnia. Exposure of *Bufo marinus* to 10% CO_2 elicits a significant behavioral reduction of selected T_b (Riedel and Wood 1988). This was not an effect of acidosis, because addition of 40% O_2 to the chamber air reversed behavioral hypothermia without affecting arterial pH. This severe hypercapnia is well outside the range of normally encountered CO_2 levels (in burrows). Exposure to 5% CO_2, a level that is encountered in burrows, has no effect on T_b selection. For hamsters Kuhnen et al. (1987) found an interaction between hypoxia and hypercapnia on thermoregulation. Fifteen percent O_2 and 5% CO_2 had no effect alone, but when combined, elicited behavioral hypothermia. More studies are needed to determine the mechanisms and significance of hypothermic responses to hypercapnia.

3.3 Dehydration

Lowering metabolic heat production to reduce water loss occurs in endotherms. For example, when drinking water is restricted, deer mice show a 25% reduction in metabolic rate and 47% decrease in evaporative water loss (Chew 1961). Water deprivation in camels decreases metabolic rate by 77% (Schmidt-Nielsen et al. 1967). The mechanism of the decreased basal metabolic rate during dehydration of camels is inhibition of thyroid function (Yagil et al. 1978).

In ectotherms, reducing heat gain from the environment should have a similar effect on water economy. However, few studies have investigated the relationship between dehydration and T_b in ectotherms. The lizard *Anniella pulchra* selects lower temperatures in a dry thermal gradient than in a moist one (Bury and Balgooyen 1976). In contrast, the desert iguana shows no effect of dehydration on upper and lower exit temperatures in a shuttle-box experiment. Salt loading, however, reduced upper and lower exit temperatures (Dupré and Crawford 1985).

Amphibians are more susceptible to dehydration and show more uniform responses to desiccating conditions. Dehydration of *Chiromantis xerampelina* and *Phyllomedusa sauvagei* to 80–90% of their original weight caused them to select a lower T_b in a thermal gradient (66). These tree frogs are unique in that their skin is nearly impermeable to water. In most amphibians skin is highly permeable to water, and T_b should have a major impact on water balance. Amphibians lose water at rates up to several hundred times that of most reptiles and waterproof frogs (Shoemaker et al. 1989). Consequently, a reduction in T_b should be helpful to terrestrial amphibians in conserving water in arid environments. Selected T_b is significantly reduced in toads (*Bufo marinus*) exposed to dry air in a thermal gradient (Malvin and Wood 1991a). This behavior occurs before there is significant dehydration, i.e. no change in plasma osmolality. The reduced T_b is not due to increased evaporative cooling, but by movement of the animals to a colder temperature. This suggests a change in the thermoregulatory set point(s), because the evaporative cooling in dry air might, with no change in set point(s), elicit movement to the warm end of the chamber. The amount of water potentially saved by this response is considerable. The reduction of T_b from 24.3 to 15.7 °C will reduce water loss by 41.3%. The factor(s) mediating the behavioral response to dry air is unknown, but arginine vasotocin is hypothesized.

3.4 Toxic Chemicals and Metabolic Waste Products

Mammals exposed to toxic substances regulate T_b at a lower value. For example, nickel chloride affects both behavioral and autonomic control of T_b in rats (Gordon and Watkinson 1989). The resulting hypothermia reduces the toxicity of most xenobiotic compounds (Gordon 1988a).

Exposure of fish to anoxia affects later T_b selection in a normoxic thermal gradient. Goldfish exposed to anoxia for 5 h select a T_b ~5 °C lower than the corresponding normoxic controls (Crawshaw et al. 1989). Ethanol, a by-product of anaerobic metabolism in goldfish, may mediate this response. When injected into the hypothalamic thermoregulatory center of goldfish, ethanol elicits an 8 °C reduction of selected T_b.

Nitrogenous excetory substances are not toxic at physiological concentrations, but can become so in renal failure. Prolonged uremia results in hypothermia (Om and Hohenegger 1980; Kluger et al. 1981). Injection of urine or

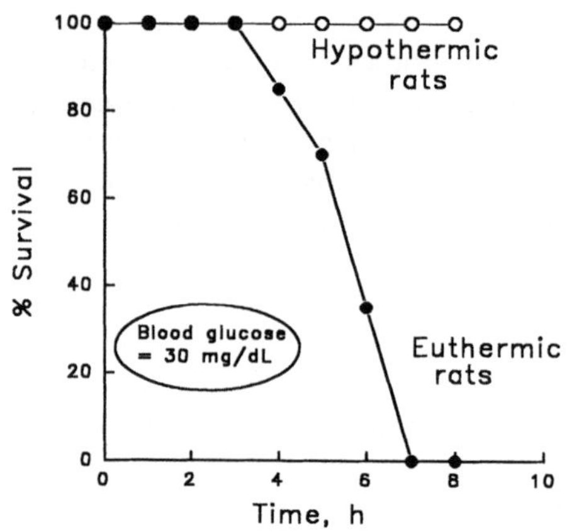

Fig. 10. Effect of insulin-induced hypoglycemia on survival of rats kept at normal body temperature (euthermic) and rats allowed to become hypothermic. (Data from Buchanan et al. 1991)

urea into experimental animals induces regulated hypothermia, suggesting the presence of an endogenous cryogen (Kluger et al. 1981; Cox et al. 1982). Mice also become hypothermic when toxic levels of other nitrogenous waste products are injected (Gordon 1988b).

3.5 Hypoglycemia

Glucose is the major fuel for brain metabolism. Consequently, hypoglycemia in mammals elicits a number of counterregulatory responses (e.g. catecholamine and glucagon release; Gerich et al. 1979) and induces hypothermia (Mayer-Gross and Berliner 1942; Kedes and Field 1964). Buchanan et al. (1991) tested the hypothesis that this hypothermia is an important adaptive response to hypoglycemia. They injected rats with insulin to lower plasma glucose to ~30 mg/dL and allowed the body temperature of half the rats to drop to ~31 °C, while the other rats were temperature-clamped to 37 °C. As shown in Fig. 10, within 7 h all of the rats kept at normal body temperature were dead, whereas all of the rats permitted to become hypothermic survived.

4 Conclusions

A common theme is that the evolution of increased locomotor and exercise capacity was linked to the evolution of endothermy, and that "warmer is better" (Bennett 1987). However, it is now clear that the preferred T_b is reduced by many ecological, environmental, and physiological factors.

Hypoxic animals have an impressive arsenal of defense mechanisms, many of which are common to other kinds of stresses. Physiological defense mechanisms, e.g. increasing cardiac output or ventilation, are effective, but also energy-demanding. An alternative to improving O_2 supply is reduction of O_2 demand. This is easily accomplished by ectothermic vertebrates by seeking a cooler environment and augmenting the rate of cooling by increased perfusion of the skin. More surprising is that endotherms respond to hypoxia in the same way. Unlike near drowning, where the hypothermia is forced (but still effective), the behavioral hypothermia appears to be regulated and analogous to fever. Clearly, both ectotherms and endotherms can take advantage of a variable T_b in both directions. By using behavioral hypothermia, they can reduce the rates of energy consumption when stressed with shortages of oxygen or substrate, or when anemic or exposed to toxins. Much more work is needed in this formative area of research to establish mechanisms, functional significance, and pathways of the thermoregulatory stress response.

Acknowledgements. Thanks to my students and colleagues who made my studies possible. My research was supported by NIH grant HL-40537.

References

Adolph EF (1951) Tolerance to cold and anoxia in infant rats. Am J Physiol 155: 366–377

Aitken PG, Schiff SJ (1986) Selective neuronal vulnerability to hypoxia in vitro. Neurosci Lett 67: 92–96

Artru AA, Michenfelder JD (1981) Influence of hypothermia or hyperthermia alone or in combination with pentobarbital on phenytoin on survival time in hypoxic mice. Anesth Analg 60: 867–870

Barber BJ, Crawford EC Jr (1977) A stochastic dual-limit hypothesis for behavioral thermoregulation in lizards. Physiol Zool 50: 53–60

Bennett AF (1987) Evolution of the control of body temperature: is warmer better? In: Dejours P, Boils L, Taylor CR, Wiebel ER (eds) Fidia Research Series, IX-Liviana Press, Padova, pp 421–431

Berman W, Wood SC, Yabek S, Dillon T, Fripp R, Burstein R (1987) Systemic oxygen transport in congenital heart diseases. Circulation 2: 360–368

Bons N (1983) Immunocytochemical identification of the mesotocin and vasotocin-producing systems in the brain of temperate and desert lizard species and their modifications by cold exposure. Gen Comp Endocrinol 52: 56–66

Boutilier RG, Lantz CJ (1989) The effects of forced and voluntary diving on plasma catecholamines and erythrocyte pH in the aquatic anuran, *Xenopus laevis*. Exp Biol 48: 83–88

Boutelier, RG, Emilio MG, Shelton, G (1986) Effects of Mechanical work on electrolyte and water distribution in amphibian skeletal muscle. J Exp Biol 120: 333–350

Brauer RW, Johnson ED, Miller CG et al (1986) Modification of temperature preference behavior by hypoxia and metabolic inhibitors. Cooper KE, Lomax P, Schönbaum E, Veale WL (eds) Homeostasis and thermal stress. 6th Int Symp Pharmacol Thermoregulation, Karger, Basel, pp 27–29

Bryan JD, Hill LG, Niell WH (1984) Interdependence of acute temperature preference and respiration in the plains minnow. Trans Am Fish Soc 113: 557–562

Buchanan TA, Cane P, Eng CC, Sipos GF, Lee C (1991) Hypothermia is critical for survival during prolonged insulin-induced hypoglycemia in rats. Metabolism 40: 330–334

Bury RB, Balgooyen TG (1976) Temperature selectivity in the legless lizard, *Anniella pulchra*. Copeia 1976: 152–155

Busto R, Dietrich WD, Globus M, Valdés I, Scheinberg P, Ginsberg M (1987) Small differences in intraischemic brain temperature critically determine the extent of ischemic neuronal injury. J Cereb Blood Flow Metab 7: 729–738
Carlsson CHM, Siesjo BK (1976) Protective effect of hypothermia in cerebral oxygen deficiency caused by arterial hypoxia. Anesthesiology 44: 27–35
Chew RM (1961) Water metabolism of desert vertebrates. Biol Rev 36: 1–31
Cowles RB, Bogert CM (1944) A preliminary study of the thermal requirements of desert reptiles. Bull Am Mus Nat Hist 83: 261–296
Cox PS, Rothenburg BA, Kluger MJ (1982) Characterization of an endogenous cryogen that appears in the urine. Am J Physiol 243: R241–244
Crawshaw LI, Hammel HT (1974) Behavioral regulation of internal temperature in the brown bullhead, *Ictalurus nebulosus*. Comp Biochem Physiol 47A: 51–60
Crawshaw LI, Wollmuth LP, O'Connor CS (1989) Intracranial ethanol and ambient anoxia elicit selection of cooler water by goldfish. Am J Physiol 256: R133–R137
Dascombe MJ (1985) The pharmacology of fever. Prog Neurobiol 25: 327–373
Dodt E (1956) Die Aktivität der Thermorezeptoren bei nichtthermischen Reizen bekannter thermoregulatorischer Wirkung. Pflügers Arch Gesamte Physiol 263: 188–200
Dunn JM, Miller JA (1969) Hypothermia combined with positive pressure ventilation in resuscitation of the asphyxiated neonate. Am J Obstet Gynecol 104: 58
Dupré RK, Crawford EC Jr (1985) Behavioral thermoregulation during dehydration and osmotic loading of the desert iguana. Physiol Zool 58: 357–363
Dupré RK, Wood SC (1988) Behavioral temperature regulation by aquatic ectotherms during hypoxia. Can J Zool 66: 2649–2652
Dupré RK, Owen L (1989) Behavioral thermoregulation by hypoxic rats. FASEB J 3: A838 (Abst)
Dupré RK, Romero A, Wood SC (1988) Thermoregulation and metabolism in hypoxic animals. In: Fedde MR, Gonzales N (eds) Oxygen transfer from environment to tissues. Plenum Press, New York, pp 347–351
Fincher K, McPherson B, Wood SC (1992) Behavioral thermoregulation of hypoxic alligators. FASEB 6: A1748
Flores G, Frieden E (1968) Induction and survival of hemoglobin-less and erythrocyte-less tapoles and young bullfrogs. Science 159: 101–103
Gautier H, Bonora M, M'Barek S, Sinclair, JD (1991) Effects of hypoxia and cold acclimation on thermoregulation in the rat. J Appl Physiol 71: 355–1363
Gellhorn E, Janus A (1936) The influence of partial pressures of O_2 on body temperature. Am J Physiol 116: 327–329
Gerich J, Davis J, Lorenzi M et al (1979) Hormonal mechanisms of recovery from insulin-induced hypoglycemia in man. Am J Physiol 236: E380–E385
Glass ML, Boutelier RG, Heisler N (1983) Ventilatory control of arterial PO_2 in the turtle, *Chrysemys picta bellii*. Effects of temperature and hypoxia. J Comp Physiol 151: 145–153
Gollan F, Aono M (1973) The effect of temperature on sanguinous rabbits. Cryobiology 10: 321–327
Gordon CJ (1988a) Temperature regulation in laboratory mammals following acute toxic insult. Toxicology 53: 161–178
Gordon CJ (1988b) Thermoregulatory responses in mice following acute administration of principal nitrogenous excretory substances. Pharmacol Biochem Behav 31: 699–703
Gordon CJ, Watkinson WP (1989) Effect of nickel chloride on body temperature and behavioral thermoregulation in rats. Fed Proc 3: A701
Hagerdal MWFA, Keykhah MM et al (1978) Protective effects of combinations of hypothermia and barbiturates in cerebral hypoxia in the rat. Anesthesiology 49: 165–169
Hammel HT, Heller HC, Sharp FR (1973) Probing the rostral brainstem of anesthetized, unanesthetized, and exercising dogs and of hibernating and euthermic ground squirrels. Fed Proc 32: 1588–1597
Heath JE (1970) Behavioral regulation of body temperature in poikilotherms. Physiologist 13: 399–410

Hennessey TM, Saimi Y, Kung C (1983) A heat-induced depolarization of *Paramecium* and its relationship to thermal avoidance behavior. J Comp Physiol 153: 39–46
Hicks JW, Malvin GM (1995) Mechanisms of intracardiac shunting in reptiles: pressure vs washout shunting. Advances in environmental and comparative physiology, vol 21. Respiration and circulation. Springer, Berlin Heidelberg New York (in press)
Hicks JW, Wood SC (1985) Temperature regulation in lizards: effects of hypoxia. Am J Physiol 248: R595–R600
Hicks JW, Wood SC (1989) Oxygen homeostasis. In: Wood SC (ed) Comparative pulmonary physiology: current concepts. Marcel Dekker, New York, pp 311–342
Hirsch H, Müller HA (1962) Funktionelle und histologische Veränderungen des Kaninchengehirns nach kompletter Gehirnischämie. Pflügers Arch 275: 277–291
Hitzig BM (1989) Effects of hypoxia on brain cell acid–base and high-energy phosphate regulation by 31P-NMR spectroscopy. In: Lahiri S, Forster RE, Davies RO, Pack AI (eds) Chemoreceptors and reflexes in breathing. Oxford Univ Press, New York, pp 255–266
Huey RB (1982) Temperature, physiology, and ecology of reptiles. In: Gans C, Pough FH (eds) Biology of the reptilia, Academic Press, New York, pp 25–92
Johnson DC, Nishimura M, Okunieff P, Kazemi H, Hitzig BM (1989) Effects of hypothermia on rat brain pHi and phosphate metabolite regulation by 31P-NMR. J Appl Physiol 67: 2527–2534
Ikonomidou C, Mosinger JL, Olney JW (1989) Hypothermia enhances protective effect of MK-801 against hypoxic/ischemic brain damage in infant rats. Brain Res 487: 184–187
Kavaliers M, Courtenay S, Hirst M (1984) Opiates influence behavioral thermoregulation in the curly-tailed lizard, *Leiocephalus carinatus*. Physiol Behav 32: 221–224
Kedes LH, Field JB (1964) Hypothermia: a clue to hypoglycemia. N Engl J Med 271: 785–787
Kluger MJ (1979) Fever in ectotherms: evolutionary implications. Am Zool 19: 295–304
Kluger MJ, Turnbull AJ, Cranston WI, Wing AJ, Gross MP, Rothenburg BA (1981) Endogenous cryogen excreted by the kidneys. Am J Physiol 241: R271–276
Krogh A (1914) The quantitative relation between temperature and standard metabolism in animals. Int Z Physik-Chem Biol 1: 491–508
Kruhøffer M, Glass ML, Abe AS, Johansen K (1987) Control of breathing in an amphibian, *Bufo paracnemius:* effects of temperature and hypoxia. Respir Physiol 69: 267–275
Kuhnen G, Wloch B, Wünnenberg W (1987) Effects of acute hypoxia and/or hypercapnia on body temperatures and cold induced thermogenesis in the golden hamster. J Therm Biol 12: 103–107
Lintzel W (1931) Über die Wirkung der Luftverdünnung auf Tiere V. Mitteilung Gaswechsel weißer Ratten. Pflügers Arch Ges Physiol 227: 673–708
Lister G (1984) Oxygen transport in the intact hypoxic newborn lamb: acute effects of increasing P_{50}. Pediar Res 18: 172–177
Lister G, Moreau G, Moss M, Talner NS (1984) Effects of alterations of oxygen transport on the neonate. Semin Perinatol 8: 192–204
Malvin GM, Wood SC (1991a) Effect of air humidity on behavioral thermoregulation of the toad *Bufo marinus*. J Exp Zool 258: 322–326
Malvin GM, Wood SC (1991b) Behavioral hypothermia enhances survival in the protozoan, *Paramecium caudatum* Science 225: 1423–1425
Malvin GM, Macias S, Sanchez M, Parkrapid A (1992) Hypoxia-induced hemoconcentration in the toad *bufo woodhousei*: role of spleen and lymph heart. FASEB 6: A 1529
Mayer-Gross W, Berliner F (1942) Observations on hypoglycemia IV body temperature and coma. J Ment Sci 88: 419–427
Mendelssohn M (1895) Über den Thermotropismus einzelliger Organismen. Pflügers Arch Gesamte Physiol Menschen Tiere 60: 1–27
Middler SA, Kleeman CR, Edwards E (1968) Lymph mobilization following acute blood loss in the toad, *Bufo marinus.* Comp Biochem Physiol 25: 335–341
Miller JA, Miller FS (1966) Interactions between hypothermia and hypoxia-hypercapnia in neonates. Fed Proc 25: 1338–1341
Minamisawa H, Nordström Smith JL, Siesjö BK (1990) The influence of mild body and brain hypothermia on ischemic brain damage. J Cereb Blood Flow Metab 10: 365–374

Misgeld U, Frotscher M (1982) Dependence of the viability of neurons in hippocampal slices on oxygen supply. Brain Res Bull 8: 993–1003
Moore FL, Miller LJ (1983) Arginine vasotocin induces sexual behavior on newts by acting on cells in the brain. Peptides 4: 97–102
Naylor AM, Ruwe WD, Veale WL (1986) Thermoregulatory actions of centrally administered vasopressin in the rat. Neuropharmcology 25: 787–794
Nishimura M, Johnson DC, Hitzig BM, Okunieff P, Kazemi H (1989) Effects of hypercapnia on brain pHi and phosphate metabolite regulation by 31P-NMR. J Appl Physiol 66: 2181–2188
Olton DS, Becker JT, Handelmann GE (1979) Hippocampus, space, and memory. Behav Brain Sci 2: 313–365
Om P, Hohenegger M (1980) Energy metabolism in acute uremic rats. Nephron 25: 249–253
Perry SF (1989) Structure and function of the reptilian respiratory system. In: Wood SC (ed) Comparative pulmonary physiology. Marcel Dekker, New York, pp 193–236
Pörtner HO, MacLatchey LM, Toews DP (1991) Metabolic responses of the toad, *Bufo marinus*, to environmental hypoxia: an analysis of the critical PO_2. Physiol Zool 64: 836–849
Rausch RN, Crawshaw LI (1990) Effect of hypoxia on behavioral thermoregulation in the goldfish, *Carassius auratus*. FASEB J 4: A551
Reynolds WW, Casterlin ME (1982) The pyrogenic response of non-mammalian vertebrates. In: Milton AS (ed) Pyretics and antipyretics, Springer, Berlin Heidelberg New York, pp 649–668
Riedel C, Wood SC (1988) Effects of hypercapnia and hypoxia on temperature selection of the toad, *Bufo marinus*. Fed Proc 2: A500
Robinzon B, Koike TI, Neldon HL, Kinzler SL, Hendry IR, el-Halawani ME (1988) Physiological effects of arginine vasotocin and mesotocin in cockerels. Brit Poultry Sci 29: 639–652
Ruth V, Fyhrquist F, Clemons G, Raivio KO (1988) Cord plasma vasopressin, erythropoietin, and hypoxanthine as indices of asphyxia at birth. Pediatr Res 24: 490–494
Schaefer K, Messier AA, Morgan C, Baker GT III (1975) Effect of chronic hypercapnia on body temperature regulation. J Appl Physiol 38: 900–906
Schmidt-Nielsen K, Crawford EC, Newsome AE, Rawson KS, Hammel HT (1967) Metabolism rate of camels: effect of body temperature and dehydration. Am J Physiol 212: 341–346
Schubring C (1986) Temperature regulation in healthy and resuscitated newborns immediately after birth. J Perinat Med 14: 27–33
Shoemaker VW, Nagy KA (1977) Osmoregulation in amphibians and reptiles. Annu Rev Physiol 39: 449–471
Shoemaker VH, Baker MA, Loveridge JP (1989) Effect of water balance on thermoregulation in waterproof frogs (*Chiromantis and Phyllomedusa*). Physiol Zool 62: 133–146
Stegner H, Leake RD, Palmer SM, Oakes G, Fisher DA (1984) The effect of hypoxia on neurohypophyseal hormone release in fetal and maternal sheep. Pepiatr Res 18: 188–191
Sutariya B, Penney D, Barnes J, Helfman C (1989) Hypothermia protects brain function in acute carbon monoxide poisoning. Vet Hum Toxicol 31: 436–441
Szelenyi Z, Donhoffer S (1968) The thermogenic function of brown adipose tissue and the response of body temperature to hypoxia and hypercapnia in the cold- and warm-adapted rat. Acta Physiol Acad Sci Hung 33: 31–39
Tamaki Y, Nakayama T (1987) Effects of air constituents on thermosensitivities of preoptic neurons: hypoxia versus hypercapnia. Pflügers Arch 409: 1–6
Wade CE, Keil LC, Ramsey DJ (1983) Role of volume and osmolality in the control of plasma vasopressin in dehydrated dogs. Neuroendocrinology 37: 349–353
Walker BA (1986) Role of vasopressin in the cardiovascular response to hypoxia in the conscious rat. Am J Physiol 251: H1316–H1323
Walker BA (1987) Cardiovascular effects of V1 vasopressinergic blockade during acute hypercapnia in conscious rats. Am J Physiol 252: R127–R133
White RJ, Austin PE Jr, Austtin JC, Taslitz N, Takoaka Y (1973) Recovery of the subhuman primate after deep cerebral hypothermia and prolonged ischaemia. Resuscitation 2: 117–122.

Wilson DA, Hanley DF, Feldman MA, Traystman RJ (1987) Influence of chemoreceptors on neurohypophyseal blood flow during hypoxic hypoxia. Circ Res 61: 94–101
Wood SC (1984) Cardiovascular shunts and oxygen transport in lower vertebrates. Am J Physiol (Reg Int Comp Physiol) 247: R3–R14
Wood SC (1990) Effect of hematocrit on behavioral thermoregulation of the toad, *Bufo marinus*. Am J Physiol 27: R848–R851
Wood SC, Malvin GM (1991) Physiological significance of behavioral hypothermia in hypoxic toads, *Bufo marinus*. J Exp Biol 159: 203–215
Wood SC, Dupré RK, Hicks JW (1985) Voluntary hypothermia in hypoxic animals. Acta Physiol Scand 124: 46
Yagil R, Etzion Z, Ganani J (1978) Camel thyroid metabolism: effect of season and dehydration. J Appl Physiol 45: 540–544

Chapter 10

Sulfide Tolerance in Marine Invertebrates

S. Völkel and *M.K. Grieshaber*

Contents

1 Introduction

Hydrogen sulfide is generally considered as a highly toxic substance. Its best known and most drastic effect concerns the binding to cytochrome aa_3, which is part of the respiratory electron transport chain resulting in the inhibition of aerobic respiration (Nicholls 1975; National Research Council 1979). In addition, sulfide inhibits several other enzymes such as uricase, succinic dehydrogenase, and catalase. The inhibitory effect is partially due to the reduction of critical disulfide bonds and to the binding of sulfide to heme-type enzymes (National Research Council 1979). Finally, sulfide is also reported to affect blood pigments by generating "sulfhemoglobin" when H_2S is bubbled through pure hemoglobin solutions (Keilin 1933), although neither "sulfhemoglobin" nor any other abnormal blood pigments were found in vivo (Sörbo 1958). Recently, Bagarinao and Vetter (1992) showed that blood of the sulfide-tolerant California killifish

Institut für Zoologie, Lehrstuhl für Tierphysiologie, Heinrich-Heine-Universität, Düsseldorf, FRG

Advances in Comparative and Environmental Physiology, Vol. 22

Fundulus parvipinnis is relatively insensitive to sulfide, and that sulfhemoglobin is not formed at environmental realistic and physiologically relevant sulfide concentrations.

Sulfide is poisonous to human beings and vertebrates at about 2 $\mu mol\ l^{-1}$ (Longwell and Pentelow 1935; Broderius et al. 1977; Burnett et al. 1977; National Research Council 1979; Torrans and Clemens 1981), whereas invertebrates are affected by concentrations of about 200 $\mu mol\ l^{-1}$ (Oseid and Smith 1974; Theede et al. 1969). Therefore, most organisms try to avoid contact with sulfide (e.g. see Meadows et al. 1981).

Nevertheless, marine environments some of which are rich in sulfide, such as the sulfide system of the sediment (Fenchel and Riedl 1970), are inhabited by a rich fauna. Some species of these habitats may be confronted with sulfide. Survival is therefore only possible when an adaptation to this toxin has evolved. In the following, the sulfide system is characterized and possible adaptations to a sulfide-rich environment with special emphasis on animals of marine sediments are discussed.

2 Sediment

Most marine sediments consist of an upper oxic layer, an underlying anoxic, sulfidic layer, and an intermediate chemocline called the redox-potential-discontinuity (RDP) layer (Fig. 1). The sulfidic layers of marine sediments are character-

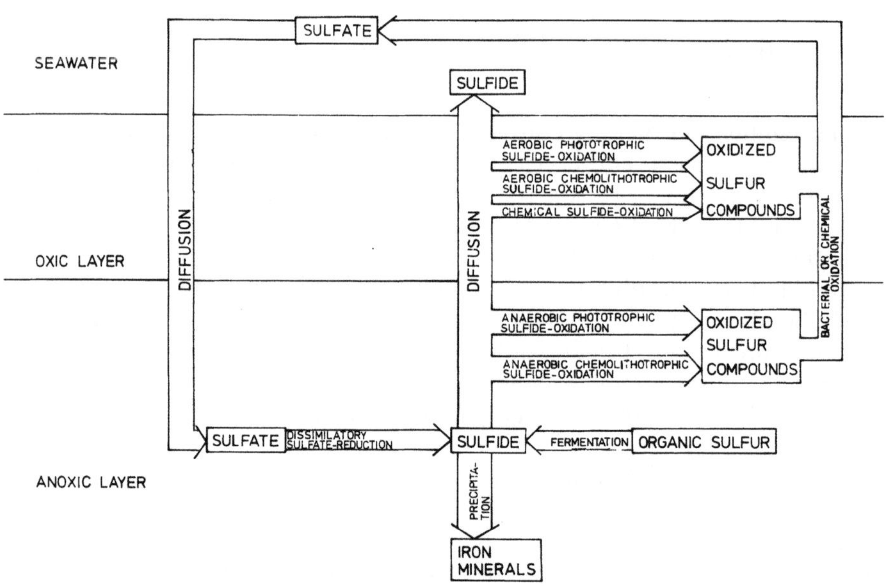

Fig. 1. Dynamics of sulfide in marine sediments

ized by anoxic, reducing conditions and various amounts of sulfide in the pore water, which can reach concentrations of 20 mmol l^{-1} (Fenchel and Riedl 1970).

The development of the chemical composition of this zone was described by Jørgensen and Fenchel (1974). In the beginning the sediment is saturated with oxygen and the constituents of the pore water are the same as in the overlaying seawater. Detritus causes an increase in the content of organic compounds, and thereby a rapid growth of heterotrophic aerobic bacteria. Oxygen becomes depleted, the redox potential decreases below – 100 mV and dissimilatory sulfate reduction starts (Nedwell 1982). Sulfate-reducing bacteria, *Desulfovibrio* and *Desulfomaculatum* are able to oxidize end-products of carbohydrate degradation such as fatty acids, alcohols, and dicarbonic acids using sulfate as an electron acceptor (Jørgensen and Fenchel 1974; Nedwell 1982; Jørgensen 1982; Kröger et al. 1988). Maximal rates of sulfate reduction occur in the upper layer of the anoxic zone (Nedwell 1982), because that is the site of the highest concentrations of electron donors and acceptors. Most sulfide in the anoxic zone results from dissimilatory sulfate reduction (Jørgensen and Fenchel 1974). Only a small amount (less than 4%) originates from the anaerobic chemotrophic degradation of thio-organic compounds such as cysteine and methionine.

Part of the sulfide (10%) is precipitated as insoluble metallosulfides, mainly pyrite (FeS_2) (Berner 1970; Jørgensen 1977; Morse and Cornwell 1987; Thode-Andersen and Jørgensen 1989). Up to 95% of the sulfide will be reoxidized (Jørgensen 1988). It diffuses in the pore water to the interface between reduced and oxidized layers, and can be oxidized in the suboxic zone (Eh 0–100 mV) in the absence of oxygen (Jørgensen 1988) by chemolithotrophic bacteria such as thiobacilli. They can use oxidized iron minerals and oxidized manganese as electron acceptors, whereas sulfide and thiosulfate as well as other reduced sulfur compounds serve as electron donors (Kelly 1988, 1989).

Sulfide seldom reaches the oxic zone, unless the sulfate-reduction rate is very high (Jørgensen 1988). Reaching the oxygen–sulfide interface, sulfide will be oxidized by aerobic bacteria such as *Beggiatoa* sp. (Nelson and Castenholz 1981; Jørgensen 1988). In the upper few millimeters of the sediment, which are penetrated by light, sulfide can be used aerobically or anaerobically as an electron donor for photosynthesis by cyanobacteria and photosynthetic bacteria (Hartmann Hansen et al. 1978; Revsbech et al. 1983; De Wit et al. 1989). In the presence of oxygen, sulfide will also be chemically oxidized to thiosulfate (Cline and Richards 1969; Chen et al. 1972; Wilmot et al. 1988; Millero 1986; Millero et al. 1987). Most of the sulfide will eventually be reoxidized to sulfate (Nedwell 1982), although thiosulfate as well as other reduced sulfur compounds are an important part of the sulfur cycle (Jørgensen 1990). Commonly, only a small amount of sulfide will be lost to the atmosphere by diffusion and bioturbation (Nedwell 1982).

The zonation described previously is characteristic for most marine sediments. The extension of the different layers and the sulfide content of the pore water depend on the influx of organic compounds and drainage of oxygen (Fenchel and Riedl 1970). In sediments with high organic influx and a high rate of sulfate reduction, e.g. salt marshes, sulfide can reach the oxic layer. Consequently, sulfide and oxygen can overlap in a distinct zone of the sediment (Jørgensen 1988). Seasonal changes of temperature influence the rate of sulfate reduction (Nedwell and Floodgate 1972) and diurnal changes of light effect the depth profile of oxygen and sulfide via the intensity of photosynthesis (De Wit et al. 1989). Therefore, abiotic conditions can rapidly change in marine sediments.

3 Living System

It is not too long ago that the sulfidic, anoxic zone was considered to be azoic, because high sulfide concentrations and the lack of oxygen supposedly excluded all metazoan life (Boaden 1975, 1977; see also Fenchel and Riedl 1970). However, Fenchel and Riedl (1970) listed a variety of organisms as representatives of the sulfide system, including Turbellaria, Gnathostomulida, Nematoda, Gastrotricha, Oligochaeta, etc. Following Fenchel and Riedl's (1970) description of a bilayered system, Boaden and Platt (1971) termed the inhabitants of the sulfide system "thiobios," whereas the living system of the overlying oxic layer was called "oxybios" by Powell et al. (1983).

It has recently been shown that marine sediments are not a simple two layer-system, but a more complex system with microhabitats generated predominantly by chemical gradients associated with the tube-building activity of the macroinfauna (Reise 1981a,b, 1984; Bell 1983). Hence, there are two gradients occurring in the sediments: firstly, the sulfide–oxygen gradient following the transition from surface-oxidized sediment to the underlying reduced sediment, and secondly, the analogous horizontal gradient around the burrows of macrofauna (see also Powell 1989).

The distribution of meiofauna within these microhabitats has been discussed (see Reise and Ax 1979, 1980; Boaden 1980; Powell and Bright 1981). According to Meyers et al. (1987) and Reise and Ax (1979), irrigated tubes are generally attractive to subsurface oxybiotic and thiobiotic meiofauna. On the other hand, the Stilbonematinae were reported to live predominantly in the interface between oxidized and sulfidic sediment (Ott and Novak 1989). Oxybios and thiobios seem to be separated by the zero-oxygen line (Meyers et al. 1987, 1988).

The microhabitats in the marine sediments are temporally variable. Macrofaunal irrigation activity changes throughout the day and according to the tide. During low tide the tube-dwelling polychaeta *Arenicola marina,* for example, stops irrigating its tube (Toulmond et al. 1983; Toulmond 1987). The oxygen content of the burrow decreases (Jones 1955; Meyers et al. 1988; Kristensen 1989; see also Watling 1991) and sulfide concentrations can increase when sulfide in the surrounding sediment is high (Völkel et al., in press). Some species relocate frequently, desert their old tubes, and create new ones (Powell 1977). Meyers et al. (1988) and Ott et al. (1991) showed that subsurface meiofauna rapidly responded to changes in sediment chemistry and tried to occupy new optimal habitats. Nevertheless, not only meiofauna, but also other members of the infauna of marine sediments, may temporarily be exposed to sulfide.

4 Sulfide Tolerance

Since Fenchel and Riedl's (1970) description of the living system in sulfidic marine sediments, many species have been shown to tolerate or even require sulfide. Several bivalves, such as *Cyprina islandica* and *Mya arenaria,* were reported to be sulfide-tolerant (von Oertzen and Schlungbaum 1972; Theede et al. 1969; Theede 1973). Powell et al. (1979) demonstrated the existence of adaptations to sulfide in a number of interstitial metazoans such as the turbellarian *Solenofilomorpha funilis* and the gastrotriche *Dolichodasys carolinensis.* The lugworm *Arenicola marina* has been shown to live in sulfide-rich sediments and to survive the exposure to high sulfide concentrations (Groenendaal 1979, 1980). The occurrence of sulfide tolerance has also been indicated in the priapulid *Halicryptus spinulosus* (Oeschger 1987) and in the Oligochaeta *Tubificoides benedii* (Giere et al. 1988; Dubilier 1993).

The knowledge of increasing numbers of sulfide-tolerant species recently sparked investigations on the mechanisms leading to sulfide tolerance. Several possibilities of physiological and biochemical adaptations to sulfide are possible (Fig. 2):

a) If an animal living in a sulfide-rich environment could exclude sulfide at its body wall, it would avoid the toxic effects completely. But if sulfide penetrates into the organism, it has to be detoxified, unless the animal is sulfide-insensitive.
b) A possible mechanism of sulfide detoxification is its oxidation to a less toxic sulfur species such as sulfite, thiosulfate, sulfate, or elemental sulfur. Sulfide oxidation can occur in the animal tissue or in bacterial symbionts, or both can participate with different compartments carrying out different steps of oxidation.
c) Immobilization by binding of sulfide to blood components, proteins, and metallic ions constitutes another possibility in the avoidance reaction.
d) If an animal is not able to detoxify or immobilize sulfide, because the available mechanisms are not efficient enough, sulfide is taken up in the body and may rise to toxic concentrations. In this case, an animal must have a sulfide-insensitive cytochrome c-oxidase to maintain its aerobic metabolism.
e) If, however, the respiratory electron transport is indeed inhibited by sulfide, survival will only be possible by switching to an anaerobic metabolism.

The mechanisms of sulfide tolerance listed above and their occurrence in several species is described in the following sections.

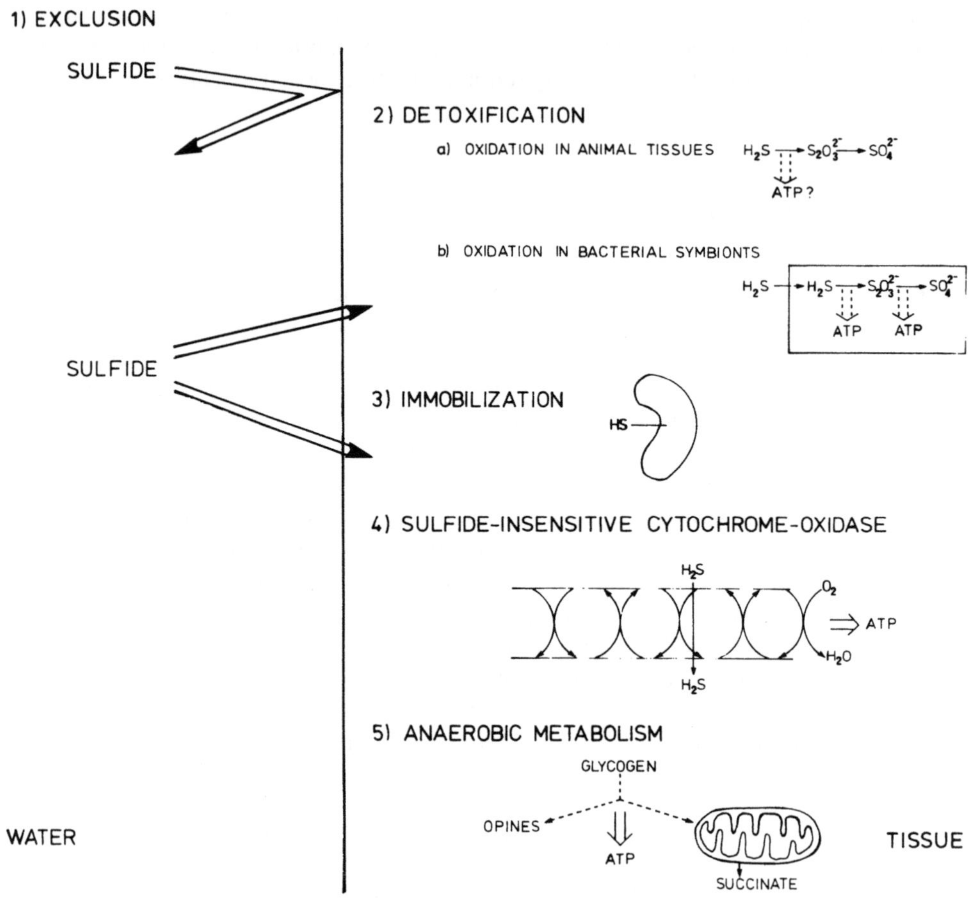

Fig. 2. Possible strategies of sulfide defense

4.1 Exclusion of Sulfide

The permeability of biological membranes for sulfide has been known for a long time (Beerman 1924). Several species show an influx of sulfide in their body during sulfide exposure, such as *Arenicola marina* (Groenendaal 1981; Völkel and Grieshaber 1992), *Sipunculus nudus* (Völkel and Grieshaber 1992), *Saduria entomon* (Vismann 1991a), *Urechis caupo* (Julian et al. 1990), and several meiofaunal species (Powell et al. 1979). The diffusion coefficient of hydrogen sulfide has been estimated as 4–4.5 10^{-5} cm^2 s^{-1} in *S. nudus* (Völkel and Grieshaber1992; see also Powell 1989). The specific permeability coefficient for H_2S and HS^- for body wall tissue of *Urechis caupo* was determined by Julian and Arp (1992) to be 0.17 and 0.063 cm h^{-1}, respectively, indicating a 2.7 times higher permeability of undissociated H_2S than of the HS^--ion.

To prevent sulfide from penetrating the body wall, an animal should have special mechanisms of defense. A highly impermeable surface, such as a thick cuticle or a shell, could perhaps exclude sulfide completely, but it would also prevent the exchange of oxygen and excretion products such as CO_2 and NH_3 with the medium. Another possibility would be a peculiar structure of the body surface, which allows the exchange of oxygen and CO_2, but not of sulfide.

Giere et al. (1988) showed that the cuticle of *Tubificoides benedii* is covered by condensed, almost solid, mucus caps. Sulfide is supposed to precipitate as iron sulfide in this surface layer (Dubilier et al. 1991; Dubilier 1993; Dubilier et al., in press). A similar mechanism may be present in the priapulid worm *Halicryptus spinulosus*. Its integument turns black in the presence of sulfide, and a considerable amount of iron was detected (Oeschger and Janssen 1991; Oeschger and Vetter 1992). However, an estimated rate of sulfide influx and the maximal rate of iron sulfide formation in the body wall of *Tubificoides benedii* calls an effective protection against sulfide by peripheral precipitation into question (Dubilier 1993; Dubilier et al., in press).

Sulfide can also be trapped on the surface of the body by epibacteria, which oxidize entering sulfide and use it for their own energy requirements (see below). This might to some extent be true in *Halicryptus spinulosus* (Oeschger and Schmaljohann 1988; Oeschger and Janssen 1991). The Stilbonematinae are likewise covered with large ectosymbiotic microorganisms that could prevent sulfide from entering the animal's body (Ott and Novak 1989; Ott et al.1991; Fig. 3). Menon and Arp (1993) discussed three possible mechanisms that may counteract the toxic effects of sulfide at the integumentary surface of the echiurid worm *Urechis caupo*: metabolism of symbiotic bacteria, dying off of peripheral, sulfide-exposed cells, and oxidation of sulfide at specialized organelles. Until now, none of these phenomena has been proven to protect the animal completely from the influx of increasing concentrations of ambient sulfide.

4.2 Mechanisms of Sulfide Detoxification

4.2.1 Sulfide Oxidation in Animal Tissue

The detoxification of sulfide by oxidizing it to a less toxic sulfur compound is a widespread defense strategy. Sulfide, as the most reduced sulfur species, can be oxidized with the participation of oxygen mainly to sulfite, thiosulfate, polythionates, and sulfate, the most oxidized compound, and to elemental sulfur. Which of them is the main oxidation product of a species depends largely on the availability of oxygen. Under limiting oxygen concentrations, it is important to have a favorable ratio of sulfide to oxygen in order to spend as little oxygen as possible to metabolize sulfide to an untoxic compound. If sulfide is directly oxidized by oxygen, the production of 1 mol thiosulfate consumes 2 mol sulfide, but only 1.5 mol oxygen, whereas only 1 mol sulfide will be detoxified by consuming 2 mol oxygen with sulfate as the end-product of sulfide oxidation.

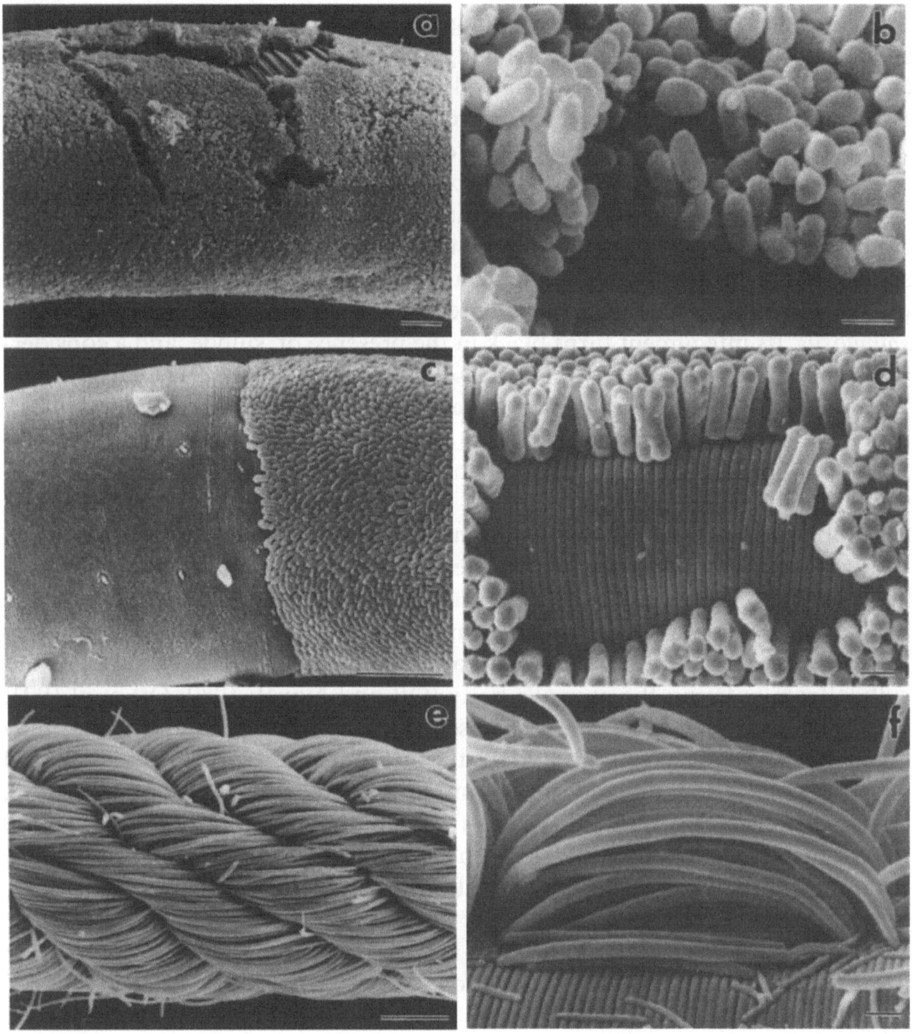

Fig. 3. Bacterial growth on different *Stilbonematinae*. **a** *Stilbonema sp.* from Belize, mid-body region completely covered by a multilayer of bacteria. **b** Coccoid bacteria at higher magnification. **c** *Catanema sp.* from the Adriatic Sea. Beginning of bacterial monolayer 220 μm from the anterior end. **d** *Catanema sp.* from Belize. In the "clearing" between the tightly packed rods standing upright the annulated cuticle of the worm is visible. **e** *Eubostrichus cf. parasitiferus* from Belize with crescent-shaped bacteria arranged in a double spiral around the body. **f** Bacteria at higher magnification, showing attachment of cells with both ends and progressive lengthening and bending of the bacteria distally. Scale *bars*: 10 μm (**a,c,e**) and 1 μm (**b,d,f**). (Ott et al. 1991)

In fact, Powell et al. (1980) found that the oxidation products of meiofauna were correlated with the oxygen availability in the animal's habitat. *Archiloa wilsoni*, a surface-dwelling turbellarian, produced sulfite and sulfate, whereas two species living in the sulfide system under low oxygen concentrations accumulated elemental sulfur as a primary end-product. *Solenofilomorpha funilis*, another

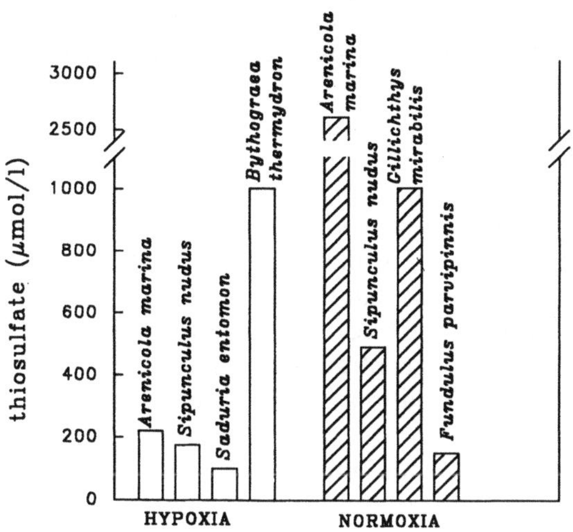

Fig. 4. Thiosulfate (μmol l^{-1}) in blood or coelomic fluid of various animals after 4 h sulfide incubation (1–1.5 mmol l^{-1}). *Open columns*, hypoxic *hatched columns*, normoxic conditions. (Data from Bagarinao and Vetter 1989; Vetter et al. 1987; Vismann 1991; Völkel and Grieshaber 1992)

sulfide-system animal, produced mainly thiosulfate, but also some sulfite and sulfate.

Because in marine sediments the exposure of animals to sulfide often occurs in the presence of low oxygen concentrations (e.g. see Powell 1989), thiosulfate will be mostly produced as an initial oxidation product. An accumulation of thiosulfate was recently shown in the sediment-burrowing isopod *Saduria entomon* (Vismann 1991a) as well as in the sipunculid worm *Sipunculus nudus* and the polychaete *Arenicola marina* (Völkel and Grieshaber 1992; Völkel, in press; Fig. 4). In *S. nudus* and *A. marina*, the production of thiosulfate could be demonstrated to depend on oxygen. The sulfide-oxidizing capacity was threefold higher under normoxia than under hypoxia in *S. nudus*, and even 15-fold higher in *A. marina*. Under normoxic conditions and in the presence of not more than 0.4 mmol l^{-1} sulfide, *A. marina* is able to oxidize all sulfide entering the body (Völkel and Grieshaber 1991; Völkel and Grieshaber 1994a).

In addition, the accumulation of thiosulfate during sulfide exposure could also be demonstrated in the shallow-water marine fish *Fundulus parvipinnis, Gillichthys mirabilis* and *Citharichthys stigmaeus* (Bagarinao and Vetter 1989 1990), *Halicryptus spinulosus* (Oeschger and Vetter 1992), and in the hydrothermal vent crab *Bythograea thermydron* (Vetter et al. 1987; Gorodetzky and Childress 1994).

The mechanism of sulfide oxidation in marine animals has been a topic for many investigations for a long time. Presently, two mechanisms appear to be mainly involved: firstly, oxidation of sulfide by blood pigments, and secondly, a mitochondrial sulfide oxidation.

Patel and Spencer (1963) suggested that the oxidation product of hemoglobin, the "brown pigment" of *Arenicola marina,* is an active catalyst of sulfide oxidation. This was confirmed by Vismann (1990), who correlated sulfide-oxidizing activity of *Nereis diversicolor* and *Nereis virens* with the concentration of brown pigment in the blood of the animals. In contrast, Wells and Pankhurst (1979) showed that another arenicolid species, *Abarenicola affinis*, did not form a brown pigment.

In the echiuran worm *Urechis caupo* hematin, a nonglobin heme compound contained in the coelomocytes, acts as a catalyst for sulfide oxidation (Arp et al. 1987a; Arp 1991; Arp et al. 1992; Eaton and Arp 1993). Powell and Arp (1989) suggested that hematin in the gill tissue of *Solemya reidi*, *Calyptogena magnifica*, and *Lucinoma annulata* may also function as a sulfide-oxidizing system. Conversely, Bagarinao and Vetter (1992) could show that the blood of the California killifish *Fundulus parvipinnis* does not catalyze sulfide oxidation to a considerable extent at environmentally realistic sulfide concentrations of 10–500 $\mu mol\ l^{-1}$, although ferric hemoglobin catalyzes the oxidation of sulfide to thiosulfate in vitro.

Other tissues, such as the hepatopancreas and muscles of *Saduria entomon* (Vimann 1991a) as well as intestinal wall and body fluid of *Nereis diversicolor* and *N. virens* (Vismann 1990), catalyze sulfide oxidation too. Powell et al. (1979) assumed a sulfide detoxification in the body wall of some meiofaunal species, although Powell (1989) suggests that in meiofauna detoxification of sulfide cannot be of physiological importance, because in small animals no sulfide oxidation system could operate at a rate sufficiently fast to oxidize all entering sulfide.

In the hydrothermal *Bythograea thermydron*, the hepatopancreas oxidizes sulfide (Vetter et al. 1987), and the same is true for the liver of some fish inhabiting tidal marshes (Bagarinao and Vetter 1989). Liver mitochondria could be demonstrated to be responsible for the sulfide-oxidizing activity (Bagarinao and Vetter 1990). Mitochondria from gill and foot tissue of the clam *Solemya reidi* were also shown to oxidize sulfide (Powell and Somero 1986a; O'Brien and Vetter 1990).

Several attempts have been made to prove whether sulfide oxidation in animal tissues is enzymatically catalyzed. Sulfide-oxidizing activity often appears to be heat- and protease-labile, and shows Michaelis-Menten kinetics (Powell and Somero 1985; Vetter et al. 1987; Vismann 1990), which provides some evidence for an enzymatic basis of oxidation. However, these measurements have been carried out according to Powell and Somero (1985) at highly unphysiological conditions, and may therefore partly be based on artifacts (see Bagarinao and Vetter 1990; Vismann, 1991a).

The sulfide-oxidizing activity of mitochondria of *Solemya reidi*, *Fundulus parvipinnis*, and *Citharichthys stigmaeus* has been followed by measuring sulfide-induced oxygen consumption (Powell and Somero 1986a; Bagarinao and Vetter 1990; O'Brien and Vetter 1990) and thiosulfate production (Bagarinao and Vetter, 1990). The oxidation process is coupled with a distinct production of adenosine trisphosphate (ATP) and is stoichiometrically linked to mitochondrial electron transport to oxygen. Oxygen consumption and ATP production were stimulated at sulfide concentrations up to 20 $\mu mol\ l^{-1}$ and decreased at higher sulfide concentrations. Similarly, respiration in *Acanthamoeba castellanii* (Lloyd et al. 1981) and mitochondrial oxygen consumption in *Halicryptus spinulosus* (Oeschger and Vetter 1992) could be enhanced by low sulfide concentrations, and was inhibited by sulfide concentrations of 25 and 50 μmol l^{-1} respectively. These findings, together with the inhibitory effect of cyanide, indicate the presence of a mitochondrial enzyme, functioning as a sulfide-cytochrome c-oxidoreductase (Powell and Somero 1986a; Bagarinao and Vetter 1990). However, the energetic significance of sulfide-stimulated ATP production needs further investigation. Sulfide oxidation seems to be an inefficient pathway to obtain ATP, because sulfide oxidation delivers less ATP than carbon substrates with both competing for oxygen as the terminal electron acceptor under hypoxic conditions (Bagarinao and Vetter 1990; O'Brien and Vetter 1990).

The oxidation of sulfide to thiosulfate was also demonstrated in the mitochondria of the body wall musculature of *Arenicola marina* (Völkel 1992; Völkel and Grieshaber 1994a). In contrast to mitochondria of other species (see above), sulfide-stimulated respiration was not inhibited by high sulfide concentrations up to 100 μmol l^{-1} and was noticeably less sensitive to cyanide than respiration with succinate as the only substrate (Völkel and Grieshaber 1994a). Further inhibitor studies indicated that mitochondrial sulfide oxidation in *A marina* is linked to the respiratory electron transport chain. At high sulfide concentrations electrons arising from sulfide oxidation are supposed to be transferred to an alternative terminal oxidase (Völkel and Grieshaber 1994b).

4.2.2 Sulfide Oxidation in Bacterial Symbionts

A special kind of symbiosis between chemoautotrophic bacteria and marine animals in sulfide-rich habitats was first discovered in the fauna living near the deep-sea hydrothermal vents. In particular, the vestimentiferan tube worm *Riftia pachyptila*, the clam *Calyptogena magnifica*, and the mussel *Bathymodiolus thermophilus* have been shown to contain sulfur bacteria as symbionts in specialized tissues (Cavanaugh et al. 1981; Cavanaugh 1983; Felbeck et al. 1981). These symbiotic bacteria are capable of oxidizing reduced sulfur compounds, such as sulfide, and use some of the energy released in the oxidation process to drive net fixation of carbon dioxide via the Calvin-Benson cycle (Cavanaugh et al. 1981; Felbeck et al. 1981; Powell and Somero 1986b; Fisher et al. 1989; Wilmot and Vetter 1990; see also Powell et al. 1987).

The gutless clam *Solemya reidi*, which is found, for example, in reduced environments near sewage outfall zones, shows a unique symbiosis with sulfur bacteria (Felbeck 1983; Powell and Somero 1985, 1986b; Anderson et al. 1987). Sulfide entering the body of the clam appears to be aerobically oxidized in the mitochondria of the animal tissue leading to the production of ATP (see above), and thiosulfate is transported to the gill bacteria for further oxidation (see also Vetter et al. 1991). Wilmot and Vetter (1992) suggested that the bacterial symbionts oxidize sulfide as well as thiosulfate in vivo. Under hypoxic conditions anaerobic sulfide oxidation by bacteria may be an important detoxification.

A number of phyla from marine sulfide-rich habitats have been shown to contain bacterial symbionts. Symbiotic sulfide-oxidizing bacteria appear to be present in several bivalve families, such as Solemyidae, Lucinidae, Thyasiridae, Vesicomyidae, and Mytilidae (Dando et al. 1985; Reid and Brand 1986; Dando and Southward 1986). Cary et al. (1989) showed that *Lucinoma aequizonata* maintains a subtoxic level of sulfide by partially oxidizing it to thiosulfate, which is utilized by endosymbionts as an energy source. The bacteria store elemental sulfur during periods when sulfide and thiosulfate are available, and utilize it when reduced sulfur compounds are limited.

Ott et al. (1982) found two mouthless interstitial animals, the namatode *Astomonema jenneri* and the turbellar *Paractenula urania*, living in reducing sediments rich in sulfide. The animals were associated with symbiotic intracellular microorganisms in their gut rudiment, but whether these bacteria are sulfide-oxidizing chemoautotrophs remains unclear. Another group of nematodes, the Stilbonematinae, has recently been shown to have ectosymbiotic bacteria, which cover the entire body surface of the worm in a highly ordered and specific pattern (Fig. 3; Ott and Novak 1989; Schiemer et al. 1990; Ott et al. 1991). These microorganisms seem to oxidize sulfide chemoautotrophically (Schiemer et al. 1990). The worms show a specific migration behavior across the chemocline between sulfidic and oxic layers, which is interpreted as a novel strategy to provide the symbionts with sulfide and oxygen (Ott et al. 1991). The bacteria may utilize sulfur stores built up in the presence of sulfide for further oxidation during the following exposure to oxygen (Schiemer et al. 1990). Ott et al. (1991) showed that the worms live exclusively off their symbionts and probably feed on them.

Epibacteria associated with the Oligochaeta *Tubificoides benedii* were described also by Dubilier (1986). The filamentous bacteria are embedded in the cuticle of the posterior region of the worm. *Tubificoides benedii* is well adapted to a sulfidic biotop, but no direct evidence is given whether the associated bacteria are sulfide oxidizers.

In the gutless Oligochaeta *Phallodrilus leukodermatus*, the endosymbiotic sulfur bacteria seem to utilize thiosulfate by oxidizing it probably to sulfate and using the energy of this process for the fixation of CO_2. Although the thiobiotic *P. leukodermatus* is reported to be sulfide-tolerant, bacterial assimilation was suppressed by the application of 75 $\mu mol\ l^{-1}$ sulfide (Giere et al. 1988).

The priapulid worm *Halicryptus spinulosus* is associated with various types of epibacteria (Oeschger 1987; Oeschger and Schmaljohann 1988). The bacteria are morphologically similar to Thiobacilli and seem to be chemoautotrophic, but no participation of the bacteria in the nutrition of the worm could be found. It is suggested that the bacteria may oxidize sulfide and protect the animal to some extent when oxygen is available (see above; Oeschger and Schmaljohann 1988).

Temara et al. (1993) has recently found symbiotic sulfide-oxidizing bacteria in the intestinal caecum of the deposit-feeding echinoid *Echinocardium cordatum*. These *Thiothrix* spp.-like bacteria build nodules by forming multilayered mats around detrital particles entering the caecum. It has not been investigated, however, whether the bacteria provide their host with chemosynthetic food.

Obviously, even less is known about the symbiosis between sulfur-oxidizing bacteria and animals of the marine sediments than about symbiosis in the hydrothermal vent fauna. In sediment-living animals, ecto- or endosymbiotic bacteria seem to function mainly as an additional protection against sulfide, whereas their role in the nutrition of the host remains questionable.

4.2.3 Immobilization of Sulfide

Another possibility to avoid toxic effects of sulfide in the body is its immobilization by binding to protein. The blood of the vestimentiferan tube worm *Riftia pachyptila* was shown to contain a sulfide-binding protein that is able to fix high amounts of sulfide (Arp and Childress 1983; Childress et al. 1984; Powell and Somero 1983; Arp et al. 1987b; Fisher et al. 1988). This protein appears to be an extracellular hemoglobin binding both sulfide and oxygen (Fisher et al. 1988). Sulfide can be bound until 60% of the protein is saturated. Low concentrations of free sulfide are maintained in the blood, and thereby tissues are protected from the toxic effects of sulfide. Sulfide is simultaneously transported to the symbiotic bacteria allowing high rates of chemoautotrophy (Fisher et al. 1988).

Similar results could be shown in the gill hemoglobin of the symbiont-containing *Solemya velum* (Doeller et al. 1988). In the presence of sulfide part of the hemoglobin is rapidly and reversibly converted to ferric hemoglobin, which can bind sulfide, while the remaining part of the hemoglobin is still able to bind oxygen. Sulfide and oxygen can thereby be transported without harmful effects for the host and delivered to the symbionts (Doeller et al. 1988). These results have been confirmed by the biochemical characterization of three different types of hemoglobin in the gill of *Lucina pectinata* (Kraus 1990), which showed different reactions to sulfide and oxygen very similar to those of the hemoglobin in *Solemya velum.* Besides hemoglobin, other proteins are reported to transport sulfide e.g. the clam *Calyptogena magnifica* (Arp et al. 1984; Childress 1987; Childress et al. 1991).

Sulfide can also be immobilized by binding to metallic ions such as iron. As mentioned previously, iron present in the cuticular mucus coat of *Halicryptus spinulosus* could serve as a short-term protection against sulfide (Oeschger and Janssen 1991). Possibly a similar mechanism works in the hepatopancreas of *Saduria entomon* (Vismann 1991a).

4.3 Sulfide-Insensitive Cytochrome c-Oxidase

The toxic effects of sulfide are mainly based on its strong, but reversible, inhibition of cytochrome aa_3 (Nicholls 1975; National Research Council 1979). A perfect strategy of adaptation to sulfide would therefore be a sulfide-insensitive cytochrome c-oxidase.

In most animals studied, the cytochrome c-oxidase appears to be highly sulfide-sensitive. In vitro the 50% inhibitory sulfide concentration (K_i) in tissue homogenates of marine fish (Bagarinao and Vetter 1989), some vent animals

(*Bythograea thermydron and Riftia pachyptila*) (Hand and Somero, 1983) as well as *Urechis caupo* (Powell and Arp 1989), and *Halicryptus spinulosus* (Oeschger and Vetter 1992) is about 1 μmol l^{-1} or below (see also Bagarinao and Vetter 1990). The inhibitory effect of sulfide on respiration is about tenfold decreased when isolated mitochondria are measured (Powell and Somero 1986a; Bagarinao and Vetter 1990), due to mitochondrial sulfide oxidation (see above).

Powell (1989) insists that thiobiotic meiofauna must be sulfide-insensitive, because no detoxification system could operate at sufficiently high speeds to be of physiological importance. Up to now only Degn and Kristensen (1981) as well as Schneider and Grieshaber (1992) described an unusually low sensitivity of *Tubifex* sp. respiration to sulfide.

4.4 Anaerobic Metabolism

If an animal is not able to exclude or detoxify sulfide at a rate sufficiently high, sulfide will increase in its body during exposure to this toxin. Lacking a sulfide-insensitive cytochrome c-oxidase, aerobic respiration is inhibited as soon as sulfide reaches a critical concentration. This situation, similar to a lack of oxygen, can only be survived if an animal is able to switch to an anaerobic metabolism (see Grieshaber et al. 1992).

Most of the inhabitants of marine sediments are able to withstand at least short periods of anoxia (Gäde 1980; Schöttler 1980; Ellington 1983; Gäde 1983; De Zwaan, 1983; Hochachka 1985; De Zwaan and Putzer 1985; Gäde and Grieshaber 1986; Bryant 1991; Nicholas 1991; Grieshaber et al 1994), because oxygen availability is limited in this habitat (see above). Intertidal animals in particular are periodically exposed to several hours of hypoxia. During low tide, e.g. water in the burrow of the lugworm, *Arenicola marina* becomes hypoxic. The worm stops ventilating and relies on anaerobic energy production (Schöttler et al. 1984a, b; Schöttler and Bennet 1991; Toulmond and Tchernigovtzeff 1984; Toulmond et al. 1983; Toulmond 1987). An anaerobic metabolism is indicated by the accumulation of end-products such as succinate, opines, and volatile fatty acids (Schöttler et al. 1984b; Schroff and Schöttler 1977).

An anaerobic metabolism has also been demonstrated in the sipunculid worm *Sipunculus nudus* (Pörtner et al. 1984, 1985), *Nereis diversicolor* (Schöttler et al. 1984b), *Halicryptus spinulosus, Astarte borealis,* and *Arctica islandica* (Oeschger 1987). Resistance to oxygen deficiency was also shown in *Mytilus edulis, Mya arenaria, Scrobicularia plana,* etc. (Theede et al. 1969) as well as in *Ctenodiscus crispatus* (Shick 1976) and others.

Although all of the hypoxia-tolerant species listed previously can be expected to survive the exposition to sulfide by relying on an anaerobic metabolism, a sulfide-dependent anaerobiosis has only been proven in a few organisms so far. Recently, Völkel and Grieshaber (1992) found that *Arenicola marina* and *Sipunculus nudus* switch to an anaerobic metabolism when exposed to 1 mmol l^{-1} sulfide, even when oxygen is available. After 8 h of normoxic sulfide incubation, as much succinate and opines have accumulated in the tissues as in hypoxic controls without sulfide, and in hypoxic sulfide incubations (Fig. 5). Because of their anaerobic capacity *Arenicola marina* and *Sipunculus nudus* survive, despite the increase in sulfide in their body (Völkel and Grieshaber 1992). Because under normoxic conditions sulfide does not appear in the body of *A. marina* up to an outer sulfide concentration of 0.4 mmol l^{-1} (see above), aerobic metabolism is continued up to this limit, and no succinate or opines accumulate (Völkel and Grieshaber 1991). Only when sulfide exceeds 0.4 mmol l^{-1} does the worm switch to an anaerobic metabolism (Völkel and Grieshaber 1994a). However, the influence of sulfide on the metabolism of *A. marina* appears to depend on the metabolic state as well as on size and population peculiarities. Völkel and Grieshaber (1994a) showed that lugworms that were in their reproductive phase were not able to maintain a completely aerobic metabolism, even at external sulfide concentrations of 0.2 mmol l^{-1}, and the same is true for much smaller specimens of *A. marina* from a different population (Hauschild and Grieshaber 1993; Grieshaber et al., in press).

The maintenance of an aerobic metabolism during normoxic sulfide incubation up to a certain threshold of the external sulfide concentration has also been shown in two other species. The symbiont-containing clam *Solemya reidi* (Anderson et al. 1990) and the oligochaete *Tubificoides benedii* (Dubilier 1993; Dubilier et al. 1994) were able to continue an aerobic respiration up to an outer sulfide concentration of 100 and 300 µmol l^{-1}, respectively. Higher sulfide concentrations overwhelmed the defense mechanisms and caused an onset of anaerobic metabolism, which could be measured by an increase of succinate.

The priapulid worm *Halicryptus spinulosus* is often exposed to long-term anoxia and may also be confronted with longer periods of sulfide conditions (Oeschger and Theede 1988). Oeschger and Storey (1990) could show that anoxia in *H. spinulosus* induced a metabolic rate depression by modifying the activity and kinetics of some glycolytical enzymes. The effect of sulfide on these enzymes were similar to the responses to anoxia, indicating that at a metabolic level the consequences of both stresses are similar. The accumulation of succinate in the tissue of *H. spinulosus* during persistent normoxic sulfide incubation gave evidence for sulfide-induced anaerobiosis in this species (Oeschger and Vetter 1992).

Nevertheless, even in species that are resistant to hypoxia, the survival capacity in the presence of sulfide is lower than under hypoxia alone (Theede et al. 1969; Levitt and Arp 1991). This may be due to the harmful effects of sulfide on other metalloenzymes (National Research Council 1979). In contrast, survival of the polychaete *Streblospio benedicti* exposed to anoxia and 60 µmol l^{-1} sulfide was not affected in comparison with anoxia without sulfide (Llansó 1991). However, the reliance on anaerobic metabolism can guarantee the survival of sulfide stress only for a limited period, because no species has been proven to be permanently completely anaerobic so far. Even parasitic helminths, where anaerobic metabolism is a steady-state condition and persists in the presence of oxygen, have been shown to possess cytochromes and to be capable of oxidative phosphorylation. They require at least low amounts of oxygen for survival and

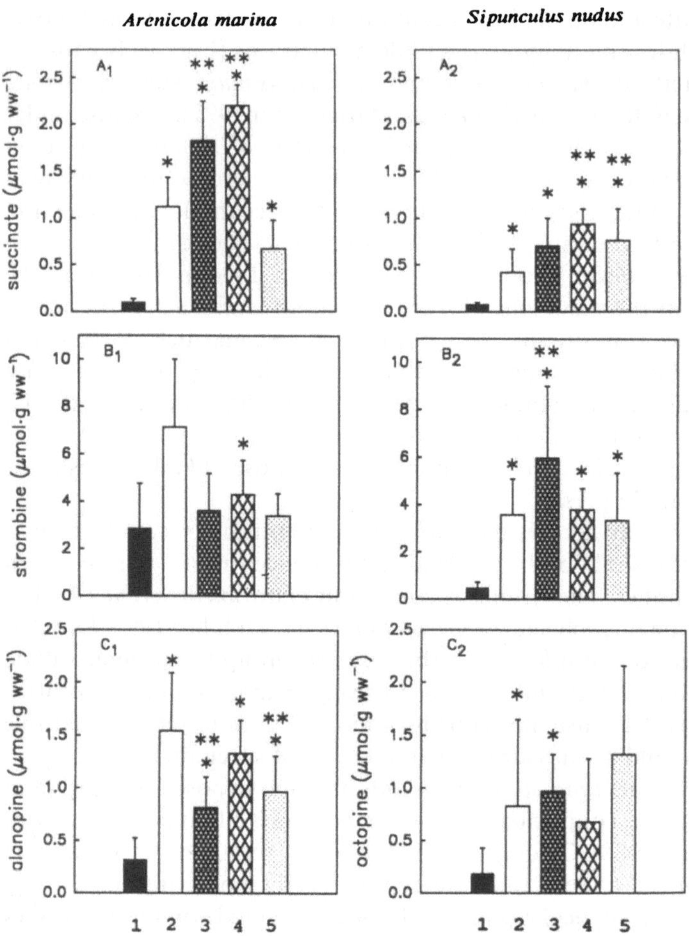

Fig. 5. Contents of succinate (**A**), strombine (**B**), and alanopine/octopine (**C**) (μmol g^{-1} wet wt.) in the body-wall musculature of *Arenicola marina* and *Sipunculus nudus* after 8 h incubations at various conditions ($n = 5$; $\times \pm$ SD; *, significant differences to the control; **, significant differences to hypoxic controls). *Column 1*, control: *column 2*, hypoxic control; *column 3*, hypoxia with 0.2 mmol l^{-1} sulfide; *column 4*, hypoxia with 1.0 mmol l^{-1} sulfide; *column 5*, normoxia with 1.0 mmol l^{-1} sulfide. (Völkel and Grieshaber 1992)

growth (see Barrett 1991). Meiofauna, which lives beyond the zero-oxygen line, does not appear to be obligatory anaerobic either. Fox and Powell (1986, 1987) demonstrated that thiobiotic species were capable of aerobic metabolism, and Powell (1989) proposed that thiobiotic meiofauna should exploit low oxygen concentrations by supplementing anaerobic with aerobic, sulfide-insensitive metabolism (see above; see also Nicholas 1991).

Because of the unusual water chemistry of deep-sea hydrothermal vents, the fauna of this habitat is simultaneously exposed to sulfide and oxygen. It has not been demonstrated yet whether vent fauna also relies on an anaerobic metabolism.

Table 1. Mechanisms of sulfide tolerance in different species 1, exclusion of sulfide; 2, oxidation(a) in animal tissue or body fluid and (b) in bacteria; 3, immobilization of sulfide; 4, sulfide-insensitive respiration; 5, sulfide-induced anaerobic metabolism; +, mechanism demonstrated;-mechanism nonexistent;?, possible, but not proved (see text)

Species	1	2		3	4	5
		(a)	(b)			
Turbellaria						
Archiloa wilsoni	–	+				
Solenofilomorpha funilis	–	+				
Paractenula urania			?			
Nematoda						
Astomonema jenneri			?			
Stilbonematinae	?		+			
Priapulida						
Halicryptus spinulosus	?	+	?	?	–	+
Bivalves						
Lucinoma annulata	–	+				
Lucinoma aequizonata			+			
Lucina pectinata				+		
Solemya velum	–			+		
Solemya reidi	–	+	+	+		+
Calyptogena magnifica	–	+	+	+		
Bathymodiolus thermophilus	–		+			
Sipunculida						
Sipunculus nudus	–	+				+
Echiurida						
Urechis caupo	–	+	?		–	
Polychaeta						
Arenicola marina	–	+				+
Nereis virens	–	+				
Nereis diversicolor	–	+				
Oligochaeta						
Tubifex sp.	–	+			?	+
Tubificoides benedii	?		?			+
Phallodrilus leukodermatus			?			
Vestimentifera						
Riftia pachytila	–		+	+	–	
Crustacea						
Saduria entomon	–	+		?		
Bythograea thermydron	–	+			–	
Fish						
Fundulus parvipinnis	–	+				
Citharhichthys stigmaeus	–	+				

An investigation of the metabolism of several vent species showed that enzymatic activities were similar to those of phylogenetically related nonvent species (Hand and Somero 1983). Only *Calyptogena magnifica* had a minimal capacity for aerobic respiration, and was suggested to rely largely on anaerobic metabolism.

5 Conclusion

The capacity to cope with sulfide as an environmental factor is not restricted to a few highly specialized species, but is widespread among the animal kingdom (see Vismann 1991b). Species of several different classes, including fish have been demonstrated to possess strategies to defend themselves against sulfide (Table 1). Even in mammals, such as rats and dogs, the existence of sulfide detoxification could be shown (Smythe 1943; Sörbo 1958; National Research Council 1979). According to Table 1, it is evident that the capacity to oxidize sulfide either in the animal tissue or via bacteria is the most abundant mechanism of sulfide tolerance. Many tissues and body fluids are more or less able to oxidize sulfide, but the adaptation to sulfide-rich environments depends on an effective oxidation rate, even at low concentrations. Sulfide detoxification and sulfide-dependent anaerobiosis may have been evolved in sulfide-tolerant organisms under the pressure to exploit new habitats.

Acknowledgments. The authors are grateful to the Deutsche Forschungsgemeinschaft (Gr 456/12–2) and Bundesministerium für Forschung und Technologie (O3F0052A) for financial support.

References

Anderson AE, Childress JJ, Favuzzi JA (1987) Net uptake of CO_2 driven by sulphide and thiosulphate oxidation in the bacterial symbiont-containing clam *Solemya reidi*. J Exp Biol 133: 1–31

Anderson AE, Felbeck H, Childress JJ (1990) Aerobic metabolism is maintained in animal tissue during rapid sulfide oxidation in the symbiont-containing clam *Solemya reidi*. J Exp Zool 256: 130–134

Arp AJ (1991) The role of heme compounds in sulfide tolerance in the echiuran worm *Urechis caupo*. In: Vinogradov SN, Kapp OH (eds) Structure and function of invertebrate oxygen proteins. Springer, Berlin Heidelberg, New York, pp 337–346

Arp AJ, Childress JJ (1983) Sulfide binding by the blood of the hydrothermal vent tube-worm *Riftia pachyptila*. Science 219: 295–297

Arp AJ, Childress JJ, Fisher CR Jr (1984) Metabolic and blood gas transport characteristics of the hydrothermal vent bivalve *Calyptogena magnifica*. Physiol Zool 57: 648–662

Arp AJ, Powell MA, Hansen BM (1987a) Sulfide tolerance and hematin levels in the mudflat-inhabiting echiuran *Urechis caupo*. Am Zool 27: 32A

Arp AJ, Childress JJ, Vetter, RD (1987b) The sulfide-binding protein in the blood of the verstimentiferan tube-worm, *Riftia pachyptlila*, is the extracellular hemoglobin. J Exp Biol 128: 139–158

Arp AJ, Hansen BM, Julian D (1992) Burrow environment and coelomic fluid characteristics of the echiuran worm *Urechis caupo* from populations at three sites in northern California. Mar Biol 113: 613–623
Bagarinao T, Vetter RD (1989) Sulfide tolerance and detoxification in shallow-water marine fishes. Mar Biol 103: 291–302
Bagarinao T, Vetter RD (1990) Oxidative detoxification of sulfide by mitochondria of the California killifish *Fundulus parvipinnis* and the speckled sanddab *Citharichthys stigmaeus*. J Comp Physiol B 160: 519–527
Bagarinao T, Vetter RD (1992) Sulfide-hemoglobin interactions in the sulfide-tolerant salt marsh resident, the California killifish *Fundulus parvipinnis*. J Comp Physiol 162: 614–624
Barrett J (1991) Parasitic helminths. In: J Bryant (ed) Metazoan life without oxygen. Chapman and Hall, London
Beerman H (1924) Some physiological actions of hydrogen sulfide. J Exp Zool 41: 33–43
Bell SS (1983) An experimental study of the relationship between below-ground structure and meiofaunal taxa. Mar Biol 76: 33–39
Berner RA (1970) Sedimentary pyrite formation. Am J Sci 268: 1–23
Boaden PJS (1975) Anaerobiosis, meiofauna and early metazoan evolution. Zool Ser 4: 21–24
Boaden PJS (1977) Thiobiotic facts and fancies (aspects of the distribution and evolution of anaerobic meiofauna). Microfauna Meeresboden 61: 45–63
Boaden PJS (1980) Meiofaunal thiobios and 'the *Arenicola* negation'; case not proven. Mar Biol 58: 25–29
Boaden PJS, Platt HM (1971) Daily migration patterns in an intertidal meiobenthic community. Thallassia Jugosl 7: 1–12
Broderius SJ, Smith LL Jr, Lind DT (1977) Relative toxicity of free cyanide and dissolved sulfide forms to the fathead minnow (*Pimephales promelas*). J Fish Res Board Can 34: 2323–2332
Bryant C (1991) Metazoan life without oxygen. Chapman & Hall, London
Burnett W, King EG, Grace M, Hall WF (1977) Hydrogen sulfide poisoning: review of 5 years' experience. CMA J Can Med Assoc J 117: 1277–1280
Cary SC, Vetter RD, Felbeck H (1989) Habitat characterization and nutritional strategies of the endosymbiont-bearing bivalve *Lucinoma aequizonata* Mar Ecol Prog Ser 55: 31—45
Cavanaugh CM (1983) Symbiotic chemoautotrophic bacteria in marine invertebrates from sulphide-rich habitats. Nature 302: 58–61
Cavanaugh CM, Gardiner SL, Jones ML, Jannasch HW, Waterbury JB (1981) Procaryotic cells in the hydrothermal vent tube worm *Riftia pachyptila* Jones: possible chemoautotrophic symbionts. Science 213: 340–342
Chen KY, Asce AM, Morris JC (1972) Oxidation of sulfide by O_2: catalysis and inhibition. J San Eng Div Proc Am Soc Civil Eng SA1: 215–227
Childress JJ (1987) Uptake and transport of sulfide in marine invertebrates In: Dejours P, Bolis L, Taylor CR, Weibel ER (eds) Comparative physiology: life in water and on land. Fidia Res Ser, Liviana Press, Padova
Childress JJ, Arp AJ, Fisher CR Jr (1984) Metabolic and blood characteristics of the hydrothermal vent tube-worm *Riftia pachyptila*. Mar Biol 83: 109–124
Childress JJ, Fisher CR, Favuzzi JA, Sanders NK (1991) Sulfide and carbon dioxide uptake by the hydrothermal vent clam, *Calyptogena magnifica*, and its chemoautotrophic symbionts. Physiol Zool 64: 1444–1470
Cline JD, Richards FA (1969) Oxygenation of hydrogen sulfide in seawater at constant salinity, temperature and pH. Environ Sci Technol 3: 838–843
Dando PR, Southward AJ (1986) Chemoautotrophy in bivalve molluscs of genus *Thyasira*. J Mar Assoc UK 66: 915–929
Dando PR, Southward AJ, Southward EC, Terwilliger NB, Terwilliger RC (1985) Sulphur-oxidizing bacteria and haemoglobin in gills of the bivalve mollusc *Myrtea spinifera*. Mar Ecol Prog Ser 23: 85–98
Degn H, Kristensen B (1981) Low sensitivity of *Tubifex sp* respiration to hydrogen sulfide and other inhibitors. Comp Biochem Physiol 69B: 809–817

De Wit R, Jonkers HM, van den Ende FP, van Gemerden H (1989) In situ fluctuations of oxygen and sulphide in marine microbial sediment ecosystems. Neth J Sea Res 23: 271–281
De Zwaan A (1983) Carbohydrate metabolism in bivalves. In: Hochachka PW (ed) The Mollusca, vol 1. Metabolic biochemistry and molecular biomechanics. Academic Press, New York
De Zwaan A, Putzer V (1985) Metabolic adaptations of intertidal invertebrates to environmental hypoxia (a comparison of environmental anoxia to exercise anoxia) In: Laverack MS (ed) Physiological adaptations of marine animals. Company of Biologists Ltd, Cambridge
Doeller JE, Kraus DW, Colacino JM, Wittenberg JB (1988) Gill haemoglobin may deliver sulfide to bacterial symbionts of *Solemya velum* (Bivalvia, Mollusca). Biol Bull 175: 388–396
Dubilier N (1986) Association of filamentous epibacteria with *Tubificoides benedii* (Oligochaeta: Annelida). Mar Biol 92: 285–288
Dubilier N (1993) Ökophysiologische und morphologische Anpassungen des marinen Oligochaeten *Tubificoides benedii* (d'Udekem) an Sauerstoffmangel und Sulfid. Dissertation Hamburg, Shaker Verlag, Aachen
Dubilier N, Völkel S, Dando PR (1991) Sulfide tolerance in annelids. J Mar Biol Assoc UK 71: 957
Dubilier N, Giere O, Grieshaber MK (1994) Concomitant effects of sulfide and hypoxia on the aerobic metabolism of the marine oligochaete *Tubificoides benedii*. J Exp Zool 269: 287–297
Dubilier M, Giere O, Grieshaber MK (in press) Morphological and ecophysiological adaptations of the marine oligochaete (*Tubificoides benedii* to sulfidic sediments. Am Zool
Eaton R, Arp AJ (1993) Aerobic respiration during sulfide exposure in the marine echiuran worm *Urechis caupo*. Physiol Zool 66: 1–19
Ellington WR (1983) Recovery from anaerobiosis. J Exp Zool 228: 431–444
Felbeck H (1983) Sulfide oxidation and carbon fixation by the gutless clam *Solemya reidi*: an animal-bacteria symbiosis. J Comp Physiol 152: 3–11
Felbeck H, Childress JJ, Somero GN (1981) Calvin-Benson cycle and sulphide oxidation enzymes in animals from sulphide-rich habitats. Nature 293: 291–293
Fenchel TM, Riedl RJ (1970) The sulfide system: a new biotic community underneath the oxidized layer of marine sand bottoms. Mar Biol 7: 255–268
Fisher CR, Childress JJ, Sanders NK (1988) The role of vestimentiferan hemoglobin in providing an environment suitable for chemoautotrophic sulfide-oxidizing endosymbionts. Symbiosis 5: 229–246
Fisher CR, Childress JJ, Minnich E (1989) Autotrophic carbon fixation by the chemoautotrophic symbionts of *Riftia pachyptila*. Biol Bull 177: 372–385
Fox CA, Powell EN (1986) Meiofauna and the sulfide system: the effects of oxygen and sulfide on the adenylate pool of three turbellarians and a gastrotrich. Comp Biochem Physiol 85A: 37–44
Fox CA, Powell EN (1987) The effect of oxygen and sulfide on CO_2 production by three acoel turbellarians. Are thiobiotic meiofauna aerobic? Comp Biochem Physiol 86A: 509–514
Gäde G (1980) Biological role of octopine formation in marine molluscs. Mar Biol Lett 1: 121–135
Gäde G (1983) Energy metabolism of arthropods and molluscs during environmental and functional anaerobiosis. J Exp Zool 228: 415–429
Gäde G, Grieshaber MK (1986) Pyruvate reductases catalyze the formation of lactate and opines in anaerobic invertebrates. Comp Biochem Physiol 83B: 255–272
Giere O, Wirsen CO, Schmidt C, Jannasch HW (1988) Contrasting effects of sulfide and thiosulfate on symbiotic CO_2-assimilation of *Phallodrilus leukodermatus* (Annelida). Mar Biol 97: 413–419
Gorodezky LA, Childress JJ (1994) Effects of sulfide exposure history and hemolymph thiosulfate on oxygen-consumption rates and regulation in the hydrothermal vent crab *Bythograea thermydron*. Mar Biol 120: 123–131
Grieshaber MK, Hardewig I, Kreutzer U, Schneider A, Völkel S (1992) Hypoxia and sulfide tolerance in some marine invertebrates. Verh Dtsch Zool Ges 85.2: 55–76

Grieshaber MK, Hauschild K, Sommer A, Völkel S (in press) Anaerobiosis and sulfobiosis in the lugworm, *Arenicola marina L.* 28th Europ Mar Biol Symp, Crete, Greece
Grieshaber MK, Hardewig I, Kreutzer U, Pörtner HO (1994) Physiological and metabolic responses to hypoxia in invertebrates. Rev Physiol Pharmacol 125: 44–147
Groenendaal M (1979) On sulfide and the distribution of *Arenicola marina* in a tidal mud flat in the Dutch wadden sea. Neth J Sea Res 13: 562–570
Groenendaal M (1980) Tolerance of the lugworm (*Arenicola marina*) to sulfide. Neth J Sea Res 14: 200–207
Groenendaal M (1981) The adaptation of *Arenicola marina* to sulfide solutions. Nath J Sea Res 15: 65–77
Hand SC, Somero GN (1983) Energy metabolism pathways of hydrothermal vent animals: adaptations to a food-rich and sulfide-rich deep-sea environment. Biol Bull 165: 167–181
Hartmann Hansen M, Ingvorsen K, Jørgensen BB (1978) Mechanisms of hydrogen sulfide release from coastal marine sediments to the atmosphere. Limnol Oceanogr 23: 68–76
Hauschild K, Grieshaber MK (1993) Hydrogen sulfide lowers the respiration of the lugworm *Arenicola marina*. L Verh Dtsch Zool Ges 86: 90
Hochachka PW (1985) Assessing metabolic strategies for surviving O_2 lack: role of metabolic arrest coupled with channel arrest. Mol Physiol 8: 331–350
Jones JD (1955) Observations on the respiratory physiology and on the haemoglobin of the polychaete genus *Nephthys*, with special reference to *N. hombergii.* J Exp Biol 32: 110–125
Jørgensen BB (1977) The sulfur cycle of a coastal marine sediment (Limfjorden, Denmark). Limnol Ocean 22: 814–832
Jørgensen BB (1982) Ecology of bacteria of the sulfur cycle with special reference to anoxic–oxic interface environments. Phil Trans R Soc Lond B 298: 543–561
Jørgensen BB (1988) Ecology of the sulphur cycle: oxidative pathways in sediments. In: Cole JA, Ferguson SJ (eds) The nitrogen and sulphur cycles. Univ Press, Cambridge
Jørgensen BB (1990) A thiosulfate shunt in the sulfur cycle of marine sediments. Science 249: 152–154
Jørgensen BB, Fenchel T (1974) The sulfur cycle of a marine sediment model system. Mar Biol 24: 189–201
Julian D, Arp AJ (1992) Sulfide permeability in the marine invertebrate *Urechis caupo*. J Comp Physiol B 162: 59–67
Julian D, Mansour S, Arp AJ (1990) The effect of hindgut inflation on oxygen uptake and sulfide diffusion in *Urechis caupo*. Am Zool 30: 414
Keilin D (1933) On the combination of methaemoglobin with H_2S. Proc R Soc Lond (Biol) 113B: 393–404
Kelly DP (1988) Oxidation of sulfur compounds In: Cole JA, Ferguson SJ (eds) The nitrogen and sulphur cycles. Univ Press, Cambridge
Kelly DP (1989) Physiology and biochemistry of unicellular sulfur bacteria In: Schlegel HG, Bowien B (eds) Autotrophic bacteria. Sci Tech Pub and Springer Verlag, Madison.
Kraus DW (1990) Oxygen and sulfide binding by clam haemoglobins: a division of labor within symbiosis. Cyclobios Newsletters 4: 25
Kristensen E (1989) Oxygen and carbon dioxide exchange in the polychaete *Nereis virens*: influence of ventilation activity and starvation. Mar Biol 101: 381–388
Kröger A, Schröder J, Paulsen J, Beilmann A (1988) Acetate oxidation with sulphur and sulphate as terminal electron acceptors. In: Cole JA, Ferguson SJ (eds) The nitrogen and sulphur cycles. Univ Press, Cambridge
Levitt JM, Arp AJ (1991) The effects of sulfide on the anaerobic metabolism of two congeneric species of mudflat clams. Comp Biochem Physiol 98B: 339–347
Llansó RJ (1991) Tolerance of low dissolved oxygen and hydrogen sulfide by the polychaete *Streblospio benedicti* (Webster). J Exp Biol Ecol 153: 165–178
Lloyd D, Kristensen B, Degn H (1981) Oxidative detoxification of hydrogen sulfide detected by mass spectometry in the soil amoeba *Acanthamoeba castellanii.* J Gen Microbiol 126: 167–170
Longwell J, Pentelow FTK (1935) The effect of sewage on brown trout (*Salmo trutta L.*) J Exp Biol 12: 1–12

Meadows PS, Deans EA, Anderson JG (1981) Responses of *Corophium volutator* to sediment sulphide. J Mar Biol Assoc UK 61: 739–748
Menon JG, Arp AJ (1993) The integument of the marine echiuran worm *Urechis caupo*. Biol Bull 185: 440–454
Meyers MB, Fossing H, Powell EN (1987) Microdistribution of interstitial meiofauna, oxygen and sulfide gradients and the tubes of macro-infauna. Mar Ecol Prog Ser 35: 223–241
Meyers MB, Powell EN, Fossing H (1988) Movement of oxybiotic and thiobiotic meiofauna in response to changes in porewater oxygen and sulfide gradients around macro-infaunal tubes. Mar Biol 98: 395–414
Millero FJ (1986) The thermodynamics and kinetics of the hydrogen sulfide system in natural waters. Mar Chem 18: 121–147
Millero FJ, Hubinger S, Fernandez M, Garnett S (1987) Oxidation of H_2S in seawater as a function of temperature, pH and ionic strength. Environ Sci Technol 21: 439–443
Morse JW, Cornwell JC (1987) Analysis and distribution of iron sulfide minerals in recent anoxic marine sediments. Mar Chem 22: 55–69
National Research Council (1979) Subcommittee on Medical and Biologic Effects on Environmental Pollutants, Division of Medical Sciences, Assembly of Life Sciences Hydrogen Sulfide. Univ Park Press, Baltimore
Nedwell DB (1982) The cycling of sulfur in marine and freshwater sediments. In: Nedwell DB, Brown CM (eds) Sediment microbiology. Academic Press, London
Nedwell DB, Floodgate GD (1972) Temperature-induced changes in the formation of sulphide in a marine sediment. Mar Biol 14: 18–24
Nelson DC, Castenholz RW (1981) Use of reduced sulfur compounds by *Beggiatoa* sp. J Bact 147: 140–154
Nicholas WL (1991) Interstitial meiofauna. In: Bryant C (ed) Metazoan life without oxygen. Champman & Hall, London
Nicholls P (1975) The effect of sulfide on cytochrome aa_3 isosteric and allosteric shifts of the reduced α-peak. Biochim Biophys Acta 396: 24–35
O'Brien J, Vetter RD (1990) Production of thiosulphate during sulphide oxidation by mitochondria of the symbiont-containing bivalve *Solemya reidi*. J Exp Biol 149: 133–148
Oertzen J-A von, Schlungbaum G (1972) Experimentell-ökologische Untersuchungen über O_2-Mangel und H_2S-Resistenz an marinen Evertebraten der westlichen Ostsee. Beitr Meeresk 29: 79–91
Oeschger R (1987) Vergleichende Untersuchungen zur biotopbedingten Langzeit-Anaerobiose und Schwefelwasserstoff-Resistenz bei einigen marinen Wirbellosen aus der Ostsee. Dissertation, Kiel
Oeschger R, Schmaljohann R (1988) Association of various types of epibacteria with *Halicryptus spinulosus* (Priapulida). Mar Ecol Prog Ser 48: 285–293
Oeschger R, Theede H (1988) Use of biochemical features of macrobenthic species as indicators of long-term oxygen deficiency. Kieler Meeresforsch 6: 99–110
Oeschger R, Storey KB (1990) Regulation of glycolytic enzymes in the marine invertebrate *Halicryptus spinulosus* (Priapulida) during environmental anoxia and exposure to hydrogen sulfide. Mar Biol 106: 261–266
Oeschger R, Janssen HH (1991) Histological studies on *Halicryptus spinulous* (Priapulida) with regard to environmental hydrogen sulfide resistance. Hydrobiologia 222: 1–12
Oeschger R, Vetter RD (1992) Sulfide detoxification and tolerance in *Halicryptus spinulous* (Priapulida) : a multiple strategy. Mar Ecol Prog Ser 86: 167–179
Oseid DM, Smith LL Jr (1974) Chronic toxicity of hydrogen sulfide to *Gammarus pseudolimnaeus*. Trans Am Fish Soc 4: 819–822
Ott JA, Novak R (1989) Living at an interface: meiofauna at the oxygen/sulfide boundary of marine sediments. In: Ryland RS, Tyler PA (eds) Reproduction genetics and distribution of marine organisms. 23rd Eur Mar Biol Symp, Swansea, UK, Olsen and Olsen, Fredenshurg, Denmark, pp 415–422
Ott JA, Rieger G, Rieger R, Enderes F (1982) New mouthless interstitial worms from the sulfide system: symbiosis with procaryotes. Mar Ecol 3: 313–333

Ott JA, Novak R, Schiemer R, Hentschel U, Nebelsick M, Polz M (1991) Tackling in the sulfide gradient: a novel strategy involving marine nematodes and chemoautotrophic ectosymbionts. Mar Ecol 12: 261–279

Patel S, Spencer CP (1963) The oxidation of sulphide by the haem compounds from the blood of *Arenicola marina*. J Mar Biol Assoc UK 43: 167–175

Pörtner H-O, Kreutzer U, Siegmund B, Heisler N, Grieshaber MK (1984) Metabolic adaptation of the intertidal worm *Sipunculus nudus* to functional and environmental hypoxia. Mar Biol 79: 237–247

Pörtner H-O, Heisler N, Grieshaber MK (1985) Oxygen consumption and mode of energy production in the intertidal worm *Sipunculus nudus* L.: definition and characterization of the critical PO_2 for an oxyconformer. Resp Physiol 59: 362–377

Powell EN (1977) Particle size selection and sediment reworking in a funnel feeder *Leptosynapta tenius* (Holothuroidea, Synaptidae). Int Rev Ges Hydrobiol 62: 385–408

Powell EN (1989) Oxygen, sulfide and diffusion: why thiobiotic meiofauna must be sulfide-insensitive first-order respirers. J Mar Res 47: 887–932

Powell EN, Bright TJ (1981) A thiobios does exist—gnathostomulid domination of the canyon community at the East flower Garden brine seep. Int Rev Ges Hydrobiol 66: 675–683

Powell EN, Crenshaw MA, Rieger RM (1979) Adaptations to sulfide in the meiofauna of the sulfide system ^{35}S-sulfide accumulation and the presence of a sulfide detoxification system. J Exp Mar Biol Ecol 37: 57–76

Powell EN, Crenshaw MA, Rieger RM (1980) Adaptations to sulfide in sulfide-system meiofauna endproducts of sulfide detoxification in three turbellarians and a gastrotrich. Mar Ecol Prog Ser 2: 169–177

Powell EN, Bright TJ, Woods A, Gittings S (1983) Meiofauna and the thiobios in the East flower Garden brine seep. Mar Biol 73: 269–283

Powell MA, Somero GN (1983) Blood components prevent poisoning of respiration of the hydrothermal vent tube-worm *Riftia pachyptila*. Science 219: 297–299

Powell MA, Somero GN (1985) Sulfide oxidation occurs in the animal tissue of the gutless clam *Solemya reidi*. Biol Bull 169: 164–181

Powell MA, Somero GN (1986a) Hydrogen sulfide oxidation is coupled to oxidative phosphorylation in mitochondria of *Solemya reidi*. Science 233: 563–566

Powell MA, Somero GN (1986b) Adaptations of sulfide by hydrothermal vent animals: sites and mechanisms of detoxification and metabolism. Biol Bull 171: 274–290

Powell MA, Arp AJ (1989) Hydrogen sulfide oxidation by abundant nonhemoglobin heme compounds in marine invertebrates from sulfide-rich habitats. J Exp Zool 249: 121–132

Powell MA, Vetter RD, Somero GN (1987) Sulfide detoxification and energy exploitation by marine animals. In: Dejours P, Bolis L, Taylor CR, Weibel ER (eds) Comparative physiology: life in water and on land. Fidia Res Ser, Liviana Press, Padova

Reid RGB, Brand DG (1986) Sulfide-oxidizing symbiosis in lucinaceans: implications for bivalve evolution. Veliger 29: 3–24

Reise K (1981a) Gnathostomulida abundant alongside polychaete burrows. Mar Ecol Prog Ser 6: 329–333

Reise K (1981b) High abundance of small zoobenthos around biogenic structures in tidal sediments of the wadden sea. Helgol Meeresunters 34: 413–425

Reise K (1984) Free-living platyhelminthes (Turbellaria) of a marine sand flat: an ecological study. Mikrofauna Marina 1: 1–62

Reise K, Ax P (1979) A meiofaunal "thiobios" limited to the anaerobic sulfide system does not exist. Mar Biol 54: 225–237

Reise K, Ax P (1980) Statement on the thiobios-hypothesis. Mar Biol 58: 31–32

Revsbech NP, Jørgensen BB, Blackburn TH (1983) Microelectrode studies of the photosynthesis and O_2, H_2S and pH profiles of a microbial mat. Limnol Oceanogr 28: 1062–1074

Schiemer F, Novak R, Ott J (1990) Metabolic studies on thiobiotic freeliving nematodes and their symbiotic microorganisms. Mar Biol 106: 129–137

Schneider A, Grieshaber MK (1992) Adaptation to hydrogen sulfide of limnic tubificids. Verh Dtsch Zool Ges 85.1: 39

Schöttler U (1980) Der Energiestoffwechsel bei biotopbedingter Anaerobiose: Untersuchungen an Anneliden. Verh Dtsch Zool Ges 1980: 228–240

Schöttler U, Surholt B, Zebe E (1984a) Anaerobic metabolism in *Arenicola marina* and *Nereis diversicolor* during low tide. Mar Biol 81: 69–73

Schöttler U, Wienhausen G, Westermann J (1984b) Anaerobic metabolism in the lugworm *Arenicola marina* L: the transition from aerobic to anaerobic metabolism. Comp Biochem Physiol 79B: 93–103

Schöttler U, Bennet EM (1991) Annelids. In: Bryant C (ed) Metazoan life without oxygen. Chapman & Hall, London

Schroff G, Schöttler U (1977) Anaerobic reduction of fumarate in the body wall musculature of *Arenicola marina* (Polychaeta). J Comp Physiol 116: 325–336

Shick JM (1976) Physiological and behavioural responses to hypoxia and hydrogen sulfide in the infaunal asteroid *Ctenodiscus crispatus.* Mar Biol 37: 279–289

Smythe CV (1943) Some reactions of sodium sulfide added to certain tissues. Arch Biochem 2: 259–268

Sörbo B (1958) On the formation of thiosulfate from inorganic sulfide by liver tissue and heme compounds. Biochim Biophys Acta 27: 324–329

Temara A, de Ridder C, Kuenen JG, Robertson LA (1993) Sulfide-oxidizing bacteria in the burrowing echinoid, *Echinocardium cordatum* (Echinodermata). Mar Biol 115: 179–185

Theede H (1973) Comparative studies on the influence of oxygen deficiency and hydrogen sulfide on marine bottom invertebrates. Neth J Sea Res 7: 244–252

Theede H, Ponat A, Hiroki K, Schlieper C (1969) Studies on the resistance of marine bottom invertebrates to oxygen-deficiency and hydrogen sulphide. Mar Biol 2: 325–337

Thode-Andersen S, Jørgensen BB (1989) Sulfate reduction and the formation of ^{35}S-labeled FeS, FeS_2, and S^0 in coastal marine sediments. Limnol Oceanogr 34: 793–806

Torrans EL, Clemens HP (1981) Physiological and biochemical effects of acute exposure of fish to hydrogen sulfide. Comp Biochem Physiol 71C: 180–183

Toulmond A (1987) Adaptations to extreme environmental hypoxia in water breathers. Comp Physiol Environ Adapt 2: 123–136

Toulmond A, Tchernigovtzeff C (1984) Ventilation and respiratory gas exchange of the lugworm *Arenicola marina* L as functions of ambient PO_2 (20–700 torr). Resp Physiol 57: 349–363

Toulmond A, Tchernigovtzeff C, Greber P, Jouin C (1983) Epidermal sensitivity to hypoxia in the lugworm. Experientia 40: 541–543

Vetter RD, Powell MA, Somero GN (1991) Metazoan adaptation to hydrogen sulfide. In: Bryant C (ed) Metazoan life without oxygen. Chapman & Hall, London

Vetter RD, Wells ME, Kurtsman AL, Somero GN (1987) Sulfide detoxification by the hydrothermal vent crab *Bythograea thermydron* and other decapod crustaceans. Physiol Zool 60: 121–137

Vismann B (1990) Sulfide detoxification and tolerance in *Nereis (Hediste) diversicolor* and *Nereis (Neanthes) virens* (Annelida: Polychaeta). Mar Ecol Prog Ser 59: 229–238

Vismann B (1991a) Physiology of sulfide detoxification in the isopod *Saduria (Mesidotea) entomon.* Mar Ecol Prog Ser 76: 283–293

Vismann B (1991b) Sulfide tolerance: physiological mechanisms and ecological implications. Ophelia 34: 1–27

Völkel S, Hauschild K, Grieshaber MK (in press) Sulfide stress and tolerance in *Arenicola marina* during low tide. Mar Ecol Prog Ser

Völkel S (1992) Ökophysiologische Untersuchungen über die Sulfidtoleranz des Wattwurmes *Arenicola marina* L. Dissertation Düsseldorf, Shaker Verlag, Aachen

Völkel S, Grieshaber MK (1991) Sulfide tolerance and detoxification in the lugworm *Arenicola marina.* Verh Dtsch Zool Ges 84: 433

Völkel S, Grieshaber MK (1992) Mechanisms of sulfide tolerance in the peanut worm *Sipunculus nudus* (Sipunculida) and in the lugworm *Arenicola marina* (Polychaeta). J Comp Physiol B 162: 469–477

Völkel S, Grieshaber MK (1994a) Oxygen dependent sulfide detoxification in the lugworm *Arenicola marina*. Mar Biol 118: 137–147

Völkel S, Grieshaber MK (1994b) Sulfide oxidation in the mitochondria of *Arenicola marina*. Physiologist 37: A–88

Völkel S (in press) Sulfide tolerance and detoxification in *Arenicola marina and Sipunculus nudus*. Am Zool

Watling L (1991) The sedimentary milieu and its consequences for resident organisms. Am Zool 31: 789–796

Wells RMG, Pankhurst NW (1979) An investigation into the formation of sulphide and oxidation compounds from the haemoglobin of the lugworm *Abarenicola affinis*. Comp Biochem Physiol 66C: 255–259

Wilmot PD, Cadee K, Katinic JJ, Kavanagh BV (1988) Kinetics of sulfide oxidation by dissolved oxygen. J WPCF 60: 1264–1270

Wilmot DB, Vetter RD (1990) The bacterial symbiont from the hydrothermal vent tubeworm *Riftia pachyptila* is a sulfide specialist. Mar Biol 106: 273–283

Wilmot DB, Vetter RD (1992) Oxygen- and nitrogen-dependent sulfur metabolism in the thiotrophic clam *Solemya reidi*. Biol Bull 182: 444–453

Subject Index

Springer-Verlag and the Environment

We at Springer-Verlag firmly believe that an international science publisher has a special obligation to the environment, and our corporate policies consistently reflect this conviction.

We also expect our business partners – paper mills, printers, packaging manufacturers, etc. – to commit themselves to using environmentally friendly materials and production processes.

The paper in this book is made from low- or no-chlorine pulp and is acid free, in conformance with international standards for paper permanency.

MIX
Papier aus verantwortungsvollen Quellen
Paper from responsible sources
FSC® C105338

If you have any concerns about our products,
you can contact us on
ProductSafety@springernature.com

In case Publisher is established outside the EU,
the EU authorized representative is:
Springer Nature Customer Service Center GmbH
Europaplatz 3, 69115 Heidelberg, Germany

Printed by Libri Plureos GmbH
in Hamburg, Germany